Hazards and Safety
Measures in Radio Stations

I. S. MEHLA

INDIA • SINGAPORE • MALAYSIA

Notion Press

Old No. 38, New No. 6
McNichols Road, Chetpet
Chennai - 600 031

First Published by Notion Press 2020
Copyright © I. S. Mehla 2020
All Rights Reserved.

ISBN 978-1-64828-703-9

This book has been published with all efforts taken to make the material error-free after the consent of the author. However, the author and the publisher do not assume and hereby disclaim any liability to any party for any loss, damage, or disruption caused by errors or omissions, whether such errors or omissions result from negligence, accident, or any other cause.

While every effort has been made to avoid any mistake or omission, this publication is being sold on the condition and understanding that neither the author nor the publishers or printers would be liable in any manner to any person by reason of any mistake or omission in this publication or for any action taken or omitted to be taken or advice rendered or accepted on the basis of this work. For any defect in printing or binding the publishers will be liable only to replace the defective copy by another copy of this work then available.

No part of this book may be used, reproduced in any manner whatsoever without written permission from the author, except in the case of brief quotations embodied in critical articles and reviews.

श्रीमद भगवद् गीता

कर्मण्येवाधिकारस्ते मा फलेषु कदाचन।
मा कर्मफलहेतुर्भूर्मा ते सङ्गोऽस्त्वकर्मणि॥

तुम्हारा (आपका) अधिकार तो केवल कर्म करने का है,
कर्म के फल पर नही।
इसलिए ना तो कर्म से भागना उचित है,
ना ही कर्म के फल की आशा रखना उचित है॥

श्रीमद भगवद्गीता; अध्याय-२, श्लोक ॥४७॥

VIEWS OF EXPERTS ON MY 1ST BOOK: AM RADIO TOWER ANTENNAS
By Mr I S Mehla, I B (E) S

The author of this radio broadcasting antenna compendium is an acclaimed broadcasting professional and a comrade-in-arms of this reviewer. So it is an interesting experience to review this well produced volume.

This book is a thesaurus on radio antennas, as it covers a vast amount of information on the subject which has been meticulously gathered by the author in his long career in broadcasting engineering. The publication covers the subject extensively, right from the fundamentals to the specialized nuances of broadcast antennas.

In my own broadcasting career of several decades, I always found the subject of antennas to be difficult and requiring specialized expertise. The way the author has approached this subject in the book and presented it amply displays his understanding, command and his keen interest in this area. Indeed, this publication is a very important reference book.

Of special interest to me is the section on Non-conventional MF Radiators which includes the Cross-Field Antennas. At the time these were developed, these appeared to be based on very complex theories on RF radiation.

To my knowledge, there is only one more exhaustive publication on radio antennas, that on HF antennas authored by Mr. C S R Rao. He produced this work many years ago, presumably in the Research Department of All India Radio. It can be said that these two reference books cover the entire gamut of radio broadcasting antennas.

Sharad Sadhu
I B (E) S
Feb 11, 2020, 8:25 AM

Sir,

I could have the opportunity to go through the book on AM RADIO TOWER ANTENNAS written by Shri I S MEHLA, I found it very useful for broadcast engineers. All the topics such as radio communication system, spectrum management, propagation related matters, strengthening of towers etc and all other points related with towers, starting from design to installation and maintenance are covered in a professional manner. In my entire carrier with AIR I have not found such detail information about towers in a single book. Design and maintenance related material is purely based on practical experience and field results. This book is providing the useful information for broadcast managers also. All the detailed information on the matter will help the broadcast Engg Professional in designing, installation and maintaining the AM RADIO TOWER ANTENNAS.

I wish, that author of the book deserves a big thank and congratulation. With best wishes

O.K.SHARMA
ADDL. DIRECTOR GENERAL (E),
AKASHWANI & DOORDARSHAN

Sir,

The book is a lucidly written master piece useful to Medium wave Broadcasters, Planners and Maintenance professionals.

The book covers all relevant topics of Medium wave antennas. More over the book has the compilation of information from different sources throughout the world which makes it very comprehensive.

All the nitty gritties of medium wave antennas are explained in easy and understandable language.

Medium wave Antennas is a subject which has not been in focus of All India Radio as most of the medium wave towers were installed around 30 years back or more. Information about these Antennas is available in bits and pieces.

The book also provides a useful insight regarding feeder lines, matching networks and earth radials and Directive Antennas. The book will be quite useful for medium wave setups throughout the country.

Regards
SANJEEV KUMAR SAXENA
DDG (E) P&D Unit, DG AIR
Sun, Apr 7, 2019, 9:42 PM

Aditya Chaturvedi

 **A great compilation**
14 January 2019

Book Format: Paper Back 14 January 2019

It is an exhaustive book covering each and every bit of AM radio antennas which can never be found in any conventional text book. Apart from minute technical details, a lot of emphasis has been focused on planning and implementation. The book is must have for broadcast stations, technical universities and broadcast professionals.

I salute the hard work put into writing this great book.

Chetna

 **Lucid understanding of concepts**
26 December 2018

Clear and informative: for beginners and experts, alike. Was very useful in clearing broadcast principles without complex jargons. Highly recommended for broadcast professionals.

Ajay Dabur

⭐⭐⭐⭐⭐ **Must -read!!**

27 December 2018

A very helpful guide to, provide clear concise and detailed understanding of all AM radio concepts –very helpful for students and professionals alike. A must read for professionals of broadcast engineering who needs a one stop book for different concepts and understanding.

 Gaurav Choudhary

⭐⭐⭐⭐⭐ **Must have for Radio Enthusiasts !!!**

27 December 2018

This book does go into great detail about all that you need to do to know about AM Radio Tower Antennas.

For me, it was quite an eye opening and also overwhelming understanding the basics and concepts in an easy to understand language.

 Dhir Singh

⭐⭐⭐⭐⭐ **Best of Luck** 😊

27 December 2018

The book is as good and interesting as its description.

Very beneficial and informative for people studying radio and broadcast engineering

⭐⭐⭐⭐⭐

A Wonderful Compendium of Information

By gRAdy Moates on 23 March 2019

While much of the information in this book is available elsewhere, spread among many different science texts and structural standards and regulations, it is very useful to have it all in one place. It is organised in a way that allows the reader to become aware of less-obvious physical relationship - -physical laws and engineering /mechanical principles are side -by –side, leading the reader to efficient and practical application of the information. Looks good in my reference library, too!

Grady Edward Moats,
II HEBA Field Engineer World Wide Antenna Systems - USA

** Review on book titled, "AM RADIO TOWER ANTENNAS"

I am a senior Radio Broadcast Professional having over 40 year experience in Radio Industry. My company is currently executing AIR MW ANTENNA renovation projects. In this context we often refer to the book titled "AM RADIO TOWER ANTENNAS" authored by Mr. I. S. Mehla, a Broadcast Engineering veteran.

While working in the Planning and Design Division of India's public broadcaster, All India Radio Head Quarters in New Delhi, we used to refer to large number of foreign authored books for getting information on AM Antenna design, yet sometimes the required information was difficult to find. Mr. Mehla has done a great service to the broadcast industry by writing a book covering every aspect of MW Antenna and connected topics of RF Spectrum and EM Wave Propagation, thus making it a comprehensive reference book not only for Radio Design Engineers but also for Radio Station Managers and Operation and Maintenance staff. It a must have reference book for every establishment concerned on MW Radio Broadcasting.

With my best wishes and complements to the author and the publisher.

Harkesh Gupta
Ex. IB (E) S
Fellow BES (I) Managing Partner,
Mark Integration & Technology Solutions, New Delhi
Formerly Chairman & Managing Director, BECIL; Dy. Director General, AIR

The book, 'AM Radio Tower Antennas' is a comprehensive treatise dealing with the theory and practice of all the elements that constitute an antenna system besides giving a detailed insight of the allied topics of radio communication starting from its basics and then going over to electromagnetic propagation and RF spectrum along with its management and related ITU standards. In nutshell, the book provides all the knowledge that is required by a broadcast engineer in doing justice to his or her profession while dealing with radio transmission, with the only exception of the transmitter. Being written by an experienced broadcast engineer, who had his hands on various transmission systems, the book, while giving the theoretical details, deals exhaustively on the practical aspects of antenna systems including their overall design, erection and maintenance, which is rare to find at a single place. The book would be useful for broadcast stations, broadcast organizations and broadcast professionals.

B. P. Srivastava
Senior Advisor
Broadcast Engineering Consultants India Ltd.

CONTENTS

Preface *15*
Acknowledgements. *17*

1. Installation and Constructional Hazards 19
    1.1   General 19
    1.2   Material, Processes and Equipment Handling 20
    1.3   Hazards in using Mechanical Tools and Aids 26
    1.4   Handling of Explosive powered tools 27
    1.5   Welding Hazards 29
    1.6   Blasting Operations 31
    1.7   Electro-explosive Devices 33
    1.8   Warning Signage 34
    1.9   Failure of Structures during erection 36
    1.10 Hazardous Installations 40
    1.11 Construction Hazards 40
    1.12 General requirements of safety Policy in hazardous installations .. 45
    References 48

2. Radiation Hazards 49
    2.1   General 49
    2.2   Radiation Classifications 50
    2.3   Harmful Biological Effects of Radiation 52
    2.4   Safe Exposure limits of RF Radiation 55
    2.5   Estimation for EM field Strength 57
    2.6   Evaluating Compliance with Guidelines for
          Human Exposure to RF in an AM Radio station 58
    2.7   General Field characteristics of electromagnetic fields 67
    2.8   Precautions at transmitting stations and in their vicinity 87
    2.9   Examples of calculated field strengths near broadcasting antennas . 92
    2.10 Limits and levels of RF hazards 94

2.11　Field-strength values to be determined 100
　　2.12　Additional evaluation methods for EM fields
　　　　　(Appendix 4 to Annex 1 of ITU-R, BS-1698) 102
　　2.13　Electromagnetic compatibility of Electro-medical devices
　　　　　(Appendix 5 to Annex 1)...107
　　2.14　Typical Safety Instructions 108
　　References 110

3. **Fire Hazards** 111
　　3.1　General...111
　　3.2　Classification and constituents of Fires 113
　　3.3　Fire Extinguishing Methods and Media................................117
　　3.4　Fire Protection Measures 129
　　3.5　Fire Protection and Safety Management for Various
　　　　　Types of Occupancies 159
　　3.6　Materials used in Radio stations 162
　　3.7　Fire Prevention, Protection Requirements and Plan 166
　　3.8　Fire Fighting Functions in a Radio Station 170
　　3.9　Design Considerations for fire measures in a Radio Station 174
　　3.10　Planning for fire Protection facilities.................................185
　　3.11　Plan for Reduction of Fire protection Needs 190
　　3.12　Typical Firefighting Equipments and facilities for a Radio Station ...194
　　References 198

4. **Lightning Hazards**...199
　　4.1　History...199
　　4.2　Formation Process, Type and Triggering of Lightning 201
　　4.3　Lightning and a broadcast Transmitter System 208
　　4.4　Lightning Characteristics and Incidences 212
　　4.5　Lightning Protection Principles 214
　　4.6　Static Charge and a Broadcast tower System 216
　　4.7　Methods for protecting a mast antenna from static discharges......222
　　4.8　Recommended measures for protection of an AM Transmittter set up ..225
　　4.9　Transmitter Building Equipment Layout 233
　　4.10　AC line Surge Protectors...237
　　4.11　Implementing some Practical Solutions 246
　　4.12　How to set Spark ball gap at tower base 251
　　References..257

5. **Electrical Shock Hazards** 259
　　5.1　General...259
　　5.2　Electrical Hazards 259

	5.3	Identifying Electrical Hazards 266
	5.4	General Electrical Safety Requirements............................. 271
	5.5	Working near Exposed Energized Overhead Lines or Parts 274
	5.6	Operating Equipment near Radio and Microwave Transmission Towers 276
	5.7	Compliance of Electrical Installation 277
	5.8	Personal Protective Grounding 279
	5.9	Training Requirements for Electrical Operations 281
	5.10	Indian Electricity Rules 1956 – an Overview 284
	References 335
6.	**Earthing and Grounding Systems** **337**
	6.1	General 337
	6.2	What is Earthing or Grounding? 338
	6.3	Classifications of Earthing on Generic Basis..................... 340
	6.4	Classification of Earthing on Function basis 344
	6.5	Characteristics of Grounding Structures and Materials......... 347
	6.6	Design features of an Earthing system 353
	6.7	Types of Earth Electrodes 357
	6.8	Initial Testing of Earthing Electrodes 364
	6.9	Review of Earthing Systems 365
	6.10	A typical Earth schematic layout of a High Power MW Transmitter ..371
	6.11	Measurement of earth Resistance and Soil Resistivity......... 371
	6.12	General points and Precautions 390
	References 396
7.	**Radio Frequency Interference and Shielding**........................... **397**
	7.1	The EM environment 397
	7.2	Electromagnetic Compatibility and Interferences 397
	7.3	EMI / RF Interferences Mechanisms 400
	7.4	Filters with Passive Components 405
	7.5	A Review of EM Field 407
	7.6	Shielding–General 409
	7.7	Components design and selection for minimising EMC 419
	7.8	Installation... 422
	7.9	Testing for EMC 439
	7.10	Rules and Regulations of EMC 441
	7.11	EMC Management 446
	7.12	RF Shielding Material – Cable and Copper 450
	References 456

8. Safety Practices in Radio stations 457
- 8.1 Introduction 457
- 8.2 Safety Engineering Philosophy 458
- 8.3 Interpretation of Safety Rules 462
- 8.4 Classifications of Accidents 463
- 8.5 Emergency Plans of the Organisation 466
- 8.6 Type of Disasters or Emergencies 473
- 8.7 Staff Responsibilities 477
- 8.8 Safety Rules 481
- 8.9 Induction of staff / employee 485
- 8.10 Safety Aids and facilities 487
- 8.11 Safety standard requirements for Radio Transmitting Equipments ... 492
- References 496

9. First-Aid and Occupational Health 497
- 9.1 History of First-aid 497
- 9.2 Occupational Safety and Health 498
- 9.3 What is First-Aid? 503
- 9.4 First-Aid Rules and Flow Chart 504
- 9.5 Protective Precautions in First-Aid 506
- 9.6 Primary assessment & basic life support for First-Aid ... 509
- 9.7 Secondary assessment for: First-Aid 513
- 9.8 Application and Categorization of: First-aid 520
- 9.9 Contents of a First-aid box 556
- 9.10 Elements of Training in First-aid 561
- References 565

Index *567*

PREFACE

Radio is an integral part of the culture, social and economic landscape of any country and India is no exception in that matter. Radio broadcasting is one of the most popular and affordable means for mass communication largely owing to its wide coverage, low setup costs, simplest receiver technology and affordability to own multiple receivers by the masses.

Terrestrial mode Radio broadcast in India is done in Medium Wave (521–1605 KHz) popularly known as AM Radio, Short Wave (6–30 MHz), and VHF (88–108 MHz), popularly known as FM. Presently FM band due to its clarity and variety of entertainment programmes as it's owned by private broadcasters also in addition to govt, is most popular and ruling the airwaves.

As to the origin of Radio in India, a private company named, Indian Broadcasting Company heralded the birth of Indian broadcasting as a commercial venture by starting broadcast from Bombay on July 27, 1927 and then a month later from Calcutta. But the company got liquidated within 3 years of its existence. However, to prevent the closure of broadcasting in India, the then British Govt of India took it over on 1st April 1930 and renamed it as "Indian State Broadcasting Service". In 1936 on 8th of June, it was renamed as "All India Radio" acronym as "AIR" and its Hindi avatar as "Akashwani" after independence. Back then radio was a rare commodity and it was considered prestigious to have a Radio set.

AIR; the public service broadcaster of India have established 467 Radio stations encompassing 662 transmitter centres (140 MW, 48 SW, and 474 FM transmitters) for providing multiple radio broadcasting services. It also provides overseas broadcasts services for its listeners in SW band across the world. Until 2000, AIR under the aegis of Ministry of Information and Broadcasting was the sole radio broadcaster in the country but in the year

2000, looking at the changing market dynamics, the government opened the FM radio broadcast to private sector also.

As of now ,in India,there are 243 pvt FM radio channels operational in 86 cities from phase-I & II schemes and with the 3rd phase in process ,total pvt FM channels may be 1000 channels across 323 cities.

Safety procedures and practices including the awareness of all types of hazards involved in project works for establishing a radio station, their operation and maintenance is one of the major subjects of considerable importance to Radio Engineering Managers. Since the productivity at work is directly affected by the health status of the workers, occupational safety and health (OSH) policy dealing with the welfare of very basic human resource, also is very important needing a detailed policy frame work.

Though AIR had a Manual detailing General administration, material / stores inventory handling, technical reports etc but there were no comprehensive source on hazards, risks involved with safety procedures and occupational health problems encountered and precautions to avoid them in a radio Station.

So during my working, I always felt a necessity of a comprehensive source detailing hazards including other occupational health problems with ways and means to tackle these. Hence a comprehensive compilation was in my view since then for referring by engineering managers, shift in-charges and all other technical staffs on the matters related to the safety of the project, operating and maintenance staff including occupational hazards encountered in a radio station.

A glance at the content list will reveal that the book is logically developed into nine chapters starting with the hazards and risks involved while establishing, operating and maintaining a radio station with the topics; Hazards of Installation and construction; RF Radiation; Fire; Lightning strikes; Electric shocks; Earthing and Grounding nuances; RF Interference and Shielding; various Safety Practices and measures in Radio Stations and lastly the; the First-Aid for various ailments while at work.

I dedicate this compilation to my parents (late) from whom I inherited the virtue of sincerity, dedication & hard work, and spouse Shashi and children – daughter Isha (Spouse Aarjav& their adorable son Vivaan), and son Ishan (spouse Chetna) for their unstinted support and affection.

– ISHWAR SINGH MEHLA

ACKNOWLEDGEMENTS

In preparing this compilation, I have referred many books, Application notes, printed documents etc from regulatory bodies. But while using them for my compilation, I have abridged, aggregated, and simplified.

To list and name all who have helped or contributed with the information, and acknowledging each individual would be a formidable task but some documents, books, handouts and pocket books, notes of veterans, manufacturer's brochures, specifications sheets, technical and application notes and project reports of organizations, and individuals in particularly need mentioning have been mentioned under reference after each chapter and herewith also I express my gratitude in using their material.

I am also indebted to publishers and authors acknowledged herewith or elsewhere in this book for deemed 'permission, in the cause of furthering the knowledge, from their published free or copyright material.

I acknowledge gratefully all the sources, whom I have not specifically mentioned here or presently not in a position to recollect, if I have used their material in this book.

I acknowledge my sincere thanks and gratitude to Engineering Management of All India Radio in position at that time who posed confidence in my sincerity and dedication and assigned me the challenging jobs. I also express my gratitude and indebt for using the name of AIR repeatedly, for the work done by me in various projects and I deem the express permission of authorities for using the material from AIR in my compilation.

– **ISHWAR SINGH MEHLA**
B.Sc.(Engg)-Electronics & Communication (Hons);
IB(E)S, F(IETE), F(BES-I), CE (IETE), F(IE);
Former: Dy. Director General(Engg)–AIR: PB; &
Director (O&M)-BECIL-a CPSE

CHAPTER 1
INSTALLATION AND CONSTRUCTIONAL HAZARDS

1.1 General

A radio station requires various types of equipments, machines, gadgets, plants, ancillaries, tools, sub systems and other accessories. The integration of numbers of equipments, subsystems and ancillaries makes a radio stations studio or transmitting centre. These equipments arrive from different sources and need to be checked immediately on arrival after opening their boxes. The installation of a big radio station studios or transmitter complex requires the assembly of very large numbers of shipping packages and considerable care must be exercised by the workmen during the handling of packages from containers or trucks. The shipping consist of proper transmitter cubicles 'discreet equipments, its accessories like heat exchangers, cooling water circulating pumps, heat exchangers dummy load, combining unit, antenna matching unit and ancillaries system like speech input equipments, other programme feeding arrangements, feeder line material, Mast and tower and other hardware as well as installation materials including the transformer oil for substation transformers. In addition material of all sub systems like air-conditioning plant, Power Supply equipments, Diesel Generating sets, misc installation hardware, fire fighting equipments, other watch and ward gadgets.

For studios center even though the size of the equipments is not large but there are huge quantity of small size materials like consoles, mikes, amplifiers, signal processors, recording, monitoring, program distribution, transmission and networking consoles and technical furniture in addition to acoustic and air-conditioning plant material. All this requires a great deal of logistic transportation, packing /unpacking, handling of fragile and costly material to distant sites through all modes and type of transport facilities.

1.2 Material, Processes and Equipment Handling

A typical high power radio transmitter consisting of cubicles may measure 6x3x2.4mtr (WxDxH) and weigh upto 6000 kgs. Similarly a single unit of an oil filled transformer for transmitter HT power supply may weigh 8000 kgs. A typical transmitter installation with spare parts and ancillaries may comprise of more than 200 crates of very big size to smallest of few feet with an aggregate weight of more than 150 tonnes transported in a convoy of 10–15 trucks and trailers. Steel structures for feeder lines and tower / masts may be much larger weight and quantity.

Lot of homework and detailed logistics planning is required for transporting so much material and equipments from so many different sources, which may be from foreign countries through diverse routes and means as well as on-site handling problems. Advance traffic negotiations and pre-planning of routes may be essential, particularly where station is located in an isolated area with poor connecting roads and light traffic bridges, or in some cases complete absence of bridges at river crossings and creek. The successful delivery of all equipments and plant may even in some cases require the reinforcement of roadbed.

Arrival of crated or containerized equipments and plant at the nearest port or railhead generally involves transfer into trucks or on trailers for transport to site. Transfer of the crated items from a ship to truck trailer is usually not much problem, but where the material arrives by rail at a siding, problems may occur due to non availability of unloading facilities. To transfer bulky and heavy items, hydraulic jacks are required to raise the crates from the floor of the rail wagon. Greased wooden planks or lubricated pipes are then placed below the crate to slide them. Then wooden planks are erected on the side of the rail bed, below the skid members and then it is extended to truck trailer. Cables are attached to Pulleys fixed around the crate, and a tractor or winch is used to slide the material on the truck trailer. If tractor or winch is not available, then hard wood or steel rollers can be used to slide the boxes to truck or trailer.

A well designed transmitter complex should have fork lift trucks, gantries, and portable chain pulley arrangements, (cranes if needed better be hired) to handle off-loading from the truck trailer at site. In a stations a gantry of about 6 to 8 ton capacity on an unloading platform of adequate length on the back side of transmitter hall are provided to handle the materials. The truck stops by the side of the platform and manual pulley chain or electrically driven gantry unloads the crates and put them on the platform.

These units then are shifted to their location in the transmitter hall or HT room with top and all side of packing or crate except the bottom removed by sliding over heavy duty GI pipes or steel rollers. It is recommended that transformers be transported without oil to avoid damage in the transit due to slapping of oil with internal parts. However immediately after receipt at site, the oil should be filled to avoid ingress of moisture in the windings.

During the unloading and shifting it is expected that a workman will carry a reasonable load. The supervisor should ensure that when heavy material and costly plant is being handled and the chances of injuries are more, then adequate staff is provided to carry the work safely. Care is required to prevent personal injury from bad or careless handling of the material. A common outcome of such attitude and working is in the overloading of material handling device and ignoring or failing to visualize the ultimate strength of the materials.

Chain pulley used on overhead gantries and other hoisting devices for handling heavy transmitter equipments or power supply transformers or switchgears, should be inspected at regular intervals. The chains being less reliable than wire ropes may break without warning and may be fatal for the workmen and materials. Small cracks and flaws are not easily visible, so the chains must be examined for soundness by an experienced inspector.

The handling of materials during the construction of radio mast tower work may require the provision of special built-in facilities to ensure safe execution of work. The provision of temporary guys, permanent guys, tensioning devices and availability of other faculties require detailed planning at initial system design stage itself for ensuring their availability at the time of execution of work at site. The execution and material lifting plan for a radio mast tower must be planned with drawing showing the location of lifting devices and the material to be lifted, on paper first and then executed in the field. Lifting devices for hauling the material to workmen on the tower masts should be laid in such a position that it does not interfere with the operations. The ground around the lifting device should be clear of all the material. Lifting gear used for hauling materials during the erection of the tower should not be used for transporting the staff up the structure as it is hazardous and dangerous.

Items like cylinders containing compressed gases should not be dragged, or rolled but transported in a hand cart or fork lifter truck after fastening in vertical position. When any such cylinder is required to be hauled up the tower for welding work it should be fitted into a safe cradle or platform in vertical position and hauled up.

Materials, tools, devices and other equipments should not be left in passageways and driveways against switchboards, or fire hydrants, or any other position which is likely to compromise the safety of workmen and others concerns. All the visitors and officials must be protected by barriers and warning signs. Best housekeeping and preventive maintenance with constant supervision of all the requirements of both are important specifications for the effective control and safe movement of materials during installation and construction works.

As per good engineering practices and work requirements Antenna materials, transmission line poles and other external plant materials must be placed in positions where they can be seen readily. If the material is hidden in the grass, then it could be a hazard to workmen or vehicles. Excavations pertaining to mast foundations or feeder line poles or earth pits must be secured and guarded by putting barricades till these are completed and remaining portion of excavation refilled lest where animal graze on the site, do not fall in it.

The stacked transmission lines poles (wooden or tubular steel) are to be handled carefully lest these do not fall on workmen when removing them for working. Measures like wedge up or otherwise secure the poles on which workmen have to work so that accidents may not be caused by the poles moving.

Handling of large dia RF co-axial or power supply cable drums can also be hazardous because of rolling problem. Some drums are very large in diameter as well are heavy and needs special attention in handling for safe working. Help of power house and utility company should be sought to handle such drums with special tools and devices available with these companies.

Orderliness and cleanliness is one of the most important aspects of safe handling of material and working. The safe working habits of a station staff should be cultivated from the inception itself for long term results. Orderliness and cleanliness are just as important to prevention of accidents as personal protection equipments. The approach of work should be, "a place for everything and everything in its place". The influence and experience of the supervisor and their capacity to manage organize efforts must be used to ensure safe working and all good habits. The problem of handling material safely, i.e. protection of the product, contains basically the same factors needed to protect the individuals who do the handling.

1.2.1 Handling Harmful substances

Many corrosive, poisonous, flammable, explosive and other harmful substances are also used in radio station installation, operation, and maintenance activities. The storage and handling of these substances should be such as it will prevent exposure of workmen to the harmful effects of these materials. So a system designer should try to search for less harmful material while designing the station. Typical list of harmful substances used in a radio station engineering works are;

(a) Sulphuric acid for diesel generator and other battery banks.
(b) Hydrochloric acid or sodium chloride recharging ion bed of deionizer for deionized water supply for transmitter and dummy load cooling system etc.
(c) Acetylene gas cylinders for brazing and other welding works.
(d) Cadmium plating or brasso or silvos for chassis and copper components polishing.
(e) Polyurethane enameled wires which give off toxic gas when soldered.
(f) Carbon tetrachloride (CCL4) for cleaning solders joints, PCBs and components.
(g) Other cleaning agents like Methyl chloroform etc.
(h) Adhesives like fevicol, dendrite or phenyl dehydrate for duct work.

(i) Resins, hardeners and other bonding agents for insulators and fiber glass.
(j) Creosote oil for anti-termite treatment of wood for use in studios false ceiling, acoustic panels, doors, observation windows and other works and duct insulation wrapping cloth.
(k) Flammable diesel, petrol and kerosene oil used in captive power generators, for lighting purpose in petromax and as general cleaning agent against rusting etc.
(l) Industrial spirit for polishing of acoustic wood paneling and other decoration work.
(m) The poisonous Beryllium oxide (Be O) ceramic used in construction of power electron valves and the mercury used in Thyristors and fluorescent tubes. Special storage cupboards and or rooms or underground storage space may be planned and provided as a permanent safety control measures for an operational station. But proper care is necessary during installation when proper storage and handling facilities are not available.

Where harmful gases, such as those in the battery rooms, are let off in open space, the accumulation of unsafe concentration limit should be prevented by means of exhaust fan or ventilation system.

Containers or Material used for storage of harmful substances should be leak and escape proof. The container should be appropriately marked and labeled in a manner consisting of colors, numerals and shape to identify the hazards of the material with respect to the health, reactivity and flammability, and the severity of these hazards. The labeling should also indicate the procedure for safe handling of the substance and the action to be taken in case a person comes in contact with it.

The compounds of Beryllium are very poisonous. Even the dust of the powdered metal or its oxide may cause serious illness and death when inhaled. Fumes are highly toxic. Technicians/workmen should therefore be extremely careful not to disassemble, grind, and pulverize, chemically clean or perform any other operation on the ceramic parts of electron tubes. This substance has high thermal conductivity, so main uses of Beryllium oxide ceramics are to conduct heat from power tubes directly from the plate of types using external plate construction.

Due to hazards involved in this material it should be disposed as per the instruction of the manufacturer. Similarly all other mercury and other harmful substances including like asbestos etc should be disposed off as per the recommendations of the manufacturers or as per the safety policy of the station.

1.2.2 Construction Site Workshop

A radio station installation and construction works always requires some type of onsite workshop facility to allow the work to be carried out safely. The extent of facilities depends on the nature and size of the project being undertaken and also on the remoteness of the site from the permanent commercial workshops. Normally an onsite construction workshop would generally be equipped with some basic items/machines/tools/facilities like bench drill, motor and hand grinder as well as polishing / buffing machine, mechanical hacksaw for wood and metal work, electric and acetylene welder, workbench, soldering and de-soldering station and comprehensive set of mechanical and electrical tools for installation and construction works. There may be additional tools and plants and it will depend on the nature and type of installation work and the complexity of the project.

For a large project involving the erection / construction of towers masts, transmission lines, matrix switches, transmitter plant and its associated equipments, it will be desirable to establish permanent workshop as soon as possession and access to the main building is given by the civil contractor, and before the commencement of proper installation work. The workshop may have major items such as lathe, sheet metal and pipe bender etc., together with a mobile gantry system to assist in handling the heavy materials in the workshop.

Where a field workshop has to be established, it would generally be of simple prefabricated construction with adequate protection for the plant, plenty of workshop rooms, good lighting and ventilation system. The area in which the workshop would be erected would be governed by the site layout of the plant to be erected, the location of the proposed buildings and the topography of the site. While establishing a site workshop facilities proper and due considerations should be given to all aspects of the safety.

1.3 Hazards in using Mechanical Tools and Aids

The use of mechanical aids such as Gin poles, Cranes, Work baskets, Hoists, JCBs etc used during construction and maintenance works, gives rise to special hazards owing to the danger of contact with high voltage antenna systems, transmission lines etc in a radio station. The special nature of some works, such as the erection of masts and towers, introduces safety hazard of the erection or rigging crew on the structure during use of these aids. Supervisor in charge of the work should ensure that the necessary precautions are known by and strictly observed by the operator.

In a large crane it is difficult for the operator to accurately gauge the clearance between the jib and conductors, from his position in the cabin. One safety device which is available to alert the operator of close feeder lines can be mounted on the crane jib. It can be a bare metal antenna fastened length wise to the jib, a screened detector unit and visual and audible alarms to alert the crane operator when the jib approaches within a fixed distance from the line.

No safety precautions are adequate without the cooperation of operators and other staff. They should remain alert from the possible hazardous situations which exist in many operations in which the mechanical aids are used. Some typical safety precautions to be exercised while using these aids are;

(a) Mechanical aids should be operated only by qualified and authorised persons.
(b) The operator for their own safety must wear safety helmet. The assistants also must wear safety helmet, safety shoes and gloves where wire ropes and slings are being handled.
(c) The operator should be responsible for ensuring that aid is in sound operational condition as per his knowledge and has been got serviced and is being serviced as per the OEM manual.
(d) The safe working limitations of each mechanical aid should be clearly marked at most appropriate location, say on control panel of the machine, and the operator to be responsible for ensuring compliance of these limits.
(e) No additional or special attachment to be fitted without the approval of OEM or Engineer-in-charge.

(f) On the conclusion of each day's work or at other times when the machine aid is to be left unattended, it is the responsibility of the operator to ensure that it is parked in clear area and left in safe condition with all the control in OFF position. Parking braking devices securely applied and if not on plane ground, then wheels blocked with wooden blockers available on the machine for this purpose.

1.4 Handling of Explosive powered tools

Explosive powered tools are finding wide application in installation work in the radio station project works. They find application in the fixing of copper sheets or mesh to the floors of transmitter rooms and in the complete shielding of antenna matching huts, test rooms, emergency studios or link rooms and the control room. They are also used for fixing attachment to the wall for equipment supports, raceways etc.

The use of these tools is fraught with danger and many deaths injuries have resulted from the use of these tools, so, extreme care is necessary in their use. Power driven tools are very versatile and are great labor savers. However, there are certain materials, encountered in the construction of radio stations that are not suitable for these tools. In fact, some materials can be extremely hazardous to the operator, and a thorough examination of the material should be done before the attempt to drive a projectile through it is made. Experiences have shown that it is dangerous to attempt to use a power driven tool to drive a projectile into;

(a) Hardened steel, high tensile steel, cast iron, and other hard and unyielding tempered materials.
(b) Material prone to shattering such as cut building stone, glass, tile, glazed or clinker brick.
(c) Positions close to the edge of the materials where there is danger that material might crack or break or the projectile escape and continue to travel to harm something or someone.
(d) Positions within 1 cm of the edge of the steel plate.
(e) Positions within 7.5 cm of the edge of a brick or concrete block.
(f) Concrete with large hard aggregate.

(g) Material where it is intended to fix the projectile to its full depth and where it may foul an imbedded object, such as a round concrete reinforcement that would deflect the projectile from a straight path.
(h) Materials of such strength that projectile may pass completely through it with the charge being used, unless the material is backed by a protecting material capable of fully absorbing the energy of the projectile.

Because of the damage which can occur to the materials and plant, and the hazards to which workmen may be exposed by improper explosive powered tools, many safety precaution instructions have been issued on their use and care to be taken. The following guidelines cover most of the issues outlined above;

(a) Only suitably trained; operators to be allowed to use the tools.
(b) The operator must wear suitable eye and ear protective devices, in addition to the safety helmet.
(c) The tools should be loaded only at the point where it is to be fired. It is not to be loaded on ground and lifted up or hauled up. Also loaded tool is not to be left unattended.
(d) A warning notice to be displayed at the place where the tool is being used to warn other workmen and visitors that such tool is in use.
(e) The gun to be retained in position for at least 30 seconds if the cartridge fails to detonate.
(f) The tool not to be fired intentionally or for test purposes in such a manner as to cause a projectile to fly free.
(g) The tools are not to be used in the area where there is a risk of fire or explosion from sparks or the discharge action of the projectile.
(h) All precautions and safety instructions given by OEM of the tools should be followed strictly.
(i) All safety requirements of local authorities and standard organizations applicable to these tools to be followed.
(j) The tool not to be used for the purpose other than that for it was designed.

1.5 Welding Hazards

In a large radio station construction project, there is huge amount of welding and brazing work for earth radials, RF shielding of feeder hut and framework for support structures. A survey has revealed that 2% of the injury on radio stations is due to welding and brazing works. So it is essential that the workmen (even contractors) are familiar with the potential hazards of the various processes of brazing and welding, and the means to controlling these. Aspects of welding work requiring particular attention are;

(a) Compressed Gases

All cylinders should be properly marked and color coded. The primary hazard of compressed gas cylinders is the possibility of sudden release of the gas by removal, or breaking off of the valves due to sudden fall of cylinder from elevated work platform. This action is responsible for explosion also. Oxygen is one of the popularly used gases in welding work. This is perhaps the most hazardous due to its ability to accelerate the combustion of normally combustible materials, and it is especially important that oxygen is handled with extreme care. The pressure of oxygen in a full cylinder is approx. 280 kPa. Oxygen in no case should be substituted for 'air'. It should not be used in any operation or equipment requiring compressed air. Pure oxygen or oxygen enriched air will accelerate burning.

Acetylene cylinder's safety and capacity is obtained by packing the cylinder with a porous material, the fine pores being filled with acetone, a liquid chemical having the property of dissolving or absorbing many times of its own volume of acetylene. In such cylinders, acetylene is perfectly safe and will not change its nature. However the cylinder should not be subjected to rough handling, dropping or knocking. The fusible safety plug provided to an acetylene cylinder acts as safety release in case the cylinder is exposed to excessive temperature. The plug will melt at about the temperature of boiling water and release acetylene from the cylinder.

Another gas which is used extensively on Radio construction project works for external use in the field for brazing work of feeder hut shielding, earth radials grid and stakes jointing is liquefied petroleum gas (LPG). It is petroleum by-product and is sold by various trade names. Liquefied petroleum gases liquefy at relatively low pressure and normal temperatures and can be stored in steel cylinders which are easily handled and transported.

The gas is non-corrosive, non-toxic, normally odorless, but to comply with some govt regulations, an added odorant gives it a distinct smell to assist in leakage detection.

There are special precautions which should always be taken with compressed gas cylinders. They should be properly stored in an upright position in approved safe places. They should be kept away from hot places and away from direct sun light, particularly in the locations where high temperature prevails. Oxygen cylinders should be stored well away from the cylinders containing combustible gases.

(b) Electric Shock Hazards

Any source of electricity is a shock hazard and it is important that all electrical machines including welding machine be periodically inspected and properly maintained by qualified electrical staff. Mobile generator welders provided on large installations requires due care by the operators when working in the wet or damp locations in the field. Moisture reduces the electrical resistance, and increases the likely hood of electric shock.

(c) Electric Arc Radiation

The rays emitted by an electric arc in welding process contain ultra-violet rays which can be painful to eyes. This discomfort is caused by the extreme intensity of the rays, and the fact that the eyes cannot withstand exposure to them without having its delicate nerves and membranes injured. Only glasses or helmets specially designed and approved for the purpose should be used during the welding operation. The filter glass has to perform two functions; it must render the operator immune from the effect of ultra-violet rays and it must reduce the glare from the arc and make the work distinguishable. Smoked or tinted glass should not be used as a substitute as they do not give sufficient absorption of the rays. It is also important to wear appropriate clothing by operator to prevent arc rays from contacting the skin as it can result into 'burn'.

(d) Air Contamination

Welding, and other many processes, contaminates the air. It is important to recognize the different type of contamination produced by the welding and brazing processes, and to take necessary steps to reduce their potential hazards. The concentration of toxic gases and dust should be measured and threshold values for different materials should be standardized. These gases and dust may come from the base metal being welded or, from electrode or from shielding gases. Many of the base metal used in radio engineering works produces toxic gases. These include, lead, zinc, cadmium, beryllium, and many others. Adequate ventilation and wearing of gas mask by the operator is essential for such works.

(e) Fire and Explosion
Three elements of fire triangle, fuel, heat, and oxygen, are normally present during welding operations. The level of each of these elements should be kept to a minimum and under control to avoid fires and explosions. The heat involved comes from the torch of the arc or from the base metal being welded. Sparks which fly from welding or cutting work are real little balls of metal oxide. They can travel a long distance from the work spot and may stay hot for a considerable time after landing. Fuel for fire may come from the gas being used or from combustibles in the welding area. Oxygen always present in the air and may be enriched by pure oxygen available from the welding apparatus. These should be kept under extreme care and strict control.

1.6 Blasting Operations

Blasting operation in connection with the excavation of foundations works associated with masts, towers, transmission lines and others facilities, are carried out in almost every transmitter project, in the hard rock area where other mechanical methods are impracticable or uneconomical. The object of the blasting is simply to shatter the rock or loosen the ground, thereby permitting the easy removal of the earth by mechanical means. On mountain tops, where TV or link towers are required to be located, foundations in rocky terrains is done by careful blasting by ensuring that that wall close to cliff faces are not fractured.

The use of explosive is also sometime necessary to remove old foundations of dismantled masts and towers, particularly the anchor blocks which remains protrude above ground and are potential danger to workers and work hazards. Also it may be necessary to remove new foundation where the strength of the concrete fails to meet the specified strength. Experiences have shown that holes should be spaced 45 to 60 cm apart with equal distance kept free space. The depth can be 15 cm to 30 cm depending upon the hardness and mass of the rock or foundation block. The charge per holes should be small and be distributed along the drill hole. A charge of 60 gm/m^3 is generally sufficient to break a foundation block into suitable pieces for removal.

Extreme care must always be exercised in the use of explosives because of the serious nature of the injuries. Some causes of the accidents have been attributed to followings;

(a) Smoking or use of naked light when handling explosives.
(b) Premature firing of the electric detonators from induced earth currents and electromagnetic picked up currents from radio transmitters.
(c) Returning to area too soon after firing.
(d) The absence of, or improper use of, blasting mats.
(e) Improper use of tamping rod when charging holes.
(f) Use of too short safety fuses length.

The use, storage, carriage, and destruction of explosives are generally covered under the regulations of local bodies, and in all cases the requirements of these should be strictly observed.

The following safety rules are recommended for use of explosives;

(a) No workman without current valid shot firer certificate or other qualification prescribed by local authority, or the organization, should handle uncased explosives, prepare a charge, charge a hole, fire a charge or investigate a misfire. A person registered as a shot firer should undergo a practical test fire every two years.
(b) Before the engineer or site supervisor of the work permits a workman to handle explosives, he must check the certificate of shot-firer and confirm that it is current, and satisfy himself that the shot-firer is registered for, and has sufficient experience, to carry out the particular work.
(c) The shot-firer should have full control of all blasting operations, and all persons in the area must obey his directions during the period in which charges are being prepared and fired, and until all, 'clear signal' is given.
(d) On completion of the work the engineer or supervisor should check the details recorded in the log-of-work book, which the shot-firer must keep, and sign in the appropriate column.

(e) Explosives must not be used in the vicinity of the station power plant fuel tanks or other restricted areas where underground construction exists and could be damaged.
(f) A vehicle with electric detonators must not be taken inside a premise having operational radio transmitters.
(g) When blasting close to a building, for example for a TV or microwave tower after completion of building, the building must be inspected before and after to ascertain the extent of damage if any.
(h) Blasting mats must be placed over the hole wherever there is a possibility of damage from flying debris to building, plant or personal.
(i) Explosives should be carried to site in unopened cases or in standard explosive boxes.
(j) Explosives should be taken to site only when everything is ready. Similarly, before firing operation, all unused explosive materials like, gelatin and detonators should be removed to a safe distance.
(k) Detonators with lead-in transit may pick up energy from nearby radio or radar transmitters, so it should be carried in closed metal container with no apertures and low resistance contacts between lid & the container.
(l) The vehicle used for carrying the explosives must comply with the requirements of all laws and instructions applicable as per local authorities.
(m) Explosives for destruction must be destroyed only by a qualified person as per approved guidelines.
(n) EM field from a high power (200 to 500 KW) MF radio transmitter can induce sufficient current in the wires of electric detonators to cause it fire. BSI in its standard BS 4992:1974, suggest a safe distance of 500 to 1000 mtr for using the blasting process in the vicinity of high power MF or HF radio transmitters.

1.7 Electro-explosive Devices

EM radiation is also considered to be a hazard under some conditions where particular items of military stores and equipments are involved. The use of electrically initiated explosive devices for booster rocket igniters & warhead detonators, and for reliable high speed operation of switches & valves has greatly increased in recent years. Some modern weapons contain more than 75 electro-explosive devices. Several instances involving ordinance have been attributed to initiation of their electro-explosive devices by EM radiation from nearby radio transmitters. Each incident occurred during operation while the ordinance item was being handled normally.

The most problematic part to determine the ordinance system's susceptibility to EM radiation is the evaluation of methods of pick-up by various electro-explosive devices used in the weapon system. The pick-up can be on antenna like couplings or probes, or by conduction into the weapon via firing leads etc, entering the weapon enclosure. The precise probabilities of actuation of electro-explosive devices are very difficult to predict because of so many variables involved. These includes field strength & frequency of the radiation in the vicinity of weapon, as well the geometric orientation with relation to the radiated field, the extent of metallic contact of the weapon with other bodies and the environment.

The most likely effects of premature actuation are duding, reduced reliability of the system or ignition of propellant. The probabilities of warhead detonation, although low, nevertheless exist.

1.8 Warning Signage

When construction is going on, suitable signs to provide an effective means for directing the attention of the staff to dangerous conditions or to places where caution should be exercised, and for indicating the location of safety aids and fire fighting equipments should be placed at appropriate locations throughout the installation. Signs must be legible, clear, and unambiguous, in the appropriate simple language with pictorial symbols and of the approved color recognized by standard bodies.

The selection of the location for sign is very important and it should be clearly visible to all concerned in a normal way. All dangerous places or processes temporary or permanent, where men are at work overhead, places where cautions should be exercised, locations of emergency exits, fire extinguishing equipments, first aids and safety equipments, should be provided with warning / danger signs. Appropriate international electrical danger symbol should be placed at all switchgears, interlocked panels, transformer enclosures, and in diesel generator room. 'Danger' and 'caution' signs should be placed sufficiently ahead of a particular hazard to allow a person ample time after seeing the sign to heed the warning. This is particularly necessary in case of a low level transmission lines, crossing pathways or approach roads by vehicles. Signs should not be placed on movable objects such as the door of the feeder antenna matching huts or transformer enclosures and other locations where a change in position would make the sign invisible and void the purpose of the sign.

Radio and electric equipments carrying high voltage, and mechanical plants, like generator or switchgear undergoing test or modification, should be temporarily removed from the system chain, area well illuminated and danger sign prominently displayed to warn of the staff.

Permanent 'caution' signs must be placed on filter capacitor banks of HT DC power supplies. Spring loaded shorting bar and fuses should be used on smoothing condensers of power supplies. The capacitor bank in a typical 250 KW transmitter supply has a large capacity of 80 μF for the requirement to provide low impedance path at audio frequencies, so strict adherence to safety procedure is essential in handling these capacitors.

The installation should have sufficient 'warning and caution' in the HT room of the transmitter and in floor mounted component type of matching and power division huts. There are instances where workmen have instantly died in the absence of these precautions and safety measures.

Special attention should be given to the hazards arising during power supply failures when workplace and building turns into complete darkness. So essential or emergency lighting system at all vintage and workplaces of sufficient intensity from battery backup should be provided to avoid hazards of accidents, safe exits, and to enable staff to switch on the essential supply generator, if it is not in auto mode or even otherwise as a generator even of small capacity requires few minutes to start.

There are four main type of warning signs as given below, even though there can be many more signs;

(a) Danger sign to warn of an immediate hazard.
(b) Warning and caution sign to warn against unsafe practices.
(c) Safety instruction signs, to indicate location of First aid and safety equipments.
(d) Direction signs, to indicate emergency exits, direction and location of safety services and equipments.

To ensure warning signs are followed and respected, temporary warning signs, installed during project work must be removed as soon as the work is completed.

1.9 Failure of Structures during erection

The erection of broadcast towers and masts requires skill of the highest order, experience, and nerves as well as courage for heights. While designing and fabricating these, the special requirements and work hazards of the erection crew must always be kept in mind. The need to work at great heights, under unfavorable weather conditions and with limited tools and devices, impose limitations on the configuration, size and weight of individual sections and components which have to be hoisted into the position to form the complete structure.

Erection of guyed masts, calls for special skill and methodical working procedure to avoid instability which may be introduced during temporary guying for attachment or replacement of permanent guys. Many collapses have taken place during this critical phase of erection. In case of MF radiators, base and sectionalizing insulators are particularly vulnerable to damage and special care is essential in handling to ensure that unnecessary load and stresses are not applied during the erection process.

During the erection of a guyed mast, temporary guys must be attached to keep the structure in plumb and maintain stability. The levels at which temporary guys are attached must be carefully calculated and followed to ensure that excessive bending or deflection of the structure does not occur. The anchor blocks should have provision for attachment of these temporary guys in the design and working drawings. As each permanent guy level is reached, the permanent guys should be attached and temporary guys removed for further use.

The fabrication, installation and tensioning of the permanent guys requires a well defined, established and approved procedure. The wire rope used for fabrication may be of several approved type but bridge strand type is preferred by design engineers. Individual strands are generally specified as being the largest size consistent with reasonable handling qualities of the rope, thereby providing the maximum safe guard against damage to, or corrosion of, individual strands. The wire is, as a rule, galvanized and treated with an appropriate preserver compound during manufacture. For large guy wires, terminations are of socketed type with socketing operation being performed carefully to avoid disturbance to the layer of the strands which may permit the entry of water to the core. As the special skill and the control is required in this operation, the work is considered a workshop function and should be done at workshop only and not in the field. As a safety measure the tucked guy splices at guy terminations are not permitted on most installations.

When construction has reached a permanent guy level, guys are hoisted in turn, fixed to mast attachment fittings and left dangling, but clear of the structure. Extreme care should be taken during lying out and handling to ensure that they are not dragged over rough ground or allowed to hit against mast structure. The lower ends should be pulled out together manually towards their respective anchorage points as far as possible. Further pulling and fixing to anchor block should be done with the aid of winches. The winches should be seated behind each anchor block and the rate of pull of all guys kept uniform to avoid unbalance loading on the structure. The guys and insulators of a tall mast structure are very heavy and require careful handling. A guy for a typical 300 m mast shall weigh up to 10 tonnes and is more than 300 mtr in length. Often additional lifting gear is necessary to hoist the guy to its attachment point on mast. Multi-drum winches may be required for simultaneous handling of guys.

After connection of the guys to the anchor block fittings, the tensioning device is fitted and by manipulations as directed in the designer's instruction and drawings, initial tensions are applied to all guys simultaneously. This work is of very critical nature and operation as many structures have fallen due to lack of proper control and cooperation during this stage.

Throughout the guy tensioning operation, and in all phases of mast erection, mast should be kept under observation continuously for verticality from two Theodolite positions viewing in planes at right angles. Final guy tensioning and vertical alignment of the structure is carried out in still air conditions. It is a well established and known fact that, a well designed and manufactured mast and erection work adequately supervised and executed with good engineering practices, damage and danger to the structure will seldom occur, except if the structure is bent due to uneven guy tensions at one level. This can be avoided by continually checking the mast for straightness as well as twist.

Structures have failed dramatically and sometimes embarrassingly during erection due to careless and other unprofessional methods of working. Had the working been careful and professional, majority of these failures should not have taken place. There are only few examples where it can be said that failure of the structure was due to sheer misfortune and no blame could be attributed to design, fabrication or erection work. A study of many structural failures has shown that the majority of these were caused by inherent defects, or inadequate control or supervision of the constructional activities, rather than from the action of some natural forces of unforeseen magnitude or other external factor.

Quite interesting and surprising to note that a large number of total or partial collapses have taken place during erection or adjustment on the structures. It is impossible to eliminate the mistakes altogether in the type of structural work encountered in broadcast engineering due to large content of human involvement in the work. But many mistakes could be eliminated, or at least reduced, by close liaison between the principal, the structural designer, and the erection contractor, and by having adequate and proper checking and the supervision of the work. In large numbers of cases it was observed that if there would have been better supervision of erection work and more coordination, then collapse could have been avoided. Occupational safety and Health (OSH) division of Industry labor department normally recommend instructions for the working conditions on such structures. Followings are the recommendations from Occupational safety point of view to reduce the likelihood of repetition of collapse of structures and loss to the life of workers;

(a) During the design phase of the project, careful considerations should be given to the erection stresses and critical areas should be mentioned and documented by the designer to be followed during working at site.

(b) Detailed documented erection procedure and rigging specifications should be agreed upon between the fabricator and erector. There should be no deviation in this until agreed by both parties in writing. Structure should conform to uniform Building code to the extent feasible for safe erection and other practices during construction.

(c) A third neutral party or at least the quality control department should review the design, erection procedure and the rigging specifications prior to the start of the project. Any change prior to incorporation also should be reviewed by this neutral party.

(d) A qualified engineer should be deputed on job site at all the time with the authority to enforce the agreed – upon erection procedures or to stop the work if there are any un—approved /authorised deviations from the agreed procedure, or any variation from the specifications, or if any welding or material defects are noted during the erection of the work at site.

(e) Fabrication of the structural components of the tower members should be done by certified welders under proper controlled process and quality control to ensure the integrity of the end product.

(f) The material must be clearly defined in the specifications and drawings including its fabrication procedures.

(g) Plans and specifications should be signed and sealed by a qualified design engineer.

(h) The plans, specifications and construction procedures should be in conformity to the guidelines of the appropriate local authority on these matters.

(i) In the event of catastrophic accident resulting into loss of life or serious injuries, an enquiry committee comprising of professional engineers, competent architects, and person experienced in the erection trade, including the labor and Industry department should be constituted to investigate the matter for ascertaining the cause and lapses in the erection work.

There are umpteen examples of failures of radio, television masts / towers during erection process due to lapses on the part of erection crew as well due to natural phenomena or due to defective fabrication or improper materials almost in every broadcasting organization or other departments using radio services like spectrum monitoring, navy and coast guards. During operation of the self radiating towers in radio stations maximum numbers of masts have failed or fallen along the coastal areas due to cyclonic winds due to rusting of guy wires or fastening hardwires.

So the construction or erection hazards are very important matter in installation, testing, commissioning and even afterward in operating and maintenance of a radio station and all good engineering practices, codes, procedures, guidelines and regulations must be followed with due respect to avoid the accident and loss to man and machine.

1.10 Hazardous Installations

An installation is defined as major hazardous installation that has VLF/LF/MF/HF/VHF/UHF/SHF/ Radio or Television or microwave frequency Radar stations radiating high intensity RF field, which exceeds prescribed limits, or produces, processes, handles, uses, disposes of, or stores either permanently or temporarily one or more hazardous substances. The criteria for defining the hazardous installations are;

(a) Storage, handling, using, manufacturing, or adapting of over 50 tonnes of Petrol or petroleum products.
(b) Storage, handling, using, manufacturing, or adapting of over 10 tonnes of asbestos or asbestos product.
(c) Storage, handling, manufacturing, or adapting of over 10 tonnes of mutagenic or teratogenic chemicals.
(d) Highly mechanized process utilizing over 2000 watts of power supplied by diesel or public generation process.
(e) Manufacture, generation and transmitting of over 2000 watts of RF power in RF range.
(f) Manufacture, storage, adopting or utilizing of any chemical recorded on the hazardous chemical list.

The ministry of labor and industry must specify the hazardous installations by adopting a policy for major hazardous installations by preparing an Occupational Safety and Health act.

1.11 Construction Hazards

There are many construction hazards involved in a construction project but major hazards of constructions are; (a) Falls, (b) Electrocution, (c) Being struck by falling objects, (d) Trapped during normal or excavation of treches. In this section "Electrocution" will not be takenup as it will be explained in a separate chapter under,"Electrical shock hazards". The material discussed in this section, was produced under grant number SH-22297-11 from OSHA. It does not necessarily reflect the views or policies of the U.S. Department of Labor, nor does mention of trade names, commercial products, or organizations imply endorsement by the U.S. Government.

(a) Falls

This is one of the most prominent hazards as fall is the most occuring mishaps on a construction site. So fall protection is needed for Walkways & ramps, Open sides & edges, Holes, Concrete forms & rebars, in Excavations, Roofs, Wall openings, Bricklaying, and Residential Construction.

Fall protection options; Falls are the leading cause of fatalities in the construction industry. When performing work that requires the use of scaffolds, cranes, ladders, scissor lifts etc, the implementation of fall protection is required as per buillding construction safety standard.

Fall prevention options; As shown in fig 1.1, Safety Nets, Hand Rails, Safety Harness (Personal fall Arrest Sytems – PFAS), and Guardrails are some of the usually used fall protection and prevention methods in the construction works. All safety equipment must be in place before work starts. Employer must provide safety equipment and periodic inspections should be conducted by competent person before work stars.

Conditions requiring use of fall protection; (a) A fall from as little as 4–6 feet as it causes death in some cases but loss of work always, (b) If an employee can fall 6 ft or more feet, fall protection measures are to be implemented by the employer or the competent person of project site.

Safety Nets; This is used to prevent workers from falling. The safety nets; Must be strong enough to support a falling employee; Must have sufficiently small mesh openings so the employee cannot fall through the net; Must be close enough to the surface of the walking/working surface so that the fall into the safety net will not still injure the employee (never more than 30 feet below the walking/working level); Must be close enough to the edge of the working surface (the outer edge of the net between 8–13 feet from the edge of the walking/working surface, depending on the distance to the walking/working surface) so that the falling employee will not slip past the net.

Handrails; These are used to asist workers for access to platforms and to prevent from falling down or sliipng while going up or coming down. The protective barriers must be strong enough to support and prevent a worker from falling. Wood, iron angles or round bars may be used for handrails. Handrails should be at least 750 mm high.

Personal Fall Arrest System (PFAS); Safety Harness or PFAS is a body harness for each worker that is fastened to a secure the anchorage so that worker does not fall (refer:1926.502(d) of OSH). PFAS should not be used until inspection by a competent personnel (foreman) is completed. Do not use it if any of the harness parts is defective, frayed or broken and if PFAS does not fit properly or it cannot be adjusted. Safety line must be able to support 5000 lbs weight (total) on fall.

Fig 1.1: Safety Nets, Hand Rails, Safety Harness (Personal fall Arrest Sytems-PFAS), and Guardrails

Guardrails; These are used to prevent workers from falling. The protective barriers must be strong enough to support an prevent an employee from falling. Wood, chain and wire rope may be used for top rails and mid-rails. If wire rope is used, slack must not exceed 3 inches from edge of work area. Height of top rail should be between 39 to 45 inches. Toe board i.e bottom rail 3 incheshigh, and inbetween shold be the middle rail..

(b) Falling Objects
All workers or superviser or visitors for their safety from fallinng objects in a construction site are always required to wear Hardhats. Other means of safety from falling objects use canopies and Barricade the area to prevent unauthorized entry.

(c) Trenching and Excavtion Hazards

Trenching and excavation is most hazardous construction operation and cave-ins are the greatest risk. Most accidents in Trenching and excavation occurres at a depth of 5–15 ft (Reference 29 CFR 1926.652 of OSH). Other excavation hazards include water accumulation, oxygen deficiency, toxic fumes, falls, and mobile equipment.

Note: 1. Trenches more than 5 feet require shoring or must have a stabilized slope, 2. Trenches less than 5 feet – a competent person must inspect to determine that a protection system is not necessary in soils where there is no indication of a potential cave-in.

Protection; Workers should be protected from caves-in by using a well designed protective system. Such protective systems must be able to support expected loads from cave-ins to protect the trapped workers. Protect workers from potential cave-ins by using, Slope or bench sides of excavation by placeing shields between the side of the excavation and work area.

Risk involved; Usually there is no warning before a cave-in of Trenching and excavation and following risk are involved in case cave-ins; Asphyxiation due to lack of oxygen; Being crushed by weight of dirt; Inhalation of toxic materials; Fire; Moving machinery near the edge of the excavation can cause a collapse, and the Accidental severing of underground utility lines

Protective system; A protective system is a method of protecting workers from cave-ins or from material that could fall or roll from an excavation face or into an excavation, or from the collapse of adjacent structures. Protective systems include support systems, sloping and benching systems, shield systems, hydraulic jacks and other systems that provide the necessary protection. In case of hydraulic jacks trench pins should be installed to take care of hydraulic failure. Some of the protection system for trenching and excavation are shown in fig1.2. Protective systems shall have the capacity to resist without failure all loads that are intended or could reasonably be expected to be applied or transmitted to the system.

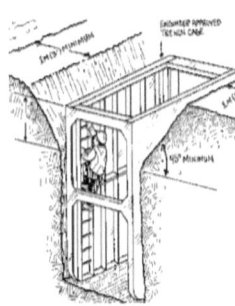  

Fig 1.2: Protection system for trenching and excavation, benching system, shilding, hydraulic jacks

Benching – excavating the sides of an excavation to form one or a series of horizontal levels or steps, usually with vertical or near-vertical surfaces between levels.

Shoring or shielding – is used when the location or depth of the cut makes sloping back to the maximum allowable slope impractical. There are two basic types of shoring, timber and aluminum hydraulic.

Trench boxes (shielding) – are different from shoring because instead of supporting the trench face, they are mostly serve to protect workers from cave-ins. The excavated area between the outside of the trench box and the face of the trench should be as small as possible. The space between the trench box and the excavation side may be backfilled (or other means may be used) to prevent lateral movement of the box. Shields may not be subjected to loads exceeding those which the system was designed to withstand. Trench boxes may be used in combination with sloping and benching.

Factors that pose hazards; Main hazards factors in trenching cave-ins are; Soil classification; Depth of cut; Water content of soil; Changes due to weather and climate; Other operations in the vicinity.

Competent person; Each Trenching and excavation construction job must have a suprvisor as competent person who had undergone specific training in and is knowledgeable about: Soils classification, use of protective systems, the requirements of the standard, and must be capable of identifying hazards, and authorized to immediately eliminate hazards. The competent person must make daily inspections of excavations, areas around them and protective systems: Before work starts and as needed, after rainstorms, high winds or other occurrence which may increase hazards, and when he can reasonably anticipate a worker is exposed to hazards.

Egress System; A stairway, ladder, or ramp must be present in excavations that are 4 or more feet deep, and within 25 feet of the employees and these must extend 3 ft above the excavation.

References for further; 29 CFR 1926: Safety and Health Regulations for construction; 29 CFR 1926. Subpart E-Personal Protective Equipment; 29 CFR 1926 Subpart L – Scaffold; 29 CFR 1926 Subpart M – Fall Protection; 29 CFR 1926 Subpart P – Excavations; 29 CFR 1926 Subpart T-Demolition.

1.12 General requirements of safety Policy in hazardous installations

Every installation having hazardous working conditions must have safety policy and other general guidelines for workers and its staff during construction and afterward for operation and maintenance as described below:

(a) Written safety Policy
A safety policy stating the intention of the establishment to uphold the safety standards and showing the commitment of the management to this policy shall be developed & signed by the management and displayed in a suitable location so that all staff members are aware of management's commitment to safety.

(b) Hazard Map
A map of the premises/ installation indicating the location of all the hazards and displayed at most prominent location for the guidance of all staff, is to be provided using following codes; (a) A Fire /explosion hazards — **red**, (b) Chemical hazards — **green**, (c) Biological hazards — **yellow**, (d) Radiation hazards — **blue**, and (e) Physical hazards — **black**

A qualitative analysis and description of the hazard should also be provided.

(c) Process Map

A line diagram showing the stages of process with appropriate labels and notes attached to outline the general processing stages of the working of the installation. This diagram is not intended to contain trade secrets or information critical to security.

(d) Establish and maintain a functioning safety committee

Under the requirements of the ILO recommendations safety committees are required for establishments with 20 or more persons in employment and in establishments where the ministry of industry and labor so order.

Hazardous installations fall under section 30 of the proposed Occupational Safety and Health Law. Therefore this requirement is crucial and integral to any Safety and Health programme in hazardous installations. The committee so established is expected to comply with the requirements of the ILO recommendations. It will be the duty of the safety committee to ensure that regular inspections are carried out.

(e) Report of regular Inspections

Regular inspections of the establishment shall be carried out. A copy of the findings of these inspections should be kept available for the ministry's inspectors to scrutinize on request. Suitable qualified persons, such as safety wardens, safety committee members or safety officer capable of identifying and assessing hazards, must carry out these inspections. In the absence of the above-mentioned persons, only persons trained in hazard identification and assessment should be allowed to carry out the inspection. Reports should also include corrective actions to deal with the defects/breaches when identified.

(f) Workplace Information Systems

A viable information system readily accessible to workers must be developed and put in use. This system can be as simple as a chemical registry (a document where chemicals are listed and workers are able to get information on safe use and handling of the chemicals used in the work place) to a computer system listing all hazards showing where they are to be found and preventative methods for reducing the effects of the hazards.

The nature of the information system will be assessed according to needs and the nature of the activities of the establishment

(g) Safety & Health officer

A safety officer capable of advising the management of the requirements of law and safety procedures for safe operations of the establishment should be appointed. Such person should have necessary training and/or experience to develop & maintain a suitable safety programme for the installation. Given the nature of hazardous installation the experience and training shall be equivalent to at least that of an industrial safety inspector.

(h) Safety & health training for workers

Suitable safety training for workers should be designed & provided to acquaint all the workers with the requirements of safety. Such training should include; (a) emergency evacuation procedures, (b) safety requirements for tools & equipment and the safe handling of chemicals, electricity, or radiation from radio transmitters as required, (c) Permit to work systems, (d) Safe systems of work procedures, (e) Safe work practices, and (f) Isolation and lockout programmes

(i) Emergency evacuation

An emergency evacuation system shall be evolved and implemented. All workers should be well acquainted with this plan. The plan should be rehearsed at least twice every year. Keeping in view the propensity for disaster in these facilities this plan should be linked to an established liaison with the fire fighting and emergency services.

(j) Registration with Disaster Management Authorities

High hazard installations should be registered with the office of national disaster management authority stating the nature of the hazard so that a national strategy can be evolved and established for taking care of disaster as & when it strikes.

(k) Mitigation and containment strategy

A suitable emergency plan, outlining procedures in case of fatal accidents, RF burns, Electric shock or lightning stroke, fire, spill containment, and hazard management should be developed and documented. Such a document must be made available to inspectors on request. All workers must be familiar with these procedures and these should be implemented in every respect without any lapse.

(l) Signs and signals

Adequate and suitable signs, signals, charts, display boards or posters should be placed so as to provide sufficient warning for workers. Such signs or posters should be conspicuously displayed in area where they are readily seen, clearly visible with glowing paint even in dark and easily read by all the staff.

(m) Demarked areas

Where heavy equipment or mechanical transporters operate in areas and where workers must also operate, suitable markings denoting the areas for mechanical transport separate for humans should be clearly marked on the ground with glowing and easily visible paint. Where ever necessary suitable guardrails and handrails should be installed to separate the areas. They should indicate among other things, High Risk Areas, safety glasses area, hearing protection areas etc.

References

1. Installation and Construction Hazards (PP 464–498) –Handbook for Radio Engineering Managers –J.F. Ross
2. The Military Public Safety and Security Division of Florida State college – Institute of Occupational Safety and Health – Safety Training Presentations Construction Hazards

CHAPTER 2
RADIATION HAZARDS

2.1 General

The developments in the field of electronics and communications have lead to the widespread use of Radio Frequency (RF) devices in various areas, including telecommunication, radio and television broadcasting, radar, industrial processing, medical applications and consumer products. There are tremendous beneficial applications of Radio Frequency waves, but concerns also has arisen about the possible health effects associated with the exposure to RF electromagnetic and microwave fields (3KHz-300 GHz) emitted by various facilities and devices. In India situation, in recent years, has got doubly compounded with the expansion in the field of Mobile telephones and FM Broadcasting networks that have led to wide spread use of towers emitting RF radiations in the middle of household and commercial localities.

These have resulted in concerns being raised about the harmful effects of radiations from these towers on the public living in the vicinity of sources of radiation. The need for specifying reasonable safety standard and measures, as an integral part of the operational practice cannot, therefore, be over emphasized especially since the perception of risk, even in the absence of full facts about the risk is an important aspect of public life. In this particular case the risk perception gets accentuated with the general knowledge about destructive effects of X-rays and other ionizing radiations on biological tissues. This has resulted in a mistaken belief that the risk involved in exposures to non-ionizing RF fields might be similar. Even though there is no reliable scientific evidence to prove that continuous exposure to low intensity RF fields results in irreversible or severe damage to biological molecules and tissues, it is necessary that these limits are specified for allaying the fears and for ensuring that care is taken by the operators so that these limits are not exceeded.

Every organization which generates radio frequency (RF) devices and gadgets (3 KHz-300 GHz) is required to have full knowledge of its harmful levels of exposure on human beings and is to be mandated to take necessary care that it not only meets the requirements of safety of their staff but also that of general public. Agencies which generates RF fields are: radio and television broadcasting, point to point microwave radio (long-distance telephone and data transmission), mobile radio including cellular telephone, ship to shore radio, amateur radio, navigation (ship and aircraft), radar (military and civilian use for detection and guidance, flight surveillance around airports, weather surveillance and prediction, traffic speed control), home (cooking), industry (heating, sealing and draying), and medicine (diagnosis and treatment).

A number of studies have been conducted by various research and medical organizations worldwide to study the adverse health effects on human bodies due to the exposure to RF fields. Much of the literature, however, seems to be limited to experimental investigations with laboratory animals, tissue preparations or cells. Many of these studies have also been carried out with low fields and may, therefore, not be conclusive for actual conditions as also for their long term effects. Consequences of exposure to RF that have been reported include effects on behavior, the central nervous system, blood parameters, the immune response, the endocrine system, metabolism and thermoregulation, reproduction, the auditory system and the eyes.

2.2 Radiation Classifications

Modern broadcasting radio transmitters with large effective radiated power, radio communications, radar transmitters and dielectric heaters have raised the problem of personal safety of staff working on or near these installations. In the recent years awareness about the biological effects on the exposure of human beings to intense radio frequency radiations has opened new research areas for setting the limits of exposures. Studies have shown that exposure to high intensity radiations for even comparatively short periods may cause damage to human tissues and structure of the body and in particular to the eye where damage can be irreversible. Radiation hazards may exist at any radio frequency, which can be absorbed by a human body. RF Spectrum on this account can be divided into;

(a) Radio Frequency waves, (10^5 Hz to 10^{12} Hz), (b) Infra-red or heat waves, (10^{12} Hz to $10^{14.5}$ Hz),(c) Visible spectrum, ($10^{14.5}$ Hz to 10^{15} Hz), and (d) Ionizing Radiations, including ultra-violet and X-rays (10^{15} Hz to $10^{20.75}$Hz)

(1) Non-ionizing Radiations

Radio equipments which produces this type of radiation includes low frequency radio telegraphy, broadcasting, radio communication, television, and radar systems. Worldwide there is no, consensus as to what levels of non-ionizing radiations constitute a hazard to human beings and due to the complexity of the subject many are ignorant of its effects or many have unreasonable fear of radiation hazards where in fact no real hazards exits. It is not possible under normal living and working conditions to avoid exposures to RF radiation. Everybody living on the earth is exposed to it from outer space emissions and emissions from manmade sources such as broadcast, television, radar, industrial radio frequency devices like industrial electronic valves dielectric heaters, microwave ovens and mobile phone towers etc.

The degree of injury from EM radiation to living tissues depends on the energy absorbed within it which in turn depends on power density, duration of exposure, type of modulation, and frequency of source. Injury may be in the form of heating or bio-chemical effects. The heating effect of microwave is well known but the biochemical effect on human beings of microwaves where no local heating occurs is generally not clearly understood.

(2) Ionizing radiations

These radiations are produced by high voltage electronic vacuum tubes and devices operating in excess of 10 KV plate voltage and by radio-active material, whereas Non ionizing radiations are caused by; (a) Radio equipment which includes VLF Marine Navigational transmitters / equipment, RF Radio Telegraphy, Radio/TV Broadcasting, Cellular Phones, Radio Communication and Radar Transmitter,(b) Industrial equipment such as ion implant equipment, drying equipment, processing and cooking of Foods, Heat Sealers, Vinyl Welders, High Frequency Welder, Induction

Heating, Microwaves Ovens and Dielectric Heating, Sputtering and Glue Curing equipment, and (c) Visible spectrum and infrared (part of sunlight).

2.3 Harmful Biological Effects of Radiation

The harmful biological effects of RF radiations have been attributed to rise in body temperature or to selective rise in certain sensitive portions of the body. It have been found that staff working continuously in close proximity to unshielded high power transmitting equipments may sustain a body temperature slightly higher than normal when the field strength is very high.

A human body in EM field acts like a dielectric and currents will be induced in it from RF radiation, which in turn will produce heat. This heating is a function of (a) the field strength of the RF, i.e. average power flow per unit area, usually expressed in milli watt per square centimeter and (b) the duration of exposures to RF field.

The most vulnerable parts of the body are those which have restricted supply of the blood. Circulating blood acts as coolant to distribute the heat developed, where supply is abundant, the damage is prevented by not permitting the rise of temperature. The extent of damage due to the radiation depends on;

(a) The frequency of the radiation
(b) Effect of Peak Powers
(c) The intensity of the incident radiation
(d) Duration of the exposure,
(e) Absorption property of the body,
(f) The distance from the source of emission.

Some parts of the body are more prone to damage by excessive heat than others and these are the parts most apt to be damaged by exposure. Major effects of radiations are;

(a) Skin burns, cataracts in the eye and also affect implants, pacemaker and rods in bones,
(b) Non thermal biological effects such as change in reaction time and memory,

(c) Long term exposure of low level RFR affects nervous system and Immune system of small animals,
(d) Prolonged use of mobile phone can result in headaches, irritation and sleep disorders.
(e) Radio Frequency Radiation (RFR) acts as cancer promoter in animals. There is, however, not sufficient evidence that it causes cancer in humans.

(1) Effects of Frequency
Tests show that at frequency below 1GHz, body absorbs about 30–40% of incident energy. Most heating takes place below surface of the skin. At 300 MHz, the penetration is between 1 to 10 cms and radiation energy is transferred deep into tissues. At this depth of penetration heat may cause damage to the brain, nervous system, lever etc. In frequency range 1 to 3 GHz, absorption is very high and may reach 100% under some circumstances. Heat is generated in the skin, fat, and muscles and a person will generally feel the rise in temperature on the skin, and act quickly to move away from danger area.

At frequency above 3 GHz, about 40 to 50% of the incident RF energy is absorbed by the body. The remaining is rejected due to the reflective property of the skin at these higher frequencies. Most of the heating takes place near the surface of skin and as skin is very sensitive to the temperature change; a temperature change is quickly felt. Living tissues act like lossy dielectrics as they contain considerable amount of conductive water and as such get heated easily. The radiation hazards produced by super power broadcasting stations in VLF, MF bands have been a matter of discussion but not much info have been published on this. Most of the biological research has emphasized the deleterious effects in the range of 3 to 30 GHz where tissue heating has been found to be a hazard.

(2) Effect of Peak Power
Radiation from transmitters may be classified as continuous or pulsed. The continuous transmission are generally associated with broadcasting and communication type systems, whereas the pulsed transmissions are generally associated with radar and other distance measuring systems which emits high intensity pulses of short duration at given intervals usually referred to as pulse repetition rate.

As far heating is concerned there is no essential difference between the modes when compared to an average power basis. The average power of a pulsed radiation is derived from the duty cycle, the pulse duration, and the pulse repetition frequency. The average power of the system is therefore much less than the peak power (usually in the range of 500 kW), since the pulse duration of normal radar is only a few micro-seconds and the pulse repetition frequency usually in the range of 200–650 per seconds.

In most non-military situation the antenna is situated some distance away from people and is in continuous rotation, so there is no biological hazard. But in case of military such as on war ships, the personnel are in close proximity to high power installations and special precautions may be necessary to automatically cut off or reduce the power when beam sweeps across exposed deck area

(3) Intensity of incident Radiations

The thermal and biological effects of electromagnetic fields on the body vary depending upon intensity of the field, the energy absorbed by the body and the part of the body directly exposed to the field. Maximum permissible dosage for human beings is precisely perhaps not yet known because of inadequate available data. It is perhaps because of this fact that maximum permissible intensities of exposures to RF energy are assigned differently by various countries; however direct human experience from therapeutic use of electromagnetic fields indicates that localized levels of up to 20 W/kg for short periods can be accommodated by the thermoregulatory mechanisms of the human body.

(4) Duration of Exposure

Among all the parameters of RF radiations, the duration of exposure is the most important factor. Higher intensities can be tolerated for smaller durations. For continuous exposures, much smaller levels are recommended. More detail shall be taken up later in the chapter.

(5) Absorption property of the body

Skin of each individual due to its molecular structure has different reflective property, so incident RF energy is absorbed by the body accordingly. Most of the heating takes place near the surface of skin and as skin is very sensitive to the temperature change; a temperature change is quickly felt. Living tissues act like lossy dielectrics as they contain considerable amount of conductive water and as such get heated up easily.

(6) Distance of source of Radiation

Intensity of RF radiation decreases as the distance from the source increases. Consequently the hazards are also less severe. In certain cases, however due to directional characteristics of the radiating source (in broadcasting field system, employing high gain antenna arrays), the radiation field may be low or high in different directions in the vicinity.

2.4 Safe Exposure limits of RF Radiation

The potential hazards of RF radiations are difficult to estimate correctly. This is because of the complexity involved in the estimation. It is also rather difficult to understand the extent of hazard involved. The type and extent of effects depend not only on the strength of the field and the exposure duration but also on various other factors such as the frequency, type of modulation, polarization and distance from the source. It would therefore, be advisable while preparing the safety standard, that limits are set much below the threshold where potentially harmful effects start manifesting as per the studies made by the scientific community. The exposures may be intentional for operators or for evaluation purpose or unintentional i.e. public exposed to such radiation.

(a) Continuous Exposure

The potential hazard that may arise from exposure to RF radiation are believed to be due to either directly or indirectly, to increase in body temperature resulting from the absorption of the electromagnetic energy. The temperature of a normal human body is constant and even a temperature rise of 5 degree centigrade above normal can be injurious or lethal, if permitted to remain for long period. The circulating blood of a living being acts as an effective distributor of heat in the body somewhat analogous to circulation of water in a transmitting tube cooling system. It has been established that the average heat dissipation of the human body under normal conditions is about 5 mW / cm² over a body surface area of about 2 m². But due to the natural ability of the body to regulate the heat loss, the body can easily handle double the amount of heat dissipation i.e.10 m W / cm² averaged over 0.1 hour (6 minutes) duration.

Safe limit for RF exposure are based upon average level of exposure and fixed time period. Generally it is; (a) 6 minutes for occupational / controlled exposure limit, and (b) 30 minutes for uncontrolled / general public limit. Minor reversible effects due to RF radiation are not considered hazardous to man.

Different country have different limit for the RF exposure and; (a) countries like UK, USA and Germany permit power intensity level up to 10 mw /cm² (field strength of 194 v/m) for limited exposure, (b) but Countries like USSR, Czechoslovakia have limits of 0.1 mw /cm2 (20 v /m) in frequency band of 1.5 MHz – 30 MHz and 0.006 mw/cm2 (5 v/m) in frequency band 30–300 MHz, and (c) Sweden has adopted a limit of 5 mw / cm2 (137.3 v/m) for continuous exposure for frequency range 10–300 MHz. All India Radio (AIR) has adopted 10 mw /cm² (field strength of 194 v/m) for limited exposure as standard. Also AIR has norm for 50 v/m field strength contours as safe limit from radiation in MF/HF band for locating the transmitter and other habitable building from radiating antenna towers. Table 2.1 summarizes recommended levels of exposure in various countries. For other general purposes 10 mW /cm² (field strength of 194 v/m) is considered as hazard.

(b) Non-continuous Exposure Level

When exposure is non-continuous, an exposure level of greater than 10 m W / cm² (field strength of 194 v/m) can be tolerated in the short period, since the important consideration is the permissible temperature rise in the body. However, the maximum permissible intensity is not easy to determine as many factors are involved. In US the standards applicable for exposure to intermittent radiation sources require that the averaged power density measured over any interval of 6 minutes should not exceed 10 m W / cm². From biological heating viewpoint, the mean power is the significant factor, and not the peak power.

Table 2.1: Typical recommended level of exposure to RF radiation

Country	Frequency range	Maximum recommended level	Conditions of exposure
Great Britain (BPO)	30 MHz – 30 GHz	10 mW/cm²	Continuous 8hr, avg power density
USA Standard C95	10 MHz – 100 GHz	10 mW/cm²	0.1 hr period
USSR	1.5 – 30 MHz	20 V/m	
	30 – 300 MHz	5 V/m	
		(a) 25 µW/cm²	Per 8 hr/day
		(b) 100 µW/cm²	Per 2 hr/day
		(c) 1 mW/cm²	Per 15 – 20 min./day
Czechoslovakia	0.01 – 300 MHz	10 V/m	Per 8 hr/day
	300 MHz	25 µW/cm²	Per 8 hr/day
Poland	300 MHz	(a) 10 µW/cm²	8 hr exposure/day
		(b) 100 µW/cm²	Per 2–3 hr/day
		(c) 1 mW/cm²	Per 15 – 20 min./day
Sweden	10 MHz – 300 MHz	5 mW/cm²	Continuous exposure
	300 MHz – 300 GHz	1 mW/cm²	
Germany	30 MHz – 30 GHz	(a) 10 mW/cm²	1 hr continuous exposure
		(b) 1 mW/cm²	continuous exposure

2.5 Estimation for EM field Strength

Before locating / installing a new transmitter installation, radiation field expected in the vicinity should be estimated. Estimation should include both direct field and indirect reflected from building, ground, passive towers and other metal objects. The RF level to which staff working in the operational or near the antenna field is likely to be exposed also should be estimated. Area where RF field (Hotspots) is substantially higher than surrounding areas should be estimated. Objects that may cause RF hotspots are metal fencing, passive towers, metal water pipe line storage, metal sheds, air-conditioning ducts etc.

RF hotspots are caused by any of the 3 factors;

(a) Intersection of multi beams of RF energy, (b) Standing wave produced by reflective ground surface, and (c) Current induced in conducting objects exposed to ambient RF fields.

Table 2.2 below lists values of radiated and induction field at 100, 200, and 400 meters distances from the tower for 100 and 1000 kW MW transmitter power operating at 1 MHz frequency with various height of towers for estimation of total field to assess the hazard level;

Table 2.2: Radiation and induction field from a MF transmitter operating on 1 MHz at 100 and 1000 kW power

Tower Height	Field strength in volt/meter at 100 KW carrier power								
	100 mtr			200 mtr			400mtr		
	Rad.	Ind.	Total	Rad.	Ind.	Total	Rad.	Ind.	Total
0.2λ	29.2	14.05	32.4	14.6	3.5	15.0	7.3	0.8784	7.3
0.3λ	31.4	15.07	38.8	15.7	3.7	16.1	7.8	0.942	7.9
0.4λ	34.2	16.4	37.9	17.1	4.1	17.6	8.5	1.027	8.6
0.5λ	38.0	18.2	42.2	19.0	4.5	19.5	9.5	1.142	9.6
0.625λ	43.2	20.7	47.9	21.6	5.18	22.2	10.8	1.29	10.8
	Field strength in volt/meter at 1000 KW carrier power								
0.2λ	92.4	44.4	103.4	46.2	11.1	47.6	23.1	2.7	23.3
0.3λ	95.0	49.4	111.3	49.6	12.4	51.1	24.8	3.1	25.0
0.4λ	108.0	51.8	119.8	54.0	12.9	55.6	27.0	3.3	27.2
0.5λ	120.52	57.8	133.5	60.1	15.0	62.0	30.0	3.7	30.2
0.625λ	136.64	65.5	151.5	68.3	17.0	70.4	34.1	4.2	34.4

Note: (1) Values shown are for radiation & induction field (2) Resultant value of field shown in 3rd column has been arrived at by root means square of both fields for deciding the safe limit as the fields are in quadrature.

2.6 Evaluating Compliance with Guidelines for Human Exposure to RF in an AM Radio station

Many Organizations have evaluated data on biological effects and fixed threshold values averaged over whole body above which adverse health effects may occur in human beings. In cooperation with Environmental health Division of World Health Organization (WHO), International Radiation Protection Association (IRPA) has developed a number of health criteria documents on Non Ionizing Radiation (NIR) as part of WHO's Environmental Health Criteria Programme, sponsored by United Nations Environment Programme (UNEP).

At Eighth International Congress of IRPA (Montreal, 18–22 May 1992), a new International Commission on Non Ionizing Radiation Protection (ICNIRP) was established as a successor to IRPA/INIRC. The functions of this commission were to investigate the hazards that may be associated with the different forms of NIR, develop international guidelines on NIR exposure limits, and deal with all aspects of NIR protection. In pursuance of these functions, ICNIRP in 1998 came out with a revised document "Guidelines for Limiting Exposure to Time Varying Electric, Magnetic and Electromagnetic Fields (upto 300 GHz.). These guidelines define the level of exposure for two groups of people: Occupational Workers (intentional) and the General Public (unintentional).

In addition inputs have also been taken from the standards of some other organizations as listed below:-

(a) IEEE Standard for Safety Levels with respect to Human exposure to Radio Frequency Electromagnetic Fields, 3KHz to 300 KHz – IEEE std C95.1,1999 Edition (ANSI)/IEEEC95.1-1992-Revision of ANSIC95.1-1982.

(b) Health Canada's, limits to Human Exposure to Radio Frequency Electromagnetic Fields in the Frequency Range from 3 KHz to 300 GHz.

(c) The ANSI/IEEE RF Safety Standard and its Rationale by Om P. Gandhi and Gianluca Lazzi, Department of Electrical Engineering, Salt Lake City, Utah 84112.

(d) IEEE Committee on Main and Radiation Home's Technical Information Statement on: Human Exposure to Microwaves and other Radio Frequency Electromagnetic Fields.

(e) Electromagnetic Fields (EMF) Protection Documents (based on WHO requirements) issued by: United States of America, Russian Federation, and United Kingdom

(f) Asia–Pacific Broadcasting Union (ABU) Guidelines for Management of Radio Frequency EM Fields

International Electro technical Commission (IEC) has a standard no: IEC 60657(1979) on Non-ionizing radiation hazards in the frequency range from 10 MHz to 300GHz. Bureau of Indian standards (BIS) who has an understanding with IEC also has a standard based on this IEC as a mirror copy IS:3293. Similarly BIS has Doc. IS: 12130 (IEC 60244–6:1976)-Methods of measurement for radio transmitters Part 9 – Cabinet radiation at frequencies between 130 kHz and 1 GHz and above 1GHz doc IS:12131 (IEC 60244–7:1979).

The following paragraphs will describe the procedures for evaluations of RF exposure limits as per Federal Communication Commission (FCC)-USA and International Telecommunication Union (ITU).

2.6.1 As per FCC standards

In determining compliance with limits for Maximum Permissible Exposure (MPE) for AM radio broadcast stations, it is normally most important to determine electric and magnetic field strength at distances relatively close to transmitting antennas. Fields from the monopole antennas (self radiating vertical towers) decreases relatively rapidly with distance, and MPE limits for AM radio frequencies are not as restrictive as those for other frequencies, such as those used for FM radio. Therefore, even for the highest powered stations, MPE limits for AM radio transmitters would normally only be exceeded relatively close to antennas. Compliance with the guidelines for AM stations typically will involve assessment of exposure potential of persons working or occupying areas in the close-in vicinity of transmitting antennas. Because such persons will always be in the near field of AM antennas, due to the relatively long wavelengths in the AM frequency band, an evaluation of both electric and magnetic field strength is necessary.

In the original version of OET Bulletin 65, staff from the U.S. Environmental Protection Agency (EPA) provided the FCC with results from a computer-based model to help determine compliance with MPE limits for AM radio broadcast stations. The EPA model used the Numeric Electromagnetic Code (NEC) computer program to predict field strength levels near AM monopole antennas. In the past several years a PC-based version of this code MININEC has also become available. FCC has used MININEC to expand and refine the predictions for electric and magnetic field-strength presented in the original version of Bulletin 65. They are included in this supplement in the form of tables and figures that can be used in evaluating compliance at these stations.

Tables 2.3: may be used to determine the minimum distance from an AM broadcast antenna to the point where electric and magnetic field strengths are predicted to correspond to MPE limit values. The tables provide compliance distances from antennas of various electrical heights transmitting at various frequencies and using various power levels. The distances specified are the distances from an antenna at which access should be restricted in order to comply with both the electric and magnetic field-strength MPE limits. For antennas that do not correspond to the specific conditions given in these tables, interpolation can be used to arrive at intermediate values, or, alternatively, the greatest distance for the range used for interpolation could be used. Since the MPE limits for the two exposure tiers are similar for most AM frequencies, and because of variability in compliance distances according to electrical height and operating frequency, one entry is given in each case that applies for both occupational /controlled and general population/uncontrolled exposures. These numbers represent the minimum worst-case distances predicted for compliance with the strictest MPE limit for each case. Note that time-averaging considerations are not taken into account in these computations. Continuous exposure is assumed in all cases. This model computes field strength values in the vicinity of single antennas. For AM stations with multiple-tower arrays a conservative "worst case" prediction could be made by assuming that all transmitted power is radiated from each antenna. Therefore, in such cases the appropriate value from the tables could be used to define a zone of restriction around the array, consisting of circles with equal radii, each of which is centered on a tower in the array. Alternatively, a more accurate prediction could be made if the power actually radiated by each tower is known.

It may be necessary to predict electric and magnetic field-strength at various locations in the vicinity of AM antennas. Therefore, Fig 2.1 to 4 for various height antennas with 1 kW power has been developed for this purpose using MININEC. These figures show conservative predictions of electric and magnetic field strength versus distance from typical AM broadcast antennas for towers with electrical heights equal to 0.1, 0.25, 0.5, and 0.625 wavelengths, respectively. Fig 2.1 to 2.4 predicts field strength for stations transmitting with 1 kW of power. For stations operating at other power levels, values be obtained from these figures by multiplied it the square root of operating power.

The example below illustrates the proper procedure. In this example a 50 KW AM station is located near an area accessible to public. It is desired to obtain an estimate of the field-strength levels in this area which is at a distance of 10 mtr from the station's single tower that has an electrical height of 0.25 wavelengths.

To arrive at the estimated field strength values proceed as follows;

1) Refer Fig 2.2 for an antenna with electrical height 0.25,
2) At 10 mtr read predicted electrical field-strength about 8 V/m, and magnetic field-strength about 0.06 A /m,
3) Multiply each value by the under root of 50, which comes out to be $\sqrt{50}$ =7.075 and
4) So Predicted values are 56.6 V/m and 0.42 A/m respectively.

Table 2.3: Predicted Distance for compliance with FCC limits for various Power and height Antennas

	Predicted Distance for compliance with FCC limits (meters)															
Power of Tx	50 kW				10 kW				5 kW				1 kW			
	Antenna Height				Antenna Height				Antenna Height				Antenna Height			
Fre (kHz)	$.1\lambda$	$.25\lambda$	$.5\lambda$	$.625\lambda$	$.1\lambda$	$.25\lambda$	$.5\lambda$	$.625\lambda$	$.1\lambda$	$.25\lambda$	$.5\lambda$	$.625\lambda$	$.1\lambda$	$.25\lambda$	$.5\lambda$	$.625\lambda$
535–740	13	4	4	4	7	2	3	3	6	2	2	2	3	1	2	1
750–940	12	4	4	4	7	2	2	2	5	2	2	2	3	1	2	1
950–1140	11	4	4	4	6	2	2	2	5	2	2	2	3	1	1	1
1150–1340	10	4	4	4	6	2	2	2	5	2	2	2	3	1	2	1
1350–1540	10	4	4	4	6	2	2	2	5	2	2	2	3	1	2	1
1550–1705	10	5	4	4	6	2	3	2	5	2	2	2	3	1	1	2

As you know that RF currents will be induced in the body of persons who climb transmitting AM broadcast antennas for maintenance or other purposes. This is a significant source of RF exposure and can be related to the limits for specific absorption rate (SAR) adopted by the FCC. Although many stations may prefer to shutdown power entirely while persons are climbing their antennas, in some cases this may be difficult or undesirable. Studies have been undertaken by the FCC and the EPA to determine appropriate operating power levels which should allow climbing on transmitting AM antennas without exceeding the SAR guidelines. The results of these studies were used to develop Figure 2.5, which shows operating power levels versus frequency for a variety of different electrical heights that are predicted to allow tower climbing without exceeding the exposure guidelines in terms of SAR. Recommended power levels are shown for tower climbing with or without the use of gloves. A study by Tell performed for the FCC indicated that certain gloves (particularly leather gloves) can significantly reduce the induction of RF currents in tower climbers.

Fig 2.5 is designed to be used to provide guidance for use by AM radio stations which finds it necessary to continue transmitting while persons are climbing their towers. It can be used to determine the levels to which operating power should be reduced before a person climbs an active tower. However, there is variability in the data, and whenever there is a question about which condition may apply in a given situation it is recommended that the most conservative power level be used or, alternatively, that power be turned off completely while the climber is on the tower. Fig 2.5 is applicable to exposure of persons climbing a transmitting AM radio tower.

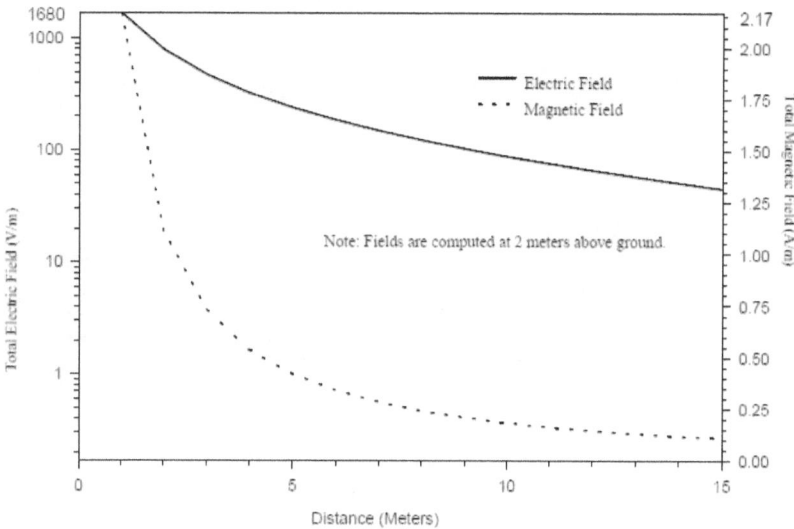

Fig 2.1: MININEC AM Model for 1 KW power, 0.1 wavelengths Tower

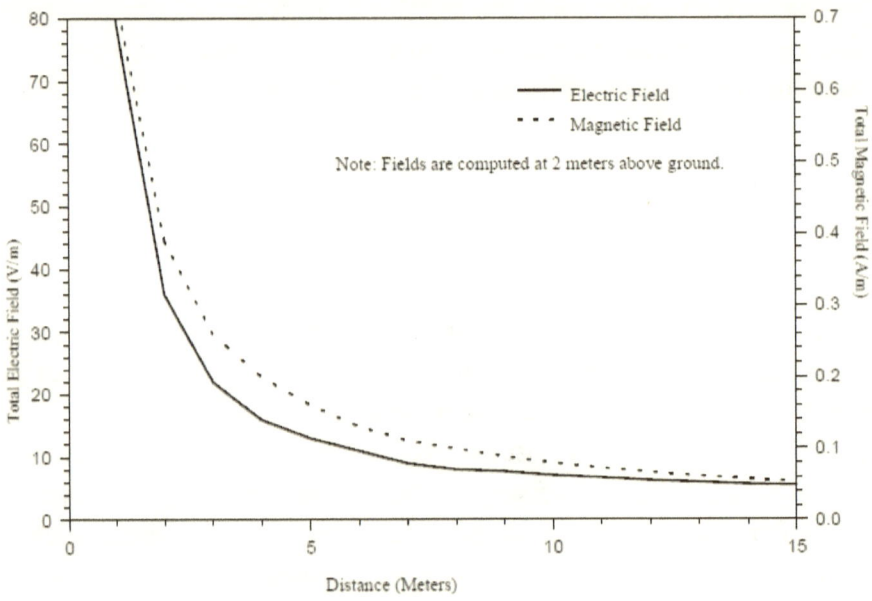

Fig 2.2: MININEC AM Model for 1 KW power, 0.25 wavelengths Tower

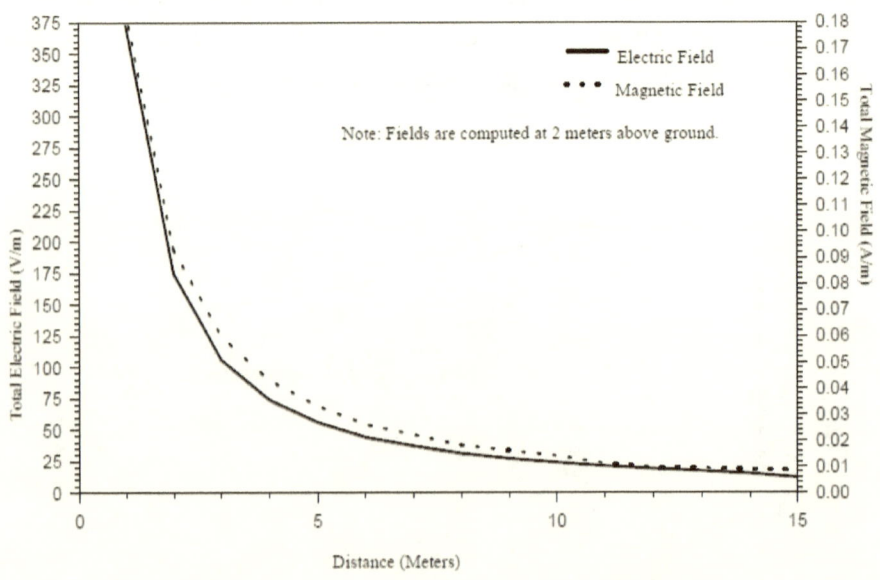

Fig 2.3: MININEC AM Model for 1 KW power, 0.5 wavelengths height Tower

Radiation Hazards 65

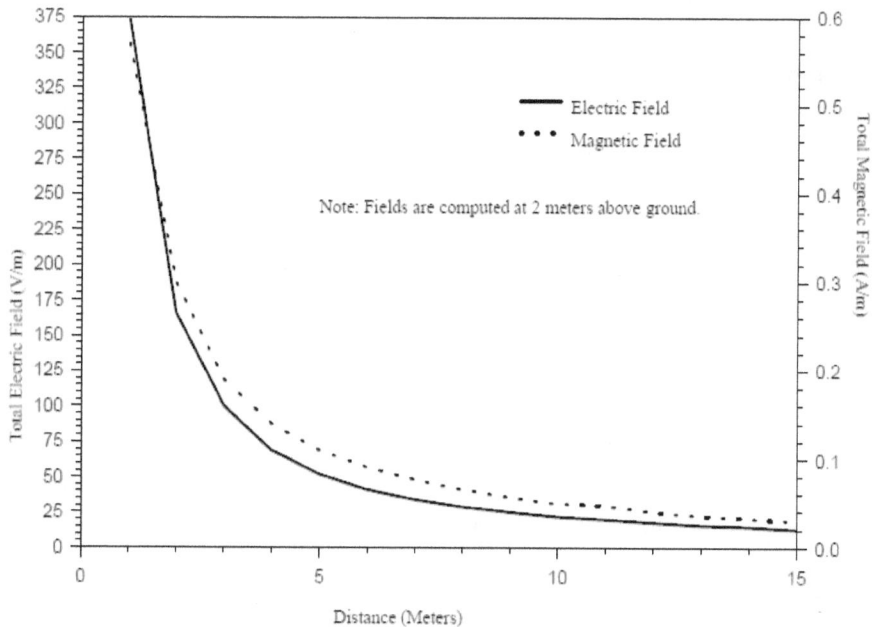

Fig 2.4: MININEC AM Model for 1 KW power, 0.625 wavelengths height Tower

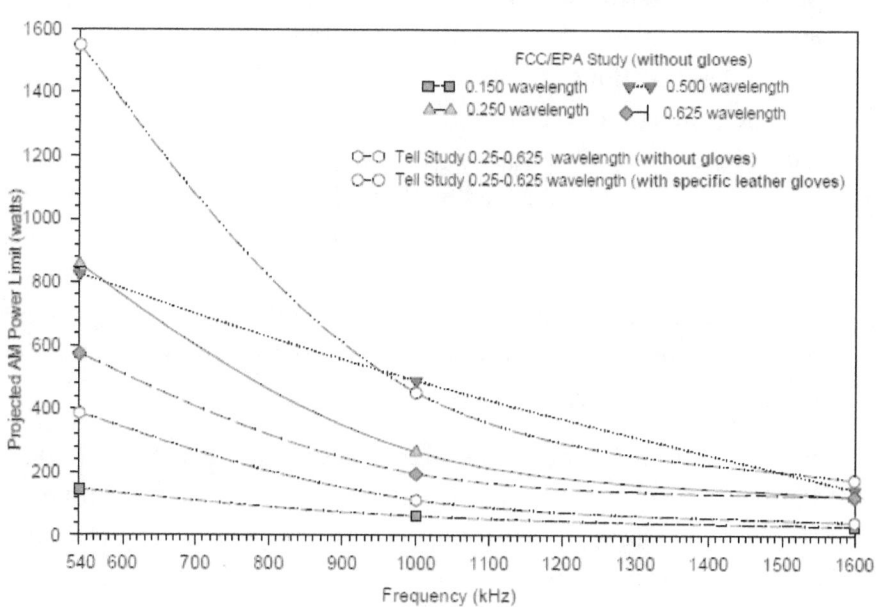

Fig 2.5: Estimated Power levels to comply with occupational / controlled limits for on-tower exposure of persons climbing AM broadcast towers

2.6.2 As per ITU Procedure

For many years the subject of the effects of electromagnetic radiation has been considered and attempts have been made to quantify particular limits that could be used to protect humans from undesirable effects. Studies in many countries by various organizations have resulted in various administrative regulations. It is noteworthy and understandable that no single standard has emerged from all the efforts in this regard.

The ITU Radio communication Assembly, considering;

a) That radio-frequency energy may have unsafe effects on the human body;
b) That radio-frequency energy may induce harmful electric potentials in conducting material;
c) That radio-frequency energy may have harmful effects on apparatus (such as Radio communication apparatus, navigation instruments, cardiac pacemakers, scientific or medical equipment, etc.);
d) That radio-frequency energy may lead to unintentional ignition of inflammable or explosive material;
e) That determination of hazardous radiation levels and electric potentials, in terms of spectrum content, intensity, cumulative effects, etc., are being made by competent authorities;
f) That identifications of areas where radio-frequency fields and electric potentials exceed safe levels are being made by competent authorities;
g) That persons not associated with such systems may be exposed inadvertently to such radiation (including travelers by air) or to such electric potentials;
h) That persons operating and maintaining radio transmitting systems may be required to work in close proximity to the source of such radio-frequency exposures;

Recommends, "That Annex 1 of ITU recommendation no; ITU-R-BS.1698 of 2005 should be used to evaluate the electromagnetic fields generated by terrestrial broadcasting transmitting systems operating in any frequency band, for assessing exposure to non-ionizing radiation".

This Recommendation is intended to provide a basis for the derivation and estimation of the values of electromagnetic radiation from a MW broadcast Radio station that occur at particular distances from transmitter site. Using such information, organizations can develop appropriate standards that may be used to protect humans or operators from undesirable exposure to harmful radiation. The actual values to be applied in any regulation will naturally depend on decisions reached by responsible health agencies, domestic and worldwide.

It is noted that this ITU-R Recommendation and ITU-T Recommendations cover similar material, but with an emphasis on different aspects of the same general subject. For example, ITU-T Recommendations K.52 (Guidance on complying with limits for human exposure to electromagnetic fields) and K.61 (Guidance to measurement and numerical prediction of electromagnetic fields for compliance with human limits for telecommunication installations) provide guidance on compliance with exposure limits for telecommunication systems. Some characteristics of electromagnetic fields shall be described in next paragraph.

2.7 General Field characteristics of electromagnetic fields

This section provides an overview of the special characteristics of EM fields that are relevant to this Recommendation, especially the distinction between the near field and the far field. Simple equations are derived for calculating the power density and the field strength in the far field, and the section concludes by defining the terms polarization and interference patterns.

2.7.1 Components of EM Field

The EM field radiated from an antenna comprises of various electric and magnetic field components, which attenuate with distance, r, from the source.

The main components are;

(i) The far field (Fraunhofer) also called the "radiation field", in which the magnitude of the fields diminishes at the rate of $1/r$;
(ii) The radiating near field (Fresnel), also called the "inductive field". The structure of this field is highly dependent on the shape, size and type of the antenna although various criteria have been established and are commonly used to specify this behavior, and
(iii) The reactive near field (Rayleigh), also called "quasi-static field", which diminishes at the rate of $1/r^3$.

As inductive and quasi-static components attenuate rapidly with increasing distance from the radiation source, they are only of significance in the vicinity of the transmitting antenna – in the so-called near-field region.

The radiation field, on the other hand, is the dominant element in the so-called far-field region. It is the radiation field which effectively carries a radio or television signal from the transmitter to a distant receiver.

(a) Far field

In the far-field region, an electromagnetic field is predominantly plane wave in character. This means that the electric and magnetic fields are in phase, and that their amplitudes have a constant ratio. Furthermore, the electric fields and magnetic fields are situated at right angles to one another, lying in a plane, which is perpendicular to the direction of propagation.

It is often taken that far-field conditions apply at distances greater than $2D^2/\lambda$ where D is the maximum linear dimension of the antenna. However, care must be exercised when applying this condition to broadcast antennas for the following reasons:

➤ It is derived from considerations relating to planar antennas,
➤ It is assumed that D is large compared with λ.

Where the above conditions are not met, a distance greater than 10λ should be used for far field.

(b) Power density

The power density vector, the Poynting vector S, of an electromagnetic field is given by the vector product of the electric, E, and magnetic, H, field components:

$$S = E \times H \qquad \rightarrow (1)$$

In the far field, in ideal conditions where no influence of the ground or obstacles is significant, this expression can be simplified because the electric and magnetic fields and the direction of propagation are all mutually orthogonal. Furthermore, the ratio of the electric, E, and magnetic, H, field strength amplitudes is a constant, Z_o, which is known as the characteristic impedance of free space and it is about 377 Ω (or 120 π Ω).

Thus, in the far field, the power density, S, in free space is given by the following non-vector equation:

$$S = E^2/Z_o = H^2 Z_o \quad \rightarrow (2)$$

The power density–at any given distance in any direction–can be calculated in the far field using the

$$\text{Power density equation: } S = P\, Gi\, /\, (4\pi r^2) \quad \rightarrow (3)$$

Where:
S: power density (W/m2) in a given direction
P: power (W) supplied to the radiation source, assuming a lossless system
Gi: gain factor of the radiation source in the relevant direction, relative to an isotropic radiator
r: distance (m) from the radiation source

The product PGi in equation (3) is known as the e.i.r.p. which represents the power that a fictitious isotropic radiator would have to emit in order to produce the same field intensity at the receiving point.

For power densities in other directions the antenna pattern must be taken into account.

In order to use equation (3) with an antenna design whose gain Ga is quoted relative to a reference antenna of isotropic gain Gr, such as a half-wave dipole or a short monopole, the gain factor Gi must be replaced by the product of Gr. Ga, as in equation (4). The relevant factor Gr is given in Table 2.4.

$$S = P\, Gr.\, Ga\, /\, (4\pi r^2) \quad \rightarrow (4)$$

{Note; as you know characteristic impedance of a medium, $z = \sqrt{(\mu/\varepsilon)}$ where μ is the magnetic permeability (=1.2566... X 10^{-6} F/m in free space), and ε is the permittivity (= 8.85418 x 10^{-12} H/m in free space)}.

Table 2.4: Isotropic gain factors for different types of reference antenna

Ref. Antenna type	Isotropic gain factor, G_r	Typical application
Isotropic radiator	1.0	Radar, satellite, terrestrial radio link system
Half wave dipole	1.64	Television, VHF and sometimes HF broadcasting
Short Mono pole	3.0	LF,MF and sometimes HF broadcasting

Thus, when the gain of the antenna G_d ($G_a = G_d$) is expressed relative to that of a half-wave dipole:

$$S = 1.64\, PG_d / (4\pi r^2) \qquad \rightarrow (5)$$

Where:
G_d: gain of the antenna relative to a half-wave dipole.

Similarly, when the gain of the antenna $Ga = Gm$ is expressed relative to that of a short monopole:

$$S = 3.0\, PGm / (4\pi r^2) \qquad \rightarrow (6)$$

Where:
Gm: gain of the antenna relative to a short monopole.

2.7.2 Field strength

If equation (2) is inserted into equation (3) to eliminate S and a factor C is introduced to take account of the directional characteristic of the radiation source, then equation (7) is obtained for the electric field strength (E) in the far field of a radiation source:

$$E = \sqrt{(Z_0 / 4\pi)} \{\sqrt{(PGi)} / r\}\, C = C / r \sqrt{(30\, P\, G_i)} \qquad \rightarrow (7)$$

Where:
E: electric field strength (V/m) – volt per meter
$Z_0 = 377\,\Omega$, the characteristic impedance of the free space
P = power fed to the radiation source (W), assuming a lossless system
C: factor ($0 \leq C \leq 1$), which takes account of the directional characteristic of the radiation source (in the main direction of radiation, $C = 1$).

If the gain of the antenna is expressed relative to a half-wave dipole or a short monopole, rather than relative to an isotropic radiator, then the factors Gd or Gm, respectively, should be used in place of Gi, as shown in equations (8) and (9).

$$E = \sqrt{(Z_0 / 4\pi)} \{\sqrt{(1.64\, PGd)} / r\}\, C = C / r \sqrt{(49.2\, P\, G_d)} \qquad \rightarrow (8)$$

$$E = \sqrt{(Z_0 / 4\pi)} \{\sqrt{(3\, PGm)} / r\}\, C = C / r \sqrt{(90\, P\, G_m)} \qquad \rightarrow (9)$$

In order to calculate the magnetic field strength in the far field of a radiation source, equation (10) is used:

$$H = E / Z_o \qquad \rightarrow (10)$$

Where:
E: electric field strength (V/m, and H: magnetic field strength (A/m)
$Z_o = 377 = (120\ \pi)$, the characteristic impedance of free space.

Equations (2) to (10) assume plane wave (far-field) conditions and are not applicable to near-field calculations.

(a) Near-Field
The field structure in the near-field region is more complex than that described above for the far field. In the near field, there is an arbitrary phase and amplitude relationship between the electric and magnetic field strength vectors, and the field strengths vary considerably from point to point.

Consequently, when determining the nature of near field, both phase and amplitude of electric and magnetic fields must be calculated or measured. In practice, however, this may prove very difficult to accomplish.

(b) Power density and field strength
It is not easy to determine the Poynting vector in the near field because of the arbitrary phase and amplitude relationship mentioned above. The E and H amplitudes, together with their phase relationship, must be measured or calculated separately at each point, making the task particularly complex and time-consuming.

Using analytical formulas, an estimation of the field strength in the near field is only feasible for simple ideal radiators such as the elementary dipole. In the case of more complex antenna systems, other mathematical techniques must be used to estimate field strength levels in the near-field region.

These other techniques allow relatively precise estimations of the field strength, the power density and other relevant characteristics of the field, even in the complex near-field region.

Measurement in the near field is even more difficult as no reference calibration method exists. The International Electrotechnical Commission (IEC) is currently working on the issue of a measurement standard for high frequency (9 kHz to 300 GHz) electromagnetic fields particularly in the near field. In addition, EN 61566 (Measurements of exposure to Radiofrequency electromagnetic field strength in the frequency range 1 kHz-1 GHz – subclause 6.1.4) gives more information on this topic.

2.7.3 Polarization

Polarization is defined as the direction of electric field vector, referenced to the direction of propagation of the wave front. In broadcasting, different types of polarization are used. The main types are vertical and horizontal
(With respect to a wave front which is travelling parallel to the surface of the Earth) although other types of polarization are used such as slant and elliptical.

2.7.4 Modulation

Modulation is a very special characteristic of the emission from a broadcasting transmitter. As certain effects of EM radiation are sensitive to the type of modulation used, it follows that the presence of modulation must be taken into consideration when making safety assessments. Modulation must also be taken into consideration when carrying out measurements or calculations to determine whether or not the limits are being exceeded. The modulation often results in a signal varying in both amplitude and frequency. For this reason temporal averaging is usually required in determining the values to be used in measurement and calculation. This requirement is also acknowledged in relevant Standards.

2.7.5 Characteristics of Radio Emission

The Radio Regulations (RR) classifies the emissions from radio transmitters according to the required bandwidths, and the basic and optional characteristics of the transmission. The complete classification consists of nine characters as follows;

- Characters 1–4 describe the bandwidth, using three digits and one letter;
- Characters 5–7 describe the basic characteristics, using two letters and one digit;
- Characters 8–9 describe any optional characteristics, using two letters.

Radiation Hazards 73

Only the three basic characteristics are relevant to the consideration of RF safety considerations. These are:

- The type of modulation of the main carrier Character 5
- The nature of the signal(s) which modulate(s) the main carrier Character 6
- The type of information to be transmitted Character 7

For sound and television broadcasts, the relevant characters are as follows:

- AM radio (LF, MF and HF double sideband) A3E, – AM radio (HF single sideband, reduced/variable carrier) R3E, – AM radio (HF single sideband, suppressed carrier) J3E, – Television pictures C3F, – Television sound F3E or A3E, – FM radio F3E or F9E, – DVB G7F, and– DAB G7E

2.7.6 Expressing transmitter power and field strength in terms of modulation type

Information about the transmitter power supplied to the antenna and the type of modulation can be obtained from the transmission authority, operating the equipment at a particular site. It is important to know whether the transmitter power is expressed in terms of the carrier power P_c, the mean power P_m, or the peak power P_p, so that the measured or calculated values can be compared accurately with the derived levels.

As an example, a MF sound-broadcasting transmitter (i.e. a type A3E emission) is considered. It is assumed that the calculations or measurements take account of the carrier power only, but the derived levels take account of the modulation components also (in terms of transmitter power, this corresponds to the mean power). Furthermore, it is assumed that only RMS values are used.

In order to compare the calculated or measured values with the derived levels, one of the following transformations must be made:

- The calculated/measured values must be modified to include the modulation components, or
- The derived levels must be modified to correspond with carrier power–only (without modulation components).

Table 3a of ITU-R BS.1698 gives multiplication factors which relate one type of power notation to another (these different notations for power are defined in the RR). In the case of an A3E transmission, shown as A*E in Table 3a, it can be seen that the mean power, Pm is 1.5 times the carrier power Pc. It should be noted that Table 3 of ITU-R BS.1698 a gives "worst-case" values, by assuming a modulation depth of 100%. In practice, the modulation depth of a broadcast transmitter will be less than 100% and hence the mean power will actually be less than 1.5 times the carrier power. For this reason Table 3b of recommendation gives the factors for a typical modulation depth (70% for an A3E transmission corresponding to a Pm/Pc ratio of 1.25 instead of 1.5). Refer ITU-R BS.1698 documents for table 3a, b, and c.

Table 3b of recommendations can also be used to convert field strength values to other notations; note, however, that the square root of the conversion factors given in Table 3b must be used when dealing with field strengths. Thus, in the above example of AM radio, the carrier-only RMS field-strength should be multiplied by 1.5 (or 1.25) to give the RMS field strength, which includes the modulation components. Conversely, the derived level (including modulation components) should be divided by 1.5 (or 1.25) to give an equivalent derived level for the carrier only.

The R.M.S. value of the field strength in the far field can be calculated from the known power, using equation (7); the appropriate type of power to use (i.e. Pm, or Pp) is shown in Table 2.5.

Table 2.5: Relationship between field-strength notations and power notations

To calculate	Use power expressed as
The effective value of field-strength	Average transmitter power, P_m
Average value of equivalent field–strength occurring in a period of peak RF oscillation	Peak Power, P_p
Peak (maximum) value of the equivalent field-strength	Peak power, $P_{p(1)}$

(1) The peak value of the equivalent field strength is determined from the peak power, Pp using the peak / rms correction factor. This factor is 2–1/2 for a sinusoidal carrier.

2.7.7 Interference patterns

Both natural and man-made structures re-radiate an EM field. The re-radiated field adds vectorially to the direct field. This results in interference patterns, which are comprised of localized maxima and minima of the field strength. The interference pattern is even more complex if there are multiple re-radiations of the field.

Interference patterns depend on the frequency of radiation source. Higher frequency of operation means smaller wavelength, spatially, the maxima and minima shall be closer. At UHF television frequencies, the local maxima and minima may be separated by only tens of centimeters. Several overlapping patterns occur in the case of multiple-radiation sources, e.g. if several radio and television channels are radiated from the same site.

2.7.8 Field-strength levels near broadcasting antennas

In this section, the field-strength levels which are found in the vicinity of typical LF/MF, HF, VHF and UHF broadcasting antennas are discussed.

(a) LF/MF bands (150–1605 kHz)

In LF and MF band, the frequencies are below whole-body resonance frequencies. In the case of direct effects of the EM field, the limit (also defined as "derived") levels for both the electric E and magnetic H field values are relatively high. However in many cases high values are present only very close to the transmitting antenna. This is especially true at the lower end of the LF/MF band, and for those standards/guidelines, which have specified higher derived levels. At the upper end of the band however the relevant distances may extend to the order of a few hundred meters. It should be realized that this increase in distance is due in part at least to the reduction in reference levels at the upper end of the MF band. During transmissions access to tower must be avoided owing to the high field-strengths and the risk of electric shock.

(b) Mixed frequency fields

It is common to have more than one transmitter (using different frequencies) located at the same site. In this case it is necessary to consider a total (combined) effect of human exposure to RF energy. On the other hand, effects are frequency dependent, and therefore, after calculation of the relevant parameters (S, E and H), the combined effect should be taken into account.

For thermal effects, exposure limits are given in terms of specific absorption rate (SAR) (see Appendix 4 of recommendation for additional evaluation methods), which means that appropriate power-densities should be determined. In the case of the multi-frequency transmitter site, the total power-density is recommended to be the sum of the power-density at each transmitting frequency:

$$S_t = \sum_{i=1}^{n} S_i \qquad \rightarrow (11)$$

Where Si is power density at the frequency fi ($i = 1, 2 \ldots n$), with the condition that:

$$\sum_{i=1}^{n} \frac{S_i}{L_i} \leq 1 \qquad \rightarrow (12)$$

Where Li is the power-density reference level, at the frequency fi ($i = 1,2, \ldots.n$).

This is the basic principle, but there are some differences in how the principle is applied (see Appendix 4 of recommendation).

(c) EMF inside buildings

The materials of a building and infrastructure inside a building have a very strong influence on the EM field, causing variations of the resulting field, from point to point, even in the same room. Spatial variations in the electromagnetic field are caused by multiple reflections of the incident wave, and therefore, the polarization of the resulting field may differ from that of the incident wave.

Metallic objects, wirings, cable and air-conditioning ducts, cause re-radiation (acting as secondary source), and change intensity of the fields in their vicinity. All these conditions make assessment of the exposure difficult. A rather large number of parameters should be taken into consideration when carrying out calculations or measurements. To have acceptable accuracy in calculation of exposure it is necessary to choose appropriate model for representing the environment.

Accuracy of measurement depends on the size and type of detection probe, as well as the location of the person who is doing measurements relative to radiation source and probe. There are no international standards for calculation and measurement methods yet. The critical issue is not simply the value of the exposure limits themselves, but the way in which calculations and measurements should be carried out and that is the main goal of this Recommendation.

2.7.9 Calculation of EM fields

Analytical and numerical calculation methods can predict the external or internal fields from a radiator. Calculations are useful to estimate the level of the field strengths in a certain exposure situation in order to determine if measurements are needed and what equipment should be used. Calculations can also be a complement to measurements and be used to verify that the results from the measurements are reasonable and acceptable.

In some situations, for example for complicated near-field exposure conditions when expensive SAR measurement equipment is not available, calculations can replace measurements.

The accuracy and quality of the calculations will depend on the analytical or numerical method used and on the accuracy of the description of the electromagnetic source(s) and physical objects between the radiator and the prediction point that may affect the fields. For SAR calculations, the accuracy of the body model will also affect the quality of the results.

To be able to make a calculation, the source parameters have to be known or estimated. Ex. of source parameters is frequency, mean & peak power, pulse width & repetition rate, antenna pattern, gain & geometry.

(a) Closed solutions

In the far-field region of a transmitting source, where the EMF are predominantly plane wave in character, analytical expressions can be used to estimate the field strengths. In the main direction of an antenna, the Friis free space equation can be used to calculate the power density:

$$S = PG/4\pi d^2$$

Here S = power density in (watt / m^2), P = mean output power (watt), G = antenna far field gain relative to an isotropic radiator, and d: distance from radiator (m).

The relation between power density and electric and magnetic field strengths is given by the following equation:

$$S = E^2/\eta = H^2\eta$$

Here: E= electric field strength (V/m) (RMS), H: magnetic field strength (A/m) (RMS), and η: The intrinsic impedance of free space, 377 Ω.

Hence, using the above formulas the field strengths can be calculated:

$$E = \sqrt{(PG\eta/4\pi d^2)} = (5.5\sqrt{PG})/d, \text{ and } H = \sqrt{PG/4\pi d^2\eta} = (\sqrt{PG})/68.8\,d$$

These relations are only valid in the far-field region of the radiating source, i.e. when $d > 2D^2/\lambda$, where D is the largest dimension of the radiating structure and λ is the wavelength. Field strength attenuation or enhancement due to reflection, material transmission, and diffraction is not taken into account. Using the relations above in the near-field region, or in directions other than the main direction, will generally give too large values unless a near field correction factor or a radiation pattern factor is introduced.

(b) Numerical procedures

Analytical procedures can only be used to calculate the electromagnetic properties for a few special cases and geometries. To solve general problems, numerical techniques have to be applied. The most common numerical procedures to calculate the EMF from a transmitting source or the internal fields and the specific absorption rate in biological bodies, are listed below. The most appropriate numerical techniques for a certain problem, depends on the frequency range considered, the geometrical structures to be modelled, and the type exposure situation (near-field or far-field). Some usual numerical modeling methods are given below;

(a) Physical optics (PO), (b) Physical theory of diffraction (PTD), (c) Geometrical optics (GO), (d) Geometrical theory of diffraction (GTD), (e) Uniform theory of diffraction (UTD), (f) Method of equivalent currents (MEC), (g) Method of moments (MOM), (h) Multiple multipole methods (MMP), (i) Finite-difference time-domain method (FDTD), (j) Finite element method (FEM), and (k) Impedance method.

An assessment must be carried out, for each application, to establish which one of the above methods is the most suitable for solving a given problem.

Each of these procedures enables the amplitude and phase of the following EMF field quantities to be determined, at every point in space, where the radiating and scattering elements may be either ideal conductors or dielectric bodies; (a) Electric field-strength; (b) Magnetic field-strength; (c) Power-density; (d) Current; (e) Voltage; and (f) Impedance

(1) Method of moments (MOM)

MOM is often used in the design of broadcast antenna systems (transmitter output power, antenna gain etc.) and in calculating their resultant electromagnetic fields. It enables calculations to be made at both the transmitting and receiving ends, as well as in the near and far-fields of the antenna.

Technical structures with up to three dimensions can be modeled, taking into account their material parameters (complex dielectric constant) as well as that of the ground. The modeling works with wires that are thin w. r. t the wavelength and, in principle, is able to represent surfaces too. The limitation of this method lies in the fact that modeling of extended & complicated structures may become time & memory consuming for the computer.

The MOM is a technique which has been extensively used to solve electromagnetic problems and to make SAR calculations in block models of biological bodies. In MOM, the electric fields inside a biological body are calculated by means of a Green's function solution of Maxwell's integral equations.

(2) Fast Fourier transforms / Conjugate gradient method (FFT/CG)

The FFT/CG method is a further development of the method of moments. Iterative algorithms based on FFT and the gradient procedure is used to solve linear equations derived from the method of moments.

(3) Finite-difference time-domain method (FDTD)

This is a numerical method to solve Maxwell's differential curl equations in the time domain. It can be used to calculate internal and external EMF and SAR distribution in biological bodies for both near-field and far-field exposures. In FDTD, both time and space are discretized, and a biological body is modelled by assigning the permittivity and conductivity values to the space cells it occupies. The computer memory required is proportional to the number of space cells. FDTD is considered the most promising SAR calculation method, but for accurate calculations very powerful computers are needed.

(4) Multiple multipole methods (MMP)

MMP is based on analytical solutions to field equations which have a multipole at one point in space, and is used in conjunction with the generalized multipole technique (GMP). The MMP procedure is especially suitable for the simulation of so-called "lossy scattering" bodies, which are near to radiation sources, i.e. within the immediate near-field.

(5) Impedance method

Impedance method has been successfully used to solve dosimetric problems where quasistatic approximations can be made. For calculations of SAR in human bodies, this method has proven to be very effective at frequencies up to 40 MHz. In the impedance method; the biological body is modelled by a three-dimensional network of complex impedances.

2.7.10 Field strength calculations

Most of the methods listed above can be used to calculate field strength levels from EM radiators. The accuracy of the results depends very much on how well the radiator (antenna) is modelled. If objects near the radiator, between the radiator and the prediction point, or close to the point of field strength prediction affect the field strength levels significantly, such objects should also be modelled.

Specific absorption rate calculations: Due to the difficulty of measuring the whole-body averaged or local peak SAR in many exposure situations, numerical calculations, several of the numerical techniques mentioned above can be used for estimation of the specific absorption rate distribution in a biological body exposed to either near or far-field EM radiation, for example the FDTD, MOM, and the MMP. Most appropriate methods for a particular problem, depends on the frequency, the exposure conditions, the size of the exposed object, the required accuracy, and the maximum tolerable calculation time. Each method requires experience in biophysics and numerical analysis.

To use any of these models, a 3-D geometric numerical model of the exposed body, or part of the body, is required. The electrical properties at the exposure frequency should be known for the different parts of the body. Depending on the required accuracy, models with different complexity may be used. In some situations, simple shapes like spheres and cylinders are appropriate to model the body. The dielectric properties of human tissues are given in the literature.

Using magnetic resonance (MR) images of a human body, very complex and accurate numerical body models can be developed. MR models with several different tissue types and a spatial resolution of less than a few millimeters have been used for FDTD calculations of the SAR distribution in humans exposed to electromagnetic fields from handheld radio transmitters.

2.7.11 Measurements of EM fields

It should be noted that measurement methods are critical, especially for near and low frequency fields. For LF bands method of measurement is a very sensitive and complex, since the distance of test point (from source of radiation) usually is much smaller than wavelength. For this reason, frequency range of 10 kHz-30 GHz is divided in four main broadcasting bands: LF/MF, HF, and VHF/UHF and SHF bands.

(a) LF/MF bands

In order to verify the theoretical results, FS measurements in near zone shall be made using special instruments (field strength meters) with three orthogonal positioned short dipoles. It is recommended not to use any instrument requiring power supply cable. To prevent influence of person doing the measurement, the measuring instrument shall be attached to insulated rod. The distance between instrument and operator should be determined by taking into account whether there are any changes on the instrument scale caused by any movement of the operator. That distance is dependent on the frequency of measured signal.

In performing this kind of measurement it is necessary to take into account the possible influences of all objects in the vicinity and, particularly, those that may create re-radiation effects. When the purpose of a measurement is to verify results obtained from theoretical computation, the test points should be selected along a radial direction and at height between 1 and 2 m.

More detailed explanation is given in Recommendation ITU-R BS.1386.

(b) HF bands
Detailed explanation is given in Recommendation ITU-R BS.705.

(c) VHF/UHF bands
Detailed explanation is given in Recommendation ITU-R BS.1195.

(d) SHF bands
Taking into account the wavelength and distances from the radiation sources, standard method of measurement shall be applied.

2.7.12 Instruments for measuring EM fields

The measurement of exposure fields, in the frequency range 10 kHz-300 GHz, requires significant effort for the spatial and time variability of the field to be determined. It is necessary to use adequate instrumentation and valid measurement set up. It is important to know the characteristics of measurement instruments because these characteristics determine the appropriate choice of instrument. Frequency dependent characteristics, such as cable interactions, out-of-band uncalibrated responses, and shaped frequency response are particularly important for broadband instruments. Other field properties need to be matched with instrument characteristics; for example, reactive or radiative, polarization and modulation, or number of field sources.

Human exposures to EM fields are commonly measured in units of power density, but other measurements such as the induced current in the body may be more relevant, and these are some of the critical aspects for protection or control that the engineer must resolve. In many cases there is no simple mathematical relation between electric and magnetic field and therefore, in such situations, each must be measured. The type of measurement instruments to use in this case is: (a) Instruments to measure the value of the field strengths E and H, and (b) Instruments to measure current.

(a) Base Equipments
Base equipments of these instruments are; (a) Probes, (b) Connection cables that transfer the signal from the probe to the reading and calculation unit, and (c) The reading and calculation unit.

(b) Probes

Most probes are isotropic, or Omni-directional in three dimensions, to measure the energy from all directions. The probes must exhibit the following characteristics:

(1) Respond to the intended fields, E or H, without responding to the unintended fields,
(2) Generally, the probe is electrically small and less than $\lambda/10$ for the maximum frequency of operation; however, special evaluations have shown that some probes may be electrically large, and
(3) Respond predictably to variations of environmental conditions, such as temperature and humidity.

It is very important that isotropic probes, during the measurement, are positioned such that the connection can decrease field perturbation at the probe by the connection cables. This field perturbation is more commonly a problem when measuring medium wave or lower frequency electric fields.

(c) Cables

The cables used to connect the probe and the reading and calculating instrument must be noise free and prevent coupling of field to the measurement unit.

It is very important to note that it is possible for the cables to act as an antenna and modify the field at the probe to cause an incorrect reading. It is sometimes possible to resolve this problem by setting the cables, during the test, perpendicular to the electric field.

(d) Characteristics of measuring instruments for electric and magnetic field

Generally a measurement of exposure to EMF is executed in the frequency domain. There are two principal groups of instruments.

(1) Wideband instruments types and specifications

With broadband instruments as shown in Fig 2.6, we can measure the total field in a given frequency range (i.e. bandwidth), but it is not possible to distinguish the contribution of a single frequency source, when several sources are radiating simultaneously.

Broadband instruments are made with sensors that can be non-isotropic to measure a single spatial component of the field, or can be isotropic to measure all three components of the field at the same time. These instruments can measure total level of instantaneous electric or magnetic field, or RMS field value or the average power density value in a time period, typically 6 minutes.

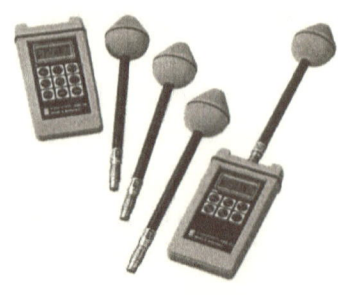

Fig 2.6: Broadband instruments

Broadband instruments can be divided in the following classes, depending on the detector used: (a) Diode, (b) Bolometer, and (c) Thermocouple

These instruments can be used in both situations, near field and far field.

(2) Narrow-band instrument types and specifications

Narrow-band instruments are selective in frequency and can measure the electromagnetic field strength at a range of different frequencies. By means of non-isotropic sensor or antenna it is possible to evaluate the direction and the polarization of the field. Care must be taken in the set up since fields can change high frequency rapidly in space relative to the antenna size, especially in the presence of reflective objects like walls, earth, metallic poles and structures.

It is important to observe that by changing the measurement point the detected field strength may be completely different. Also the measurement can be influenced by the antenna position and connecting cables.

When the measurement of the EMF in high frequency is executed in the time domain, it is necessary to use instruments with appropriate characteristic of analysis (for frequency and resolution answer) to obtain good results in the spectral analysis by Fourier's transformation. Fig 2.7 shows the block diagram of a narrow-band measurement system.

The system consists of the following basic components:

(a) A calibrated antenna that converts the electric field for a dipole antenna or magnetic field for a loop antenna to a wave on the transmission line.
(b) A calibrated connecting transmission line or coaxial cable.

(c) A selective receiver, typically a spectrum analyzer (Fig 2.8), that measures using a tuning circuit, the signal strength received as a function of frequency. The spectrum analyzer gives the values of the voltage or power in the frequency domain. It is very important to use care during these measurements, so that the measuring instruments do not disturb the field being measured.

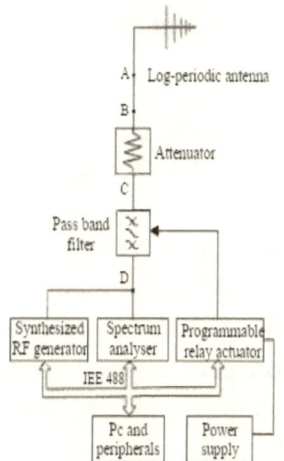

Fig 2.7: Block diagram of the measurement chain

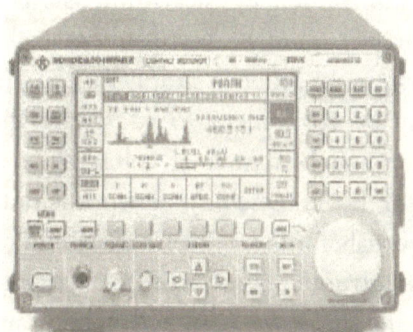

Fig 2.8: Selective receiver with spectrum analyzer

(e) Comparison between predictions and measurements
The comparison between predictions and measurements indicate that results of measurements are in good agreement with results obtained by theoretical computation. For more details, please refer to Appendix 2 to Annexure 1 of ITU-R BS.1698 recommendation.

The comparison between measured and predicted values of the E and H are, generally, not in full mutual agreement, as would seem to be expected; however there are no substantial differences among them, when compared with uncertainty of instrumentation, uncertainty in the planarity of the Earth's surface near the antenna (not taken into account in simulations, even when it is clear the field's intensity has a strong dependence on height of the measurement point) and the imposed simplicity of the model. Differences are bigger near the antenna (i.e. the first maximum at about 10 m), up to 50%, and decrease with increasing distance; at 250 m the differences are quite small.

The reasons for these differences are to be found in the difficulties in near field measurements, the uncertainty inherent in the instrumentation, the simplicity of the model, the presence of some objects near the antenna (metallic structures, the two lattices supporting dipoles and matching network, a little house) that have not been taken into account in the model, as well as the matching network and its radiation. Finally, terrain has been modelled with its typical electrical values.

In order to best approach the activity of E and H field prediction using a model, we recommend the following:

Antenna model: Physical dimensions of radiating and passive elements need to be carefully investigated as well as the complex input impedance of the system. In order to simplify a complex system, i.e. an array of radiators, it may be convenient to substitute the matching and feeding network with an equal number of voltage generators applied at the input of each radiator. If the matching network is not considered, then it is necessary to compensate for the: – Eventual mismatch between generators and radiators by the introduction of artificial matching elements or simple networks, or adjusting the power assigned to the transmitter.

The final result is quite insensitive to the presence of small mismatches that do not require modeling.

Segment subdivision: It is sufficient to represent filer antenna systems with segments not longer than $\lambda/20$.

Terrain model: it is necessary to give the exact values of permittivity and conductivity, especially in the case of a horizontally-polarized E field.

Transmitter power: It is important to take into account losses in transmission lines, matching network, resistance of junctions, mismatches to the load. In some cases it is opportune to artificially adjust the exact value of power in order to take into account various causes of losses without complicating the model of the antenna system.

Height of measurement points referred to ground: In many cases this parameter is very important if the scope is a comparison between measurements and predictions. In fact, the strong influence of height may be noted in the value of fields, and, if the terrain is modelled as a plane, large errors may be encountered, by comparison with measurement. In all these cases in which radiators are in the vicinity of terrain and terrain is not quite plane, the results of prediction must be taken with prudence.

Selection of code: the simulation based on MOM seems to be useful and easy to use in cases of filer antennas, of which physical and electrical characteristics are well known. It is possible and quite easy to simplify the model according to a minimum number of rules, without loss in precision of the results.

2.8 Precautions at transmitting stations and in their vicinity

This section outlines the precautions that should be taken at high-power transmitter stations to control the potential risks due to RF radiation. These risks fall into two main categories; the first being the direct risk to health due to human exposure to high levels of RF radiation, including shocks, burns and the possible malfunctioning of medical Implants; the second category comprises indirect risks where RF radiation could cause explosions, fires or interfere with the safe working of machines, cranes, vehicles, etc.

2.8.1 Precautions to control the direct health effects of RF radiation

Two categories of people considered for precautions are; (1) First category is employees or regular officials ,who are required to visit as part of their duty, and (2) Second category is the general public who are exposed indirectly being in the vicinity of transmitter site or on casual visit.

2.8.2 Occupational (Employee) precautionary measures

(a) **Physical measures**
Provide some form of protective barrier to restrict access to any area where either the exposure limits are exceeded or contact with

exposed RF conductors is possible. Access to such areas must be with an interlock or some physical barriers only. Mechanical or electrical interlocking or both should be provided to enclosures where access for maintenance is needed. Other physical measures such as warning lights or signs should also be used in addition to, but not instead of, protective barriers.

The risk of shock or burns from induced voltages on conducting objects, like fences and support structures, should be minimized by proper earthing. Particular attention should be paid to earthing of any temporary cables or wire ropes, winch etc. Where such objects need to be handled in an RF field, additional protection from shocks or burns should be provided by wearing heavy-duty gloves and through effective labeling.

(b) Operational procedures

RF radiation risk assessments must be carried out by suitably trained and experienced staff at construction site and also when any significant changes are made to a transmitting station. The initial objective must include the identification of the following:

(1) The areas where people may be exposed to "derived" or "investigation" levels;
(2) The different groups of people, e.g. employees, site sharers, general public etc., who may be exposed;
(3) The consequences of fault conditions, leakage from RF flanges, antenna misalignment or operational errors.

An initial check on the RF radiation levels can be done by calculation or mathematical modeling, but some sample measurements should also be carried out for verification purposes. In most cases, however, measurements will be needed to determine RF radiation levels more accurately. The actual quantities to be measured (E&H-field, power density, induced current) should be determined based on the specific circumstances. These include station frequencies, field region (near/far-field) being measured and whether it is proposed to check compliance with basic restrictions of SAR or only "derived/investigation" levels. These circumstances will also largely determine whether the three individual field components should be measured separately or whether an isotropic instrument should be used. RF radiation surveys should then be carried out by staff trained in the use of such instruments, following prescribed measurement procedures, and recording results in a specified format.

A competent person should be made responsible for the identification and provision of suitable instruments within any organization. Such measuring instruments must always be used in accordance with manufacturer's instructions and be subject to regular functional (operation with a check source) and calibration. Labels showing expiration dates must be fixed to instruments following such tests or calibration. Records of calibration should be kept, including whether adjustments and/or repairs were needed on each occasion. This information should then be used to determine the interval between calibrations. Work procedures should be implemented that ensure that RF radiation limits are not exceeded.

Staff should be trained on appropriate RF safety procedures. Maintenance work, in areas subject to access restrictions due to high RF radiation levels, should be planned around scheduled transmission breaks or radiation pattern changes where possible. However there should always be a balance between exposure to RF radiation and other risks, such as working on masts at night, even when floodlit. Where necessary, transmitters should be switched to reduced power or turned off to allow safe access for maintenance or repair work.

Prohibited areas on transmitting stations must be clearly defined and marked, and "permit to work" sytem should be implemented. Appropriate arrangements should be put in place for any systems, antennas, combiners or areas shared by other organizations. All staff who regularly works in areas with high levels of RF radiation should be issued with some form of personal alarm or RF hazard meter. Records must be kept of exposure above specified RF radiation levels. Organizations responsible for operating transmitting stations should monitor the health of staff who regularly works in areas with high levels of RF radiation and take part in epidemiological surveys, where appropriate.

Details of general policies and procedures relating to RF radiation safety should be included in written safety instructions and given to all appropriate staff. In addition, local instructions for each transmitting station should be issued to ensure compliance with such policies and procedures. Safety training should also include the nature and effects of RF radiation, the medical aspects and safety standards.

2.8.3 Precautionary measures for General Public

(a) Physical measures
Similar considerations apply to general public, as those detailed above for staff. Particular attention should be given to areas where RF radiation limits could be exceeded under fault conditions. Protective barriers should be provided in the form of perimeter fencing, suitably earthed where needed. Additional hazard warning signs will probably be necessary.

(b) Operational procedures
Risk assessments, carried out under Para above, must take into account the possibility of members of the public having medical implants. A procedure for providing health hazard information to such potential visitors should be adopted with appropriate restricted access procedures. Basic RF safety instructions should be provided for regular site visitors.

The need to carry out RF radiation surveys beyond site boundaries must be considered, in particular where induced voltages in external metallic structures (cranes, bridges, buildings etc.) may cause minor burns or shocks. In carrying out such surveys the possibility of the field strength increasing with distance, usually due to rising terrain, should be taken into account. Where necessary, a procedure for monitoring planning applications or other development proposals should be implemented.

2.8.4 Precautions to control indirect RF radiation hazards

Indirect effects of RF radiation, such as ignition hazards to flammable substances, may occur at levels well below the "derived/investigation" levels particularly at MF/HF. This is because flammable substances may be stored on a site having conducting structures, such as pipe work, expended metals that could act as a fairly good receiving antenna. Actual risks are however rare but may include industrial processing plants, fuel storage facilities and petrol filling stations. Detailed evaluation is however far from simple. The general procedure recommended below is therefore based on progressive elimination. The detailed precautions adopted will however need to take account of any National Standards or legislation in the country concerned.

An initial assessment should be carried out based on practical, worst case estimates of the minimum separation needed between a particular type of transmitter and a conducting structure to avoid such a hazard. The 1st step in doing this is to determine the minimum field strength that might present an ignition hazard for the particular transmitter frequencies in use. This is a function of the type of flammable substance and the perimeter of any loop formed by metallic structures, usually pipe work, and can most easily be determined from tables or graphs. The vulnerable area should then be determined from this minimum field strength by calculation, mathematical modeling or from tables/graphs.

If the vulnerable area, as determined above, contains any such sites on which flammable substances are stored, or if any are being planned, a more detailed assessment should then be made. This should be based on the actual dimensions of any metallic structures, the gas category of the flammable substance(s) being stored and the measured field strength. This detailed assessment should be carried out by calculation of the extractable power from the metallic structure to determine whether this exceeds the minimum ignition energy of the flammable substance. Should this be the case, then the extractable power should be measured and any necessary modifications to the structure and/or other safeguards implemented.

In a similar category to ignition hazards, is the possible detonation of explosive materials. This will very rarely be encountered but detailed guidance is available from national standards, such as BS: 6657 in the United Kingdom. Other indirect effects that should be considered include interference to the safety systems of vehicles, machines, cranes etc. close to or within the boundaries of transmitting stations. The immunity of these systems is covered by electromagnetic compatibility (EMC) regulations (see Appendix 3).

Where necessary, precautions similar in principle to those described in Para 2.10.1 may need to be applied.

2.9 Examples of calculated field strengths near broadcasting antennas

(1) Example A: Electric and magnetic field-strength plots

Numerical calculations of electric & magnetic field-strength distribution near broadcast transmitting antennas can be done in order to determine how field strengths are at certain points. This includes the near-field zone where field structure is generally very complicated. Calculations can also be done to verify the field contours (lines with constant field strength) where relevant limiting values of EMF restrictions are kept. In this way it is possible (for planning purposes) to estimate how extended relevant zones may be where protection measures may or must be performed.

In a technical document of the European Broadcasting Union (EBU) many calculation results are given. In Figures 2.9 (a) and (b) results of these examples (MF broadcasting transmitting antennas) are given as plots.

Electric Field: Fig 2.9 (a) # 1. ICNIRP (general publication) = 60v/m. # 2. 158 v / m, E_{max} =449.8 v / m, Frequency; 1422 kHz, Power; 600 kW, Evaluation height; 1.5 m above ground level (AGL)

Magnetic Field: Fig 2 (b) #1. ICNIRP (general publication) =0.16 A/m, 2. 1.3 A/m, H_{max} =1.6A/m, Frequency; 1422 kHz, Power; 600 kW, Evaluation height; 1.5 m above ground level (AGL)

(2) Example B: Determination of magnetic field strength in near-field zone of high-power MF/LF antennas

This example has the aim to determine magnetic field strength in the near-field zone of MF and LF mast antennas (monopoles), solving Hallen's integral equation.

In frequency bands below 10 MHz physical relations in the EMF are much more complex. In contrast to microwave frequencies, where the EMF has characteristics of the field in the far zone even at very short distances from the radiation source, and where the concept of the radiated power density (Poynting vector intensity) is very useful, in the MF/LF band the field in antenna vicinity is very complex. In fact, in near – field zone, simple relationship between electric and magnetic fields no longer exists: the two fields are not in phase and their ratio is not 377 Ω. That fact additionally complicates the relationships in the EMF below 10 MHz.

Clearly, the measured field strengths will depend on the type of transmitting antenna, transmitter power and distance from the transmitting antenna. For example, in case of high-power transmitter E-component, field strengths on a typical LF/MF site may range from a few V/m to over 250 V/m. Very close to the transmitting antenna the field strength may be of the order of 1000 V/m.

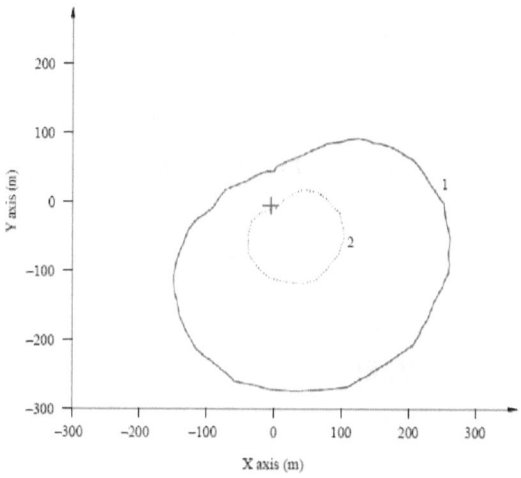

Fig 2.9: (a) MF Monopole; RMS electric field-strength contours representing certain levels

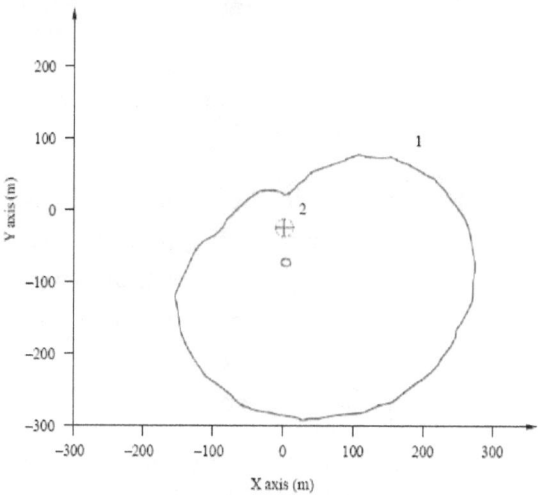

Fig 2.9: (b) MF Monopole; RMS magnetic field – strength contours representing certain levels

2.10 Limits and levels of RF hazards

A number of international and national authorities are concerned with the health aspects of RF hazards. Some of them are listed below.

2.10.1 Regulatory and advisory authorities on health aspects

A non-exhaustive list of some Regulatory and Advisory bodies is;

1. Electronic Components Committee, Comité des composants électroniques, CENELEC
2. German Commission for Electrical and Electronic Information technologies, DKE
3. European Union, EU–Recommendation from the Council of the European Union, L 199, 12 July 1999
4. Institute of Electrical and Electronics Engineer/ American National Standards Institute (IEEE/ANSI)
5. International Commission on non-ionizing radiation protection (ICNIRP)
6. International Electrotechnical Commission (IEC)
7. World Health Organizations (WHO)
8. National Radiological Protection Board (NRPB)-UK
9. European Telecommunications Standards Institute (ETSI)

2.10.2 Comparison of basic limits and derived levels from widely used regulations

Present standard guidelines differ in the definition of groups of persons potentially at risk "exposure groups" (e.g. general public, occupationally exposed workers, age) and/or in the place of locations of interest (e.g. public places, houses, fenced enclosures and restricted buildings). Differences also appear on the parts of the body considered.

Compared in this write up herewith are:

(a) IEEE Standard for safety levels w.r.t. human exposure to radio frequency EM fields,3 kHz to300 GHz.

(b) ICNIRP guidelines for limiting of exposure to time-varying electric, magnetic and EM fields (up to 300 GHz).
(c) UK NRPB restrictions on human exposure to static and time varying electromagnetic fields and radiation.

(a) Comparison of basic biological limits

Standards "basic limits" & guideline for contact current; current density and SAR are indicated in Table 2.6, together with measuring conditions. A common guideline for compared standards is that all of them do not allow SAR over the entire body to exceed the value of 4 Watt /kg.

By applying a safety factor of 10, the basic SAR limit (whole-body-averaged SAR) is reduced to 0.4 Watt/kg. NRPB basic limit of 0.4 Watt/kg is proposed as maximum level allowed for core temperature increase of 1 K.

It is to be noted that the basic whole body limit for SAR is averaged over a time of 6 minutes. There are also limits applicable to specific part of the body, referred as "local SARs"; since resonance effects may give rise to local "hot spots", these limits are higher than the whole-body-averaged SAR.

(b) Comparison of derived limit levels for E, H and power flux limits at various frequencies

Direct measurement of current density and specific absorption rate is very difficult and not possible in practice. Hence, derived levels are given in addition to basic limits in the standards and guidelines considered here.

On the other hand measurement of EM field quantities E and H are of more practical feasibility. SAR is frequency dependent, while the basic limits as in Table 2.6 are assumed constant. By means of the relevant data the SAR limit may be transformed into free field corresponding quantities E, H and S.

Tables 2.7 to 9 show the maximum level respectively of E, H and power density in various bands in the frequency range 1 kHz-300 GHz. The limits are calculated under the conservative assumption of optimum electromagnetic coupling between the RF field and the body.

Table 2.6: Comparison of basic restrictions (SAR limits) and reference level (current parameters)

Parameter	IEEE/ANSI		ICNIRP		NRPB
	Controlled	uncontrolled	Occupational	General Public	
RMS induced or contact current (mA)	1000f (13a)	450 f (13a)	100 (2)	45 (2)	
	100 (1), (13b)	45 (1), (13b)	40 (3)	20 (3)	
Current density RMS (A/m^2) Averaging Area (cm^2) Averaging time (s)	350 f 1 1 (4)	15.7 f 1 1 (4)	10f 1 -(4)	2f 1 -(4)	10 f
Whole-body-average (W/kg), SAR	0.4, (5a)	0.08, (5b)	0.4, (5a)	0.08 (5a)	0.4 (5c)
Local SAR (W/kg) Averaging mass (kg)	8, 13 (c) 0.001, (6)	1.6, 13 (c) 0.001, (6)	10, 13 9d) 0.001, (5a),(7)	2, 13 (d) 0.01, (5a),(7)	10 0.01 and 0.1, 5 (a),(10)
Local SAR $^{(7)}$ (W/kg) Averaging mass (kg)	20, (13c) 0.010$^{(8)}$	4, (13c) 0.010$^{(8)}$	20, (13c) 0.01$^{(5a),(9)}$	4, (13c) 0.01$^{(5a),(9)}$	20 0.1$^{(5a),(9)}$
Power Density (W/m^2) Averaging time (minute)	-	-	50 68/f$^{1.05}$, (12), (13)	10 68/f$^{1.05}$, (12),(13)	100 68/f$^{1.05}$, (12)

Legends: (1) Current through each foot, f: is frequency in MHz, (2) Current induced in any limb (10–110 MHz), (3) Contact current from conductive objects (100 kHz-110 MHz), (4) Current density over any 1 cm^2 area of tissue, (5) a) The SAR limits relate to an averaging time of 6 min, b) The SAR limits relate to an averaging time as given in Table 2.7, c) The SAR limits relate to an averaging time of 15 min, (6) Localized SAR except for the hands, wrists, feet and ankles (100 kHz-6 GHz), (7) Localized SAR for head and trunk (100 kHz-10 GHz), (8) Localized SAR for the hands, wrists, feet and ankles (100 kHz-6 GHz), (9) Localized SAR for limbs (100 kHz-10 GHz), (10) Localized SAR for head, neck, trunk and fetus (10 MHz-10 GHz), (11) 10 grams for the head and fetus, 100 grams for the neck and trunk, (12) For frequencies between 10 and 300 GHz, f: frequency (GHz), (13) Averaged over any 20 cm^2 of exposed area; a) 3 kHz < f < 100 kHz, b) 100 kHz < f < 100 MHz, c) 100 kHz < f < 6 GHz, and d) 100 kHz < f < 10 GHz.

Table 2.7: Comparison of derived levels; E field (R.M.S. values V/m)*

Frequency Range	IEEE/ANSI		ICNIRP		NRPB	
	Controlled	Uncontrolled	Occupational	General Public	Adults only	Adults only & children
0.6 -3 KHz	614		610.. (1)	87		1000
3 -30 KHz						
30 -38 KHz						
38 -65 KHz						
65 -100 KHz						
100 -410 KHz						
410 -600 KHz						
600 -610 KHz						600 / f
610 -680 KHz						
680 -920 KHz						
0.92 -1 MHz						
1 -1.34 MHz						
1.34 -3 MHz	614	823.8 / f	610 / f	87 / f$^{0.5}$		
3 -10 MHz						
10 -12 MHz	1842 / f	823.8 / f				600 / f
12 -30 MHz					60	
30 -60 MHz	61.4	27.5	61	28	f...(2)	50.....(2)
60 -100 MHz						
100 -137 MHz						
137 -200 MHz						
200 -300 MHz						0.25 f.... (2)
300 -400 MHz					137...(2)	
400 -800 MHz						100...(2)
0.8 -1.1 GHz			3 f$^{0.5}$	1.375 f$^{0.5}$		
1.1 -1.55 GHz					0.125 f....(2)	0.125 f....(2)
1.55 -2 GHz						194....(2)
2 -3 GHz						
3 -15 GHz			137	61		
15 -300 GHz						

Legends: a) $f^2/0.3$, and b) 30 min.
(1) This value is in the range 0.82 kHz to 1 MHz, (2) Plane wave equivalent value of the E field
  f: frequency (MHz, unless otherwise stated)
  * Values should be averaged over 6 minutes, except as shown in tables

Table 2.8: Comparison of derived levels; H field (RMS values A/m)[1],[2]

Frequency Range	IEEE/ANSI Controlled	IEEE/ANSI Uncontrolled	ICNIRP Occupational	ICNIRP General Public	NRPB Adults only	NRPB Adults only & children
1-3 KHz						
3-30 KHz						
30-38 KHz	163		5....(3)		64	
38-65 KHz						
65-100 KHz						
100-140 KHz						
140–150 KHz						
150-535 KHz						
535-610 KHz			1.6/f			
610-680 KHz						
0.68-1 MHz	16.3/f			0.73/f		
1-1.34 MHz						18/f^2
1.34-3 MHz						
3-10 MHz						
10-12 MHz						
12-30 MHz			0.16	0.073		
30-60 MHz	16.3/f	158.3/$f^{1.668}$ (1a)			0.16....(2)	0.13
60-100 MHz					f/377....(2)	
100-137 MHz						
137-200 MHz	0.163	0.0729 (1b)				
200-300 MHz						
300-400 MHz					0.36.....(2)	0.66x 10^{-3} f
400-800 MHz						0.26
0.8-1.1 GHz			0.008 $f^{0.5}$	0.0037 $f^{0.5}$		0.33 x 10^{-3}f
1.1-1.55 GHz						0.33 x 10^{-3} f.......(2)
1.55-2 GHz						
2-3 GHz						0.52
3-15 GHz			0.36	0.16		
15-300 GHz						

Legends: f: frequency (MHz, unless otherwise stated).
(1) Values should be averaged over 6 min, except as shown below: a) 0.0636 $f^{1.337}$ min, b) 30 minutes, (2) Plane wave equivalent value of the H field, based on power density values given for adults. (Note – these values are not given explicitly in the same way as the E field and power density values are specified), (3) This value is valid in the range 0.8 kHz to 150 kHz.

Table 2.9: Comparison of derived levels; power density (W/m²) [3]

Frequency Range	IEEE/ANSI[1] Controlled E field	IEEE/ANSI[1] Controlled H field	IEEE/ANSI[1] Uncontrolled E field	IEEE/ANSI[1] Uncontrolled H field	ICNIRP Occupational	ICNIRP General Public	NRPB Adults only	NRPB Adults only & children
< 100 Hz								
0.1–1 KHz								
1–3 KHz								
3–30 KHz	1000	10×10^6	1000	10×10^6				
30–100 KHz	1000	10×10^6	1000	10×10^6				
100–410 KHz								
0.41–1 MHz	1000	$10^5/f^2$	1000	$10^5/f^2$				
1–1.34 MHz	1000	$10^5/f^2$	1000	$10^5/f^2$				
1.34–3 MHz	1000	$10^5/f^2$	$1800/f^2$ [2],[3a]	$10^5/f^2$ [2]				
3–10 MHz	$9000/f^2$	$10^5/f^2$	$1800/f^2$ [2],[3a]	$10^5/f^2$ [2]				
10–12 MHz	$9000/f^2$	$10^5/f^2$	$1800/f^2$ [2],[3a]	$10^5/f^2$ [2]				
12–30 MHz	10	$10^5/f^2$	2 [3a],[3b]	$(9.4\times10^6)/f^{8.366}$ [2],[3c]	10	2	10	
30–60 MHz	10	$10^5/f^2$	2 [3a],[3b]	$(9.4\times10^6)/f^{8.366}$ [2],[3c]	10	2	10	6.6
60–100 MHz	10	$10^5/f^2$	2 [3a],[3b]	$(9.4\times10^6)/f^{8.366}$ [2],[3c]	10	2	10	$2.7\times10^{-3}\,f^2$
100–137 MHz	10		2 [3b]		10	2	10	$2.7\times10^{-3}\,f^2$
137–200 MHz	10		2 [3b]		10	2	10	$2.7\times10^{-3}\,f^2$
200–300 MHz					10	2	10	$0.165\times10^{-3}\,f^2$
300–400 MHz					10	2	50	26
400–800 MHz	f/30		f/150 [3b]		f/40	f/200	50	$41\times10^{-6}\,f^2$
0.8–1.1 GHz	f/30		f/150 [3b]		f/40	f/200	50	$41\times10^{-6}\,f^2$
1.1–1.55 GHz	f/30		f/150 [3b]		f/40	f/200	$41\times10^{-6}\,f^2$	$41\times10^{-6}\,f^2$
1.55–2 GHz					f/40	f/200		100
2–3 GHz								100
3–15 GHz	100		f/150 [3d]		50	10		100
15–300 GHz	100		100 [3e]		50	10		100

Legends: f: frequency (MHz, unless otherwise stated).
(1) Below 100 MHz, plane-wave equivalent values are given for the E and H fields, (2) As given by some commercially available meters; (3) Values should be averaged over 6 minutes, except as shown below:

a) $f^2/0.3$ min, b) 30 min, c) $0.0336\,f^{1.337}$ min, d) $90\,000/f$ min, and e) $616\,000/f^{1.2}$ min.

2.11 Field-strength values to be determined

There are two procedures to determine values of Field-strength as below:

(a) Graphical Method

The range of electrical and magnetic field strengths can be arrived at by using data given in the tables of previous para and are also as shown in Figs. 2.10 and 2.11 below, respectively. These curves/graphs however should not be used as a basis for an administration's regulatory requirements. They represent a composite view of the limits currently depicted and are certain to evolve over time. As such, they are merely illustrative of the methodology that could be applied to develop useful standards within an administration.

Also, it must be recognized that results of independent studies of the subject are not entirely consistent and, as a result the interpretation, of the results by responsible authorities has in the past, and will continue in the future, to result in differing requirements in different countries.

The curves "a" and "b" represent the upper and lower boundaries respectively of some known, existing recommendations for RF exposures levels (presented as example). All curves from authorities making such recommendations lie between these boundaries, and any curve between curves "a" and "b" should allow adequate broadcasting services.

The differences between the suggested maximum levels at the same frequency (Fig 2.10 and 2.11) depend on the different conditions considered by the various sources suggesting the limits.

(b) Numerical procedures and calculation of EMF quantities

In few relatively simple cases, electromagnetic radiation and scatter problems can be solved by using analytic procedures in a closed form. However, the solution of general problems, with random geometries, requires the application of numerical calculation procedures, running on powerful computers.

The numerical procedures, depending on the frequency range under consideration and the size of the geometrical structures used, are available for the calculation of EMF field quantities.

Among these different methods we have decided to use the MOM, used in the design of broadcast antenna systems and in calculating their resultant electromagnetic fields.

Radiation Hazards 101

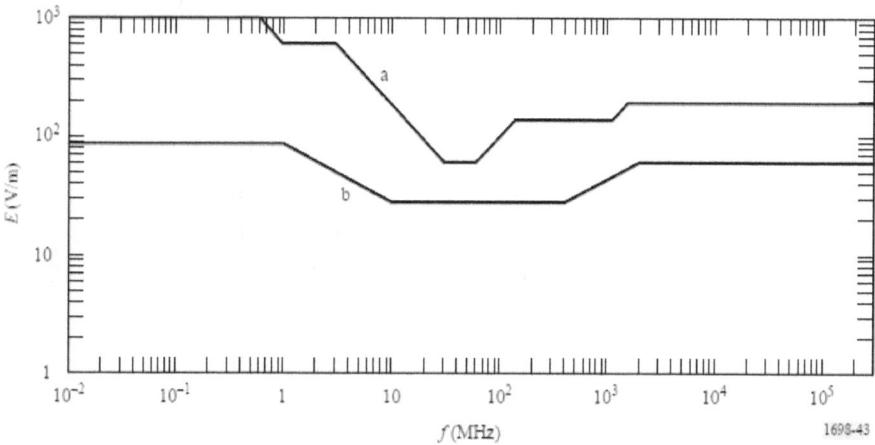

Fig 2.10: Range of the electrical field strengths derived from the Tables shown in previous Para

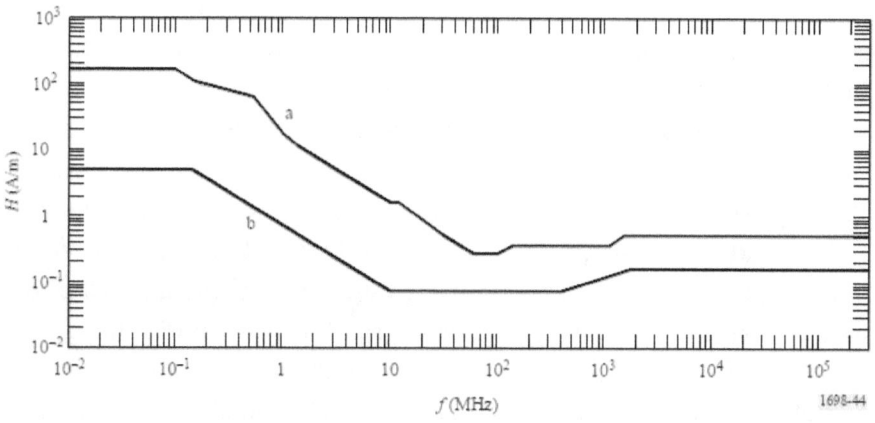

Fig 2.11: Range of the magnetic field strengths derived from the Tables shown in previous Para

The MOM is used intensively for calculating the SAR distribution in so-called /block model/.

The electric field intensities within the body are calculated from the solution of the integral equation of the electric field by using Maxwell's equations.

The software used is: NEC – WIN Professional V 1.1 (1997) by Nittany Scientific, Inc. – http://www.nittany-scientific.com/.ic.com/.

2.12 Additional evaluation methods for EM fields (Appendix 4 to Annex 1 of ITU-R, BS-1698)

(1) Dosimetry Method

The application of dosimetric concepts enables the link to be established between external (i.e. outside the body) field strengths and internal quantities of electric field strength, induced current density and the energy absorption rate in tissues. The development of experimental and numerical Dosimetry has been complementary. Both approaches necessitate approximations to the simulation of human exposure; however the development of tissue equivalent materials and minimally disturbing probes in the experimental domain and the use of anatomically realistic models for computational purposes have improved the understanding of the interaction of RF fields with the body.

Whereas current density is the quantity most clearly related to the biological effects at low frequencies, it is the SAR, which becomes the more significant quantity as frequencies increase towards wavelengths comparable to the human body dimensions.

In most exposure situations the SAR can only be inferred from measured field strengths in the environment using dosimetric models. At frequencies below 100 MHz non-invasive techniques have been used to measure induced current, and in extended uniform fields, external electric field strengths have been related to induced current as a function of frequency. In the body resonance region, exposures of practical significance arise in the reactive near field where coupling of the incident field with the body is difficult to establish owing to non-uniformity of the field and changing alignment between field and body. In addition, localized increases in current density and SAR may arise in parts of the body as a consequence of the restricted geometrical cross-section of the more conductive tissues.

Dosimetric quantities can be calculated by use of suitable numeric procedures and calculational models of the human body. On the other hand such quantities can be measured using suitable physical models (phantoms).

(2) SAR measurement

The SAR (W/kg) is the basic limit quantity of most RF exposure regulations and standards. SAR is a measure of the rate of electromagnetic energy dissipated per unit mass of tissue.

The SAR may be specified as the value normalized over the whole body mass (sometimes referred to as the "whole body averaged SAR") or the localized value over a small volume of tissue (localized SAR).

SAR can be ascertained from the internal quantities in three ways, as indicated by the following equation:

$$SAR = \sigma E^2 / \rho = C_i\, dT/dt = J^2/\sigma\rho$$

Where:
E: value of the internal electric field strength in the body tissue (V/m⁻¹)
σ: conductivity of body tissue (S/m⁻¹)
ρ: density of body tissue (kg/m⁻³)
Ci: heat capacity of body tissue (J/kg⁻¹ °C⁻¹)
dT/dt: time derivative of temperature in body tissue (°C/s⁻¹)
J: value of the induced current density in the body tissue (A/m²).

The local SAR in an incremental mass (dm) is defined as the time derivative of the incremental absorbed energy (dW) divided by the mass: SAR = d/dt (dW/dm)

This quantity value is important from two standpoints; the resulting non-uniform distribution of energy absorption when exposed to a uniform plane wave, and the localized energy absorption arising from non-uniform fields in close proximity to a source of exposure.

Exposure regulations or standards contain derived electric and magnetic field limits. The underlying dosimetric concept assures that compliance with the (external) derived levels will assure compliance with the basic SAR limits. However, external or internal SAR measurements can also be used to show compliance. For partial-body near-field exposure conditions, the external electromagnetic fields may be difficult to measure, or may exceed the derived limits although the local SAR is below the basic limits. In these cases internal SAR measurements in body models have to be conducted. The most important methods to measure SAR will be described below.

(3) Electric field measurement

The SAR is also proportional to the squared RMS electric field strength E (V/m) inside the exposed tissue:

$$SAR = \sigma E^2/\rho$$

Where:

σ (S/m): conductivity
ρ (kg/m3): mass density of the tissue material at the position of interest.

Using an isotropic electric field probe, the local SAR inside an irradiated body model can be determined. By moving the probe and repeating the electric field measurements in the whole body or in a part of the body, the SAR distribution and the whole body or partial-body averaged SAR values can be determined. A single electric field measurement takes only a few seconds, which means that three-dimensional SAR distributions can be determined with high spatial resolution and with a reasonable measurement time (typically less than an hour).

(4) Temperature measurement

The specific absorption rate (SAR) is proportional to initial rate of temperature rise dT/dt (C/s) in the tissue of an exposed object:

$$SAR = c \, \Delta T/\Delta t$$

Where c is the specific heat capacity of the tissue material (J/ kg C)

Using certain temperature probes, the local SAR inside an irradiated body model can be determined. One or more probes are used to determine the temperature rise ΔT during a short exposure time Δt (typically less than 30 sec to prevent heat transfer). The initial rate of temperature rise is approximated by $\Delta T/\Delta t$, and the local SAR value is calculated for each measurement position. By repeating the temperature measurements in the whole body or in a part of the body, the SAR distribution and the whole-body or partial-body averaged SAR values can be determined.

Three-dimensional SAR-distribution measurements are very time consuming due to the large number of measurement points. To achieve a reasonable measurement time the number of points has to be limited. This means that it is very difficult to measure strongly non-uniform SAR distributions accurately. The accuracy of temperature measurements may also be affected by thermal conduction and convection during measurements, or between measurements.

(5) Calorimetric measurement

The whole-body average SAR can be determined using calorimetric methods. In a normal calorimetric measurement, a full-size or scaled body model at thermal equilibrium is irradiated for a period of time. A calorimeter is then used to measure the heat flow from the body, until the model is at thermal equilibrium again. The obtained total absorbed energy is then divided by the exposure time and the mass of the body model, which gives the whole-body SAR. The calorimetric twin-well technique uses two calorimeters and two identical body models. One of the models is irradiated, and the other one is used as a thermal reference. This means that the measurement can be performed under less well-controlled thermal conditions than a normal calorimetric measurement.

Calorimetric measurements give rather accurate determinations of whole-body SAR, but do not give any information about the internal SAR distribution. To get accurate results a sufficient amount of energy deposition is required. The total time of a measurement, which is determined by the time to reach thermal equilibrium after exposure, may be up to several hours. Partial body SAR can be measured by using partial-body phantoms and small calorimeters.

(6) Body current measurement
Measurement devices for body current may be carried out in two categories:

- Measurement devices for body to ground current;
- Measurement devices for contact current.

(6.1) Induced body currents
Internal body currents that are induced in persons occur from partial or whole-body exposure of the body to RF fields in the absence of contact with objects other than the ground.

The two principal techniques used for measuring body currents include clamp-on type (solenoidal) current transformers for measuring current flowing in the limbs, and parallel plate systems that permit the measurement of currents flowing to ground through the feet.

Clamp-on current transformer instruments have been developed that can be worn.

The meter unit is mounted either directly on the transformer or connected through a fiber-optic link to provide a display of the current flowing in a limb around which the current transformer is clamped. Current sensing in these units may be accomplished using either narrow-band techniques, e.g., spectrum analyzers or tuned receivers (which offer the advantage of being able to determine the frequency distribution of the induced current in multi-source environments, or broadband techniques using diode detection or thermal conversion).

Instruments have been designed to provide true R.M.S. indications in the presence of multiple frequencies and/or amplitude-modulated waveforms.

The upper frequency response of current transformers is usually limited to about 100 MHz however air cored transformers (as opposed to ferrite-cored), have been used to extend the upper frequency response of these instruments. Whilst air-cored transformers are lighter and therefore useful for longer term measurements, they are significantly less sensitive than ferrite cored devices.

An alternative to clamp-on device is parallel plate system. In this instrument, the body current flows through the feet to a conductive top plate, through some form of current sensor mounted between the plates, and thereby to ground. The current flowing between the top and bottom plates may be determined by measuring the RF voltage drop across a low impedance resistor. Alternatively, a small aperture RF current transformer or a vacuum thermocouple may be used to measure current flowing through conductor between two plates. There are several issues that should be considered when selecting an instrument for measuring induced current.

Firstly, stand-on meters are subject to the influence of electric-field induced displacement currents from fields terminating on the top plate. Investigations have shown that apparent errors arising in the absence of a person are not material to the operation of the meters when a person is present.

Secondly, the sum of both ankle currents measured with clamp-on type meters tends to be slightly greater than the corresponding value indicated with plate type meters. The magnitude of this effect, which is a function of RF frequency and meter geometry, is not likely to be material. Nonetheless, more accurate method of assessing limb currents is the current transformer. The precise method of measurement may depend upon the requirements of protection guidelines against which compliance assessments are made.

Thirdly, the ability to measure induced currents in limbs under realistic grounding conditions as found in practice need to be considered. In particular, the differing degree of electrical contact between ground and bottom plate of the parallel plate system and the actual ground surface may affect the apparent current flowing to ground.

Measurements can be made using antennas designed to be equivalent to a person. This enables a standardized approach to be used and permit current measurements to be made without the need for people to be exposed to potentially hazardous currents and fields.

(6.2) Contact current measurement

The current measurement device has to be inserted between hand of the person and the conductive object. The measurement technique may consist of a metallic probe (definite contact area) to be held by hand at one end of the probe while the other end is touched to the conductive object. A clamp-on current sensor (current transformer) can be used to measure the contact current which is flowing into the hand in contact with the conductive object. Alternative methods are:

- The measurement of the potential difference (voltage drop) across a non-inductive resistor (resistance range of 5–10Ω) connected in series between the object and the metallic probe held in the hand;
- A thermocouple milliammeter placed directly in series.

The wiring connections and the current meter must be set up in such way that interference and errors due to pick-up are minimized. In the cases where excessively high currents are expected an electrical network of resistors and capacitors can simulate the body's equivalent impedance.

(7) Touch voltage measurement

The touch voltage (no-load-voltage) is measured by means of a suitable ac voltmeter or oscilloscope for the frequency range under consideration. The measurements devices are connected between the conductive object charged by field induced voltage and reference potential (ground). The input impedance of the ac voltmeter must not be smaller than 10 kΩ.

2.13 Electromagnetic compatibility of Electro-medical devices (Appendix 5 to Annex 1)

Electromagnetic compatibility (EMC) is a general concern for electronic equipments and particularly electronic medical devices. These devices, as described below, if worn by operators or general staff or public visiting radio transmitting station should not have adverse effect on their functioning in the RF field and hence study must be conducted for the effect on the functioning of these devices for their electromagnetic compatibility and safety guidelines prepared and followed accordingly.

(1) Electronic Equipments & Electronic-medical devices

If Electromedical devices are used in the presence of strong EM fields they may malfunction. The risk of such malfunction increases if the field strengths are great enough. The risk of malfunction depends upon several variables, such as the level of field strength, which is dependent on distance between the radiating antenna and the device, the transmitter power, the frequency of the waves, the type of modulation of the radiated signal, the effect of cable coupling as well as the electronic device's own RF immunity

RF interference to Electromedical devices can usually be reduced or eliminated by suitable RF screening or electronic filtering. Special limits, that may be significantly lower than general limits for population, may apply to medical devices, implanted or not, and to medical instrumentation.

(2) Implanted and portable devices

EMF can cause RF interference to active implanted or portable medical devices. Insulin pumps and cardiac pacemakers belong to this class and, in future, there may be an increasing number of these devices. Also the range and the number of different new devices appear to be increasing, e.g. portable monitors, prosthetic aids for sight and motion. Generally speaking, pacemakers and other medical devices could suffer interference from radiated EMF. However in the case of Electromedical implanted devices, RF interference problems have not yet been completely solved due to the lack of full awareness of the problem by manufacturers and suppliers.

2.14 Typical Safety Instructions

Instructions as defined in this paragraph should be issued by the competent authority for the staff that may be required to work on or near the equipments which produces EM radiation of high intensity.

(a) General Instructions

(1) This is intended to warn personnel engaged in installation, maintenance and testing of radio equipments, to exercise extreme care to avoid exposure to intense RF radiation especially in r.o. eyes and reproductive organs which are more susceptible to damage.

Radiation Hazards 109

(2) For normal environmental conditions and for incident EM energy of fre 10MHz to 100MHz, the radiation protection guide is 10 mW /cm² as averaged over any possible 0.1 hour period (6 minute) i.e., Power density;10 mW / cm² for period of 0.1 hour or more and Energy Density; 1mWh /cm² during any 0.1 hour period or more.

(3) For non-continuous exposure, involving power density higher than 10 mW /cm², for short duration, the important consideration is the permissible temperature rise within the body.

(4) The level considered safe has been under discussion, so a lower level needs to be specified.

(5) To specify intense local field like a wave guide if any, even though there may be a little or no danger from total body say a microwave dish antenna.

(6) There is a little evidence of cases of injury or damage to personnel since the first installation, but there have been few cases where staffs have been doubtful whether radiation might have caused temporary discomfort to body and eye. Unfortunately it has been difficult to obtain firm evidence that radiation has been the cause.

(b) Precautions

(1) Inspection of parts like open ends of waveguides, lines etc, through which EM radiation exceeds the limit shall not be carried until the equipment have been made safe for the purpose of inspection.

(2) Staff tuning /adjusting transmitters or matching network etc. are required to wear personal safety equipment such as eye protectors, head protectors and overalls fabricated from silverised wire knitted mesh material.

(3) Notice, Danger board, and Warning Signage shall be prominently displayed on all radio equipments which are subject to exceeding radiation limit.

(4) Access to area above permissible limit will be to authorised staff only.

(5) Any increase in body temperature or loss of acuity of vision resulting from RF radiation exposure shall be reported immediately.

(6) Staff or visitors who have metal plate or wire or bolts in the bone for fracture repair or plastic implants or who wear cardiac pacemakers shall be specially warned against exposure.

(C) Measurements locations for RF fields

Measurements of RF fields at any installation must be done at all the strategic locations near the sources of radiation or where operational staff or public is likely to be exposed. Locations of hazardous spots may differ from installation to installation. The distances of these points also vary depending upon the frequency and intensity of radiated field, type of service being provided, antenna characteristics and availability of reflecting structures in the vicinity. The locations become more critical when a number of transmitters are operating at a common site.

Sources of Hazards are typically very common at high power AM Broadcasting centers. For example; some of the common points near the intended/unintended sources of hazards at most of the transmitting stations are:

(a) Antenna /Aerial Field, (b) Mast base enclosures, (c) Standby or parasitic antennas, (d) Guy wire Anchor blocks, (e) Transmission lines, (f) Antenna Coupling enclosures, (g) RF Transmitter Cubicles, (h) Transmitter Inspection Windows,(i) Transmitter space openings at top of cabinets, (j) Dummy loads,(k) Control rooms and other places where operational staff remains for most of the time,(l) Steel structures in the vicinity, and (m) Platforms of Towers used for mounting FM, TV antennas and Microwave dishes.

An equivalent of IEC-60244-6 First edition 1976-01 spec – Methods of measurement for radio transmitters Part 6: Cabinet radiation at frequencies between 130 kHz and 1 GHz is being issued with BIS number by LITD committee.

References

1. Evaluating fields from terrestrial broadcasting transmitting systems operating in any frequency band for assessing exposure to non-ionizing radiation –Rec. ITU-R BS.1698 – (PP – 1to 77) – (2005)
2. Evaluating Compliance with FCC Guidelines for Human Exposure to Radiofrequency Electromagnetic Fields, OET BULLETIN 65, Edition 97–01 August 1997, Authors: Robert F. Cleveland, Jr. David M. Sylvar Jerry L. Ulcek, Standards Development Branch Allocations and Standards Division Office of Engineering and Technology, Federal Communications Commission Washington, D.C. 20554.
3. Radiation Hazards – Chapter 25, (PP 372-387) –Handbook for Radio Engineering Managers –J.F. Ross

CHAPTER 3

FIRE HAZARDS

3.1 General

Highest standard of reliability is required from most of the modern radio broadcast equipments and setups. So design as well as project engineers give considerable emphasis on reducing the risk of failure due to fire to the lowest possible level in radio stations. Fire hazards in radio installations arise from the use of various flammable organic materials such as acoustic tiles, wooden technical furniture, and wooden framework for supporting acoustic treatment on walls & ceiling, insulation for acoustics & air-conditioning duct work, dielectrics in the supporting and housing of all types of radio and electronics equipments. Even in a well designed facility fault conditions and environmental factors may cause overheating of wires and cables resulting in very high temperatures or the breakdown of insulation material due to aging, physical damage, lightning, or other factors may cause an electric arc. Any of these conditions may result in ignition of flammable materials or the generation of explosive or toxic gases.

For protection of radio equipments against fire, the nature of the materials likely to be used has to be taken into account. The ignition of the insulating materials, transformer oil, capacitor wax, rubber, pitch, wooden coil formers, cotton fibers, printed circuit boards are a common occurrence. In a radio installation with tall towers particular concern is the protection of equipments from fire due to lightning strike on these structures.

Inspite of care to avoid fires, the occurrence of a fire can highlight the importance of care to be needed. It is interesting to analyze some cases of fire which occurred under broad categories of transmitting stations, Radio studios and fire caused by lightning. The financial losses due to fire incidences worldwide in broadcast & TV installations and other radio communication stations are staggering. This necessitates the need to develop sound and continuing educational precautionary & safety program for prevention and protection from the fire.

Fire can strike anywhere there is a fuel (any combustible material) and igniter (any source of heat able to raise the temperature to ignition point) and the circumstances to sustain it. Fire can strike unexpectedly with dramatic suddenness and a radio station is no exception. In fact large radio transmitters, in particular old generation type with not so elaborate protection system, are potentially self-destructing. A large numbers of radio, electronics and electric equipments installed at transmitting centers, and studios are potentially self-igniting, either due to heating of conductors on account of overloading beyond their safe designed capacity, or by arching at contact points or through insulation or by excessive heat loss in a dielectric in the absence of poor ventilation.

Majority of the inter-unit wiring cables in a radio station have either a rubber or thermoplastic insulation protected with thermoplastic Jacket, in most cases polyvinyl chloride (PVC). Due to circuit requirements these are normally grouped in large numbers in bunches in a confined space. Such cables bunches running in troughs, trenches, raceways and below floors in confined, unventilated space are a potential fire hazard. Although so-called self extinguishing insulations help to retard burning, but are still a fire hazard due to large numbers bunched together.

From the records of fire in radio installations it has been observed that most of these are of common type or ordinary nature. These can be grouped into due to electric faults, lightning strikes, heating system, hot devices, poor ventilation, and human carelessness. Combustible construction materials, lack of automatic fire detection devices and non availability of automatic fire extinguishing facilities are responsible for large number of losses. Radio transmitter installations suffers most and the fire cases are devastating as these facilities are situated in remote and far off areas lacking adequate water supply and fire fighting facilities to combat the fire.

Sometimes it has been observed that equipment design and environmental conditions are main factor for aggravating the fire. Fire in a cable may place an earth on the start lead of the cubicle blower motor and make it operational resulting into supply of more fresh air to the fire to aggravate it further. Environmental conditions due to installation and maintenance features appear to have been major contributory factor in major fires, causing the equipment to be operated beyond specified limits. Some unattended transmitters involved in fires contained overload protection circuits which would normally be adequate under a staffed operation but the fire which occurred showed that a new approach is necessary in the design of protective circuits for unattended operation.

In a normal office situation, fire protection facilities are aimed at limiting the fire to a single area to give significant time for evacuation of personnel. These also apply to a broadcast studios located in single or multi storey buildings. In buildings housing the machine, plants and equipments, the facilities are arranged to prevent fire damage to the technical installation which, because of its constructions cannot be removed.

3.2 Classification and constituents of Fires

Internationally accepted classifications of the fires are as below;

(a) Class A
Fires involving solid materials normally of an organic nature (compounds of carbon), in which combustion generally occurs with the formation of glowing embers. Class A fires are the most common. Effective extinguishing agent is generally water in the form of a jet or spray.

(b) Class B
Fires involving liquids or liquefiable solids. For the purpose of choosing effective extinguishing agents, flammable liquids maybe divided into two groups; (i) Those that are miscible with water, and (ii) Those that are immiscible with water. Depending on (i) and (ii), the extinguishing agents may be water spray, foam, vaporizing liquids, and carbon dioxide and chemical powders.

(c) Class C
Fires involving gases or liquefied gases in the form of a liquid spillage, or a liquid or gas leak, and these include methane, propane, butane, etc.

Foam or dry chemical powder can be used to control fires involving shallow liquid spills. (Note; Water in the form of spray is generally used to cool the containers.)

(d) Class D

Fires involving metals and extinguishing agents containing water are ineffective, and even dangerous. Carbon dioxide and the bicarbonate classes of dry chemical powders may also be hazardous if applied to most metal fires. Powdered graphite, powdered talc, soda ash, limestone and dry sand are normally suitable for class D fires. Special fusing powders have been developed for fires involving some metals, especially the radioactive ones. Presently special dry chemical powders have been developed for extinguishing metal fires.

(e) Electrical fires

Now-a-days, it is not considered a separate class, since any fire involving; electrical equipment must be a fire of class A, B or D. The normal procedure in such cases is to cut off the electricity and use any extinguishing method appropriate to what is burning. Only when this cannot be done with certainty, special extinguishing agents will be required which are non-damaging to the equipment. These include vaporizing liquids, dry powder carbon-dioxide, and other gaseous extinguishing agents.

3.2.1 Constituents of Fire

Three basic constituents of the fire are;

(a) Combustion Process

An understanding of the basic principles of combustion or fire, causes and sources of ignition, fire growth and fire spread is necessary for understanding the principles of fire control and extinguishment. Combustion usually involves an exothermic chemical reaction between a substance or fuel and oxygen. Unlike slow oxidation, a combustion reaction occurs so rapidly that heat is generated faster than it is dissipated, causing a marked increase of temperature, even up to a few hundreds of degrees. Very often, the temperature reaches so high that visible light or flame is generated.

(b) Fire Triangle

Fire or combustion can be described in terms of "the triangle of fire" or "combustion" as in fig 3.1. It has been seen that for combustion to occur three factors are essential; (a) heat, (b) oxygen (or air), and (c) fuel or a combustible substance. Fire or combustion will continue as long as these three factors are present. Removal of one of them leads to the collapse of the triangle and the combustion process stops.

(c) Nature of flame

As is known, burning of most materials produces a flame. A flame front stemming from a local ignition source is established in a flammable medium. A form of chemical reaction is set-up in the layer of gas adjacent to this source with the result that heat and chain carriers pass into the next layer of gas

Fig 3.1: A Triangle of Fire, the three constituents

and continue the cycle of the operations, as runners in a relay race. Chain carriers, known as free radicals, believed to be atoms or part of molecules, are extremely reactive. Combustion, therefore, is a type of chain-reaction.

The flame temperature is very important because the rate of a key combustion reaction ($H+O_2 = OH+O$) is very sensitive to temperature. A small decrease in temperature causes a disproportionately large decrease in the rate of the reaction. A single H atom, when introduced into an H_2-O_2 mixture at an elevated temperature will be transformed in a fraction of a millisecond to form 2 molecules of H_2O and 3 new H atoms. Each of these H atoms can immediately initiate the same sequence, resulting in a branching chain reaction, which continues until the reactants are consumed. The remaining H, O and OH species recombine according to the reaction;

$$H+O = OH \text{ and } H+OH = H_2O$$

Similar chain reactions occur in flames of any H containing fuel. H is present in the vast majority of combustibles except for metals and pure carbon. H atoms or other active species (radicals) may also be removed from the flame by purely chemical means that is by an extinguishing agent capable of chemical inhibition. Hence, there are two fundamental ways of reducing combustion intensity in a flame, ultimately causing extinguishment:

(i) Reducing the flame temperature, and (ii) Adding a chemical inhibitor to interfere with the chain reaction.

The flame temperature is generally around 1900°C and above. By bringing the flame temperature to below around 1200°C to 1300°C, flame is unable to continue burning. Thus, below a critical temperature (1200°C to

1300°C), the chain breaking reactions will dominate and the flame can no longer burn. In the absence of oxygen within certain zones of the flames, the organic (carbon containing) materials are decomposed by heat, giving rise to tarry and sooty decomposition products. In other words, smoke is generated. Besides, carbon monoxide is also formed due to incomplete combustion.

3.2.2 Spread of Fire

It is not easy to predict where and when a fire may start or the conditions by which it may spread. The broad patterns, however, a fire may follow in the early stages will be determined largely by whether the equipment, the building itself, or both becomes involved initially. Since many of the material used in Radio equipments, including papers and other stationery type documents, can be readily ignited or can be set alighted by a high temperature source or sparking in the relays etc. In some favorable circumstances which normally occur in broadcasting like presence of forced air movements from air cooling blowers, rapid flaming can be induced. Unless blowers are shut down immediately a fire starts, the effects can be disastrous. This can produce enough heat to ignite the building materials and other items which might not have hazardous situation had they been ignited directly.

In rack type of equipments installations and configurations, heat can usually rise freely and flame will spread vertically in the equipments rather than spreading horizontally. In the case of enclosed cubicle the fire may be reasonably contained but if there is air flow it may spread through air-duct. In rack type of equipment, the fire will move towards ceiling and may ignite the overhead cabling and ceiling materials.

The use of cubicles with equipment mounted on horizontal shelves at various levels can introduce unusual situations. In case of a transmitter, a fire started amongst transformers, capacitors, and a blower motor and miscellaneous equipments mounted in the lower bottom compartment. The fire burnt slowly, consuming a large quantity of insulation materials but flames did not spread to the top compartment housing the radio and audio frequency units of the system. But this slow fire in the bottom compartments resulted in the intense accumulation of very hot combustible distillate in the top compartment and when the fire flames entered top compartment, they ignited the distillate, resulting into very fast ignition of all the combustible contents.

A fire will spread very fast and violently when fed by oil or other flammable liquids from exploding capacitors, coils, transformers, and fuel tanks. The situations from exploding of high voltage power or modulation transformers or smoothing or modulation chokes can be disastrous, as burning oil will spill to nearby components or on workers and will prove to be uncontrollable and devastating. So to contain the spread of the fire should be one of the main concerns once the fire had been detected.

3.3 Fire Extinguishing Methods and Media

It is seen from the triangle of fire of fig 3.1 that three factors are essential for combustion, namely; (i) The presence of a fuel (A), or combustible substances; (ii) The presence of oxygen (B), usually as air or other supporter of combustion; and (iii) The attainment and maintenance of a certain minimum temperature or heat (C).

3.3.1 Extinguishing Methods

Fire extinguishing, as shown in fig 3.2, in principle, consists in limiting or eliminating one or more of the factors, and accordingly the methods of extinguishing fire may be classified under following headings: (a) Starvation (or cut off the fuel); (b) Smothering / Blanketing (or the limitation of oxygen); and (c) Cooling (or reduction of temperature). In practice, specific methods of fire extinction often embody more than one of these principles, but it will be convenient to consider them according to the main principle involved.

(a) Starvation Method

The extinction of fire by starvation is applied in three ways; i) By removing combustible material from the neighborhood of the fire. Examples of these are, the drainage of fuel from burning oil tanks; the working out of cargo at a ship fire, the cutting of trenches

Fig 3.2: Fire Extinction Methods

in peat, heath, and forest fires; the demolition of buildings to create a fire stop; counter-burning in forest fires; ii) By removing the fire from the neighborhood of combustible material as, for instance, pulling apart a burning haystack or a thatched roof; and iii) By sub-dividing the burning material, when the smaller fires produced may be left to burn out or to be extinguished more easily by other means. A typical example is the emulsification of the surface of burning oil, whilst the beating out of a heath fire owes much of its effectiveness to this.

(b) Smothering Method

If the oxygen content of the atmosphere in the immediate vicinity of burning material can be sufficiently reduced combustion will cease. The general procedure in methods of this type is to prevent or impede the access of fresh air to the seat of the fire, and allow the combustion to reduce the oxygen content in the confined atmosphere until it extinguishes itself.

An important practical application of the smothering method is the use of foam, which forms a viscous coating over the burning material and limits the supply of air. It also tends to prevent the formation of flammable vapour.

Another method of smothering is by the application of a cloud of finely divided particles of dry powder, usually sodium bicarbonate, from a pressurized extinguisher.

A further development in the smothering method has been the discovery of a powdered compound for use on metal fires, such as uranium and plutonium, thorium and magnesium. This powder (ternary eutectic chloride) is applied by means of a gas cartridge pressurized extinguisher. As the fusing temperature of the powder is in the region of 580^0 C, it forms a crust over the burning metal and this excludes the oxygen of the air. The vigorous discharge of an inert gas in the immediate vicinity of the fire so reduce the oxygen content of the atmosphere for the time being that combustion cannot be maintained. Carbon-dioxide and nitrogen are familiar examples of this.

A group of extinguishant consisting of volatile liquids based on the halogenated hydrocarbons is also in use. These evaporating liquids act partly as inerting blankets similar to those mentioned in the preceding section, and partly by chemical interference with the chain reaction of flame propagation.

(c) Cooling Method

If the rate at which heat is generated by combustion is less than the rate at which it is dissipated through various agencies, combustion cannot persist.

The application of a jet or spray of water to a fire is invariably based on this simple but fundamental principle. There are many variations. Another example is the emulsification of the surface of oil by means of the emulsifying type of spray nozzle producing an oil-in-water-emulsion. The cooling principle in fire extinction is the one most commonly employed form as it is done on the application of water and other liquids to burning materials.

The action of water depends predominantly on its thermal capacity and latent heat of vaporization, the latter being by far the more important. Thus it takes about six times as much heat to convert a certain weight of water at its boiling point into steam as is required to raise the temperatures of the same amount of water from the usual atmospheric temperature to its boiling point. In fact, while changing from liquid (water) to vapour state (steam) water expands about 1760 times which also contributes to its smothering effect.

To be highly effective, water should be applied to a fire in the liquid condition and in such a way that as much as possible it is converted to steam. The smothering effects of the steam produced at fire spot are thought to play a part in assisting in the extinguishing process. On the basis of thermal capacity and latent heat of vaporization, water is an excellent fire extinguishing agent since both figures are high. For instance, the thermal capacity or specific heat of water is 4.2 kJ/kg /°C and latent heat of vaporization is 2260 kJ/kg. This fact, combined with its availability in large quantities, makes it the most useful fire extinguishing agent for general purposes.

(d) Tetrahedron of Fire (Fourth factor contributing to fire)

The triangle of fire representing three basic constituents of fire is the conventional concept. Fire scientists have now found that there is a fourth constituent in all flaming fires which play a vital part in the fire growth and sustenance. This is the unbroken or uninhibited chain reaction. Thus, as per modern concept, the previous triangle of fire has been transformed into a

Fig 3.3: Tetrahedron of Fire

tetrahedron of fire as in fig 3.3, each of its four sides representing one of the four basic requirements: fuel, temperature, oxygen and unbroken chain reaction. This last factor comes into play only in flaming mode of combustion which is normally applicable in case of flammable liquids and gases.

In the flame front, due to chemical reaction, active free radicals of OH*, H* and O* species are produced which act as chain carriers which help to sustain the flame. Extinguishment of fire by flame inhibition or breaking the chain reaction is achieved when these active free radicals or chain carriers are inhibited or eliminated. This principle of fire extinguishment is known as breaking the chain reaction which is achieved by removal / suppression of the free radicals. The extinguishing agents used for this purpose are halogenated hydrocarbons or Halons / Halon alternatives and several types of dry chemical powders. On application of these agents, the flame becomes inhibited and extinguishment is achieved.

3.3.2 Fire Extinguishing Agents/Media

Main media /agents in decreasing order of importance/efficiency for extinguishing the fires are as below;

(a) Water
(1) Despite that many new techniques have come to the assistance of firemen, but water is still the most efficient, cheapest and most readily available medium for extinguishing fires of a general nature. The method of applying water to a fire varies according to the size of the fire.
(2) For major fires, greater quantities of water are necessary, and the built-in pumps driven by the vehicles engines are often capable of pumping 4500 litres (1000 gallons) per minute (or more) giving the necessary energy to the water to provide adequate striking power.
(3) A variation in the application of water can be made by means of nozzles that produce jets or sprays ranging from large sized droplets down to atomized fog effects. Judicious use of this type of application can not

only cut down the amount of water used, minimizing water damage, but will ensure that it is used to greater effect.
(4) Few special properties of water which makes it most efficient and generally accepted extinguishing agent are:
   (a) Water has a high specific heat capacity of: 4.2 kJ / kg / per º C.
   (b) Water has a high latent heat per unit mass, at least 4 times higher than that of any other non flammable liquid.
   (c) It is outstandingly non-toxic.
   (d) Its B.P (100ºC) is well below 250 to 450ºC range of pyrolysis temperatures for most solid combustibles.
   (e) Water extinguishes a fire by a combination of mechanisms cooling the combustible substance, cooling the flame itself, generating steam that prevents oxygen access, and as fog blocking the radiative transfer of heat.
(5) In practical fire fighting water has to be applied at 10 to 100 times the rates prescribed in laboratory tests because of the difficulty of ensuring that the bulk of it reaches the burning surfaces.

(b) Foam and Foam-making Compounds
(1) Foam as used by fire brigades is usually generated by the mechanical agitation of a diluted foam compound solution in the presence of air.
(2) The desirable characteristics of foam are resistance to radiant heat, to fuel vapors and to loss of water content by drainage. The most satisfactory measure of the efficiency of the foam as a firefighting agent is the minimum rate of application at which a fire is controlled by the agent. As per conventional standards, it was usual to allow 50 litres per square meter (1 gallon of foam per square foot) of surface area per minute as the ideal rate, although in most cases it would be rather less than this.
(3) Classification: Foam concentrates can be classified in two ways:
   (i) Classification by Expansion; (a) Low expansion (LX); Low expansion ratio up to 50:1 (Usually between 5:1 and15:1),(b) Medium expansion (MX); expansion ratio between 50:1 and 500:1 (Usually between 75:1 and 150:1), and (c) High expansion (HX): expansion ratio between 500:1 and 1000:1 (Usually between 750:1 and 1000:1).

(ii) Classification by Constituents;
  (a) Protein Foam Concentrate: Generally used at 4% concentration for low expansion foam production. Expansion Ratio about 8:1. This is effective on most hydrocarbon fuels but not on water miscible liquids. It makes stiff foam with good resistance to burn back.
  (b) Fluoroprotein Foam Concentrate: Generally used at 4% concentration for low expansion foam production having an expansion ratio of about 9:1. More fluid than protein foam gives quicker control and extinction of fires. Good resistance to burn back and resistant to fuel contamination, making it the most suitable type for sub-surface injection for oil tanks. Fluoro-Chemical Foam Concentrate Generally used at 3 to 6% concentration for low expansion foam production having an expansion ratio of about 10:1. It is effective on hydrocarbon fuels and some water miscible liquids. Particularly effective on low boiling point hydrocarbon fuels. Gentle surface application will give quick control and extinction, but burn back resistance is not good, and it is susceptible to fuel/foam mixing and breakdown by radiant heat and hot fuel. Undiluted concentrate may strip paint and care should be taken not to allow contact with the skin. (c) Alcohol Resistant Foam Concentrate; Usually protein foams with additives used at 4 to 6% concentration for low expansion foam. Has the ability to resist water miscible liquids, and is the only practical choice for fires in many polar solvents, like acetone.
  (c) Fluoro-Chemical Foam Concentrate: Generally used at 3% to 6% concentration for low expansion foam production having an expansion ratio of about 10:1. Effective on hydrocarbon fuels and some water miscible liquids. Very fluid foam, gives rapid control and extinction of fire, but burnback resistance not as good as the protein and fluoroprotein types. Undiluted concentrate may strip paint, and care should be taken not to allow contact with the skin. Commonly known as 'Aqueous Film – Forming Foam' (AFFF), as it creates a film over the liquid surface which prevents vapour formation.

(d) Synthetic Foam Concentrate: Normally used at 2 to 3% concentration for low & medium expansion foams, & 1.5 to 2% for high expansion foam, (Expansion Ratio 11:1, 75 to 150:1, & 750 to 1000:1 respectively). Particularly effective on low boiling point hydrocarbon fuels. Gentle surface application will give quick control and extinction, but burnback resistance is not good, and it is susceptible to fuel/foam mixing and breakdown by radiant heat and hot fuel. Undiluted concentrate may strip paint and care should be taken not to allow contact with the skin.

(e) Alcohol Resistant Foam Concentrate: Usually protein foams with additives used at 4% to 6% concentration for low expansion foam. Has the ability to resist water miscible liquids, and is the only practical choice for fires in many polar solvents, like acetone.

(c) Halogenated Agents (Halon and Halon alternatives)

(1) Halogenated extinguishing agents, though a relatively recent innovation in fire protection, are already being phased out, since they have very high Ozone Depletion Potential or ODP (power for depleting the ozone layer above the earth which acts as a shield for protecting mankind from the harmful ultra-violet rays from the sun). They have been phased out already in the developed countries from 1st Jan 1994, and will be phased out in the developing countries like India by 2010.

(2) The Halons are chemical derivatives of Methane (CH_4) or Ethane (C_2H_6), in which some or all H atoms are replaced with Fluorine (F), Chlorine (Cl) or Bromine (Br) atoms, or by combinations of these halogen elements.

(3) Of various Halons, Halon 1301 (Bromo-trifluoro-methane, CF_3Br) is most commonly used, since it has lowest toxicity and highest extinguishing efficiency. Most of fires can be extinguished with 4 to 6% by volume of H1301.

(4) Due to phasing out of Halons, during last decade its several alternatives have been developed. In so far as India is concerned, 12 new Indian Standards on Halon alternatives are in the process of publication by BIS. They are: (i) Gaseous Fire Extinguishing Systems—General

Requirements for design, Installation and Commissioning; (ii) Inert Gaseous Total Fire Protection (Total flooding) Systems— Inergen, Argonite, Nitrogen, Argon; (iii) HFC-227 ea (FM-200) Total Flooding System; (IV) NAF S-III (HCFC Blend A) Total Flooding System; (v) Water Mist Fire Protection Systems; (VI) Specification for Powdered Aerosol System; (vii) Gaseous Fire Extinguishing System—Regular Maintenance; (viii) Test Methods for determining fire extinguishing and inerting concentrations for flammable liquids and gases; (ix) Specification for Halon 1211 and Halon 1301 for essential use (ISO7201–1:1989); (x) Code of Practice for Safe Handling and Transfer Procedures of Halon 1301 and 1211; (xi) Carbon dioxide systems, including high and low pressure and in cabinet sub floor system; and (xii) Fire Protection-Extinguishing Media, Carbon-dioxide-Quality Assurance Test for Fire Extinguishing CO2 Gas.

(d) Water Mist
(1) It is a comparatively recent development as a Halon Alternative. Fine Water Mist technology relies on relatively small (less than 200 microns) droplet sprays to extinguish fires. The three methods of application of Water Mist are: (i) Fixed installation – in a compartment / room for total flooding (ii) Fixed spray nozzles, for local application, and (iii) In portable extinguishers.
(2) Water Mist extinguishes a flame by adopting the following mechanisms: (i) Mist droplets evaporate removing heat and producing cooling (gas phase cooling, which acts as the primary fire suppression factor), (ii) The fine droplets evaporate in the hot environment even before reaching the flame, generating steam and effecting smothering (oxygen depletion), and (iii) Mist blocks radiative heat transfer between the fire and the combustible.
(3) Analytical studies have indicated that water liquid volume concentrations of the order of 0.1 lit of water per m3 of air is sufficient to extinguish fires in the gas phase. This represents a potential of two times effectiveness in extinguishment over application rates for conventional sprinklers.
(4) There are currently two basic types of water mist systems: single and dual fluid systems-both promising. Single fluid systems utilise water

stored at 40–200 bar pressure and spray nozzles which deliver droplets of size 10–100 microns dia range. Dual systems use air, Nitrogen (N) or other gas to atomise water at the nozzle.

(5) Water mist system using pure water does not present a toxicological and physiological hazard and are safe for use in occupied areas. Also there are no concerns regarding ozone depletion or atmospheric lifetime potentials.

(e) Fine Solid Particulate Technology:

(1) This has also been developed recently relating to fine solid particulates and aerosols. These take advantage of the well established fire suppression capability of solid particulates. One principle of these aerosol extinguishants is in generating solid aerosol particles and inert gases in the concentration required and distributing them uniformly in the protected space. Aerosol and inert gases are formed through a burning reaction of the pyrotechnic charge having specially proportioned composition.

(2) Extinguishment is achieved by combined action of two factors such as flame cooling due to aerosol particles heating and vaporising in the flame front as well as a chemical action on the radical level. Solid aerosols must act directly upon the flame. Gases serve as a mechanism for delivering aerosol towards the seat of a fire.

(3) However, toxicity problems about this new technology pose potential concerns.

(f) Carbon Dioxide

(1) Carbon-dioxide possesses a number of properties which make it a good fire extinguishing agent. It is non-combustible, does not react with most substances and provides its own pressure for discharge from storage container. Being a gas, it can easily penetrate and spread to all parts (including hidden) of the fire area. It will not conduct electricity and can be used on energised electrical equipment. Also it leaves no residue.

(2) At normal temperatures, carbon-dioxide is a gas. 1.5 times as dense as air. It is easily liquefied and bottled, where it is contained under a pressure of approximately 51bars (1bar=14.7psi) at about 15°C. As the fire extinguisher is discharged, the liquid boils off rapidly as a gas, extracting

heat from the surrounding atmosphere. The gas, however, extinguishes fire by smothering, or reducing the oxygen content of the air.

(3) As regards toxicity, a concentration of 9% in air is the maximum, most persons can withstand without losing consciousness within a few minutes.

(4) CO_2 concentration to extinguish various types of fuels as shown in Table 3.1, vary from approx. 30 to 62%.

(5) On a volume basis, CO_2 is substantially more effective than N. However, on weight basis, both have nearly equal effectiveness as CO_2 is 1.57 times heavier than N.

Table 3.1: Minimum CO_2 Concentration required for extinguishment of fire in various materials

S. No	Materials	Concentration required	S. No	Materials	Concentration required
1	Acetylene	55	15	Ethylene Dichloride	21
2	Acetone	26*	16	Ethylene Oxide	44
3	Benzol, Benzene	31	17	Gasoline	28
4	Butadiene	34	18	Hexane	29
5	Butane	28	19	Hydrogen	62
6	Carbon Disulfide	55	20	Isobutane	30*
7	Carbon Monoxide	53	21	Kerosene	28
8	Coal Gas or Natural gas	31*	22	Methane	25
9	Cyclopropane	31	23	Methyl Alcohol	26
10	Dowtherm	38*	24	Pentane	29
11	Ethane	33	25	Propane	30
12	Ethyl Ether	38*	26	Propylene	30
13	Ethyl Alcohol	36	27	Quench, Lub. Oils	26
14	Ethylene	41	28		

Note: Apart from above, add safety factor concentration. *May vary depending upon the concentration.

(6) It is actually the depletion of the O_2 level in the air which is responsible for extinguishment in the case of inert gases. A reduction of the O_2 percentage in the air from 21 to 10% by volume would make fires and

explosions impossible, except for a few special gases like H, C2H2, or CS2 which would require greater dilution.

(g) Steam
(1) Steam is the oldest among the smothering agents. Now extinguishing systems based on steam are rarely used. Only in certain ships holds and occasionally in industries involving flammable liquids they are used. These systems are not effective for total flooding, but only for local application by hand held branches or lances. Steam is taken from boilers through fixed piping. The control valves are opened slowly. A by-pass is opened first to warn occupants. Manual systems with flexible tubing and lances are more common. These systems may still be seen in some of the Benzol plants, refineries, oil quenching tanks etc.

(h) Inert Gases
(1) Four inert gases or gas mixtures are developed as clean total flooding fire suppression agents. Inert gases are used in concentrations of 35 to 50% by volume to reduce the ambient oxygen concentration from 14 to 10% by volume. It is known that for most typical fuels oxygen concentrations below 12 to 14% will not support flaming combustion.
(2) The inert gas mixtures developed so far contain Nitrogen and / or Argon; and one blend contains CO2 (approx 8%). They are not liquefied gases, but are stored as high pressure gases. Hence they require high pressure storage cylinders. These systems use pressure reducing devices at or near the discharge manifold. Discharge times are of the order of one or two minutes. The physical properties of the inert gas agents may be studied from any reference for further knowledge on the fire extinguishing properties of inert gases.

(i) Dry Chemical Powders
(1) On most fires involving burning metals, the result of applying water can be explosively disastrous, and so new methods of extinction have been evolved.
(2) The base chemical of most dry chemical powders is sodium bicarbonate. This, with the addition of metallic stearate as a water proofing agent, is widely used as an extinguishant, not only in portable extinguishers, but also for general application in large quantities. Apart from stearate,

other additives like silicones are also used to decrease the bulk density, and to reduce packing in the cylinder.

(3) Dry chemical is expelled from containers by gas pressure and by means of specially designed nozzles and is directed at the fire in a concentrated cloud. This cloud also screens the operator from the flames, and enables a relatively close attack to be made. Dry chemical powder can also be supplied in polythene bags for metal fires, as it is more effective to bury the fire under a pile of bags which melt and allow the contents to smother the fire.

(4) Special powders have been developed for some metal fires, especially for the radioactive metals such as uranium and plutonium. These are known as the ternary eutectic chloride group, (Chlorides of Sodium) (Na), Potassium (K) and Barium (Ba) (in the proportions of 20%, 29% and 51% respectively for the three chlorides). These powders contain an ingredient which melts, then flows a little and forms a crust over the burning metal, effectively sealing it from the surrounding atmosphere and isolating the fire. Dry chemical powders are also tested for their compatibility with foam, as it was discovered that the early powders tended to break down foam, and the two should complement each other on fires where foam is the standard extinguishant.

(5) These powders which are 10 to 75 microns in size are projected on the fire by an inert gas (usually CO_2 or N) the commonly used dry chemical agents are listed in table 3.2 below:

Table 3.2: Common type of dry chemical powder

S. No	Chemical Name	Formula	Common Name	S. No	Chemical Name	Formula	Common name
1	Sodium bicarbonate	$NaHCO_3$	Baking soda	6	Mono ammonium-phosphate	(NH_4) H_2PO_4	ABC or Multipurpose Powder
2	Sodium chloride	$NaCl$	Common salt	-			
3	Potassium bicarbonate	$KHCO_3$	Purple K	7	Urea + potassium bicarbonate	NH_2CONH_2 $+KHCO_3$	Monnex
4	Potassium chloride	KCl	Super K	-			
5	Potassium sulfide	K_2SO_4	-	-	(Pot. Carbamate)		

(6) Only one among the above is effective against deep-seated fires because of a glassy phosphoric acid coating that forms over the combustible surface on application, and that is mono-ammonium phosphate (MAP).
(7) Any dry chemical powder can cause some degree of corrosion or other damage, but MAP, being acidic, corrodes more readily than other dry chemicals which are neutral or slightly alkaline. These dry chemicals, especially MAP, can damage delicate electrical/electronic equipment.
(8) The powders act on a flame by some chemical mechanism, like breaking of chain reaction, presumably forming volatile species that react with H atoms or hydroxyl radicals. They also absorb heat by blocking radiative heat transfer, and in the case of MAP, by forming a surface coating.
(9) Potassium bi-carbonate based agent, often known by the name, Purple K, is approx. twice as effective, on unit weight basis, as conventional soda bi-carb based dry chemical.

3.4 Fire Protection Measures

In previous three sections, the constituents of fire, methods / principles of fire extinguishment, and the extinguishing agents / Media used for putting out a fire have been described. Here this section will deal with the under mentioned important component items of fire protection requirement (suppression equipments and installation) of a building or structure or facility, all of which come under the active fire protection sector:

(a) Fire Detection and Alarm Systems (Automatic Fire Alarm Systems); (b) Fixed Fire Extinguishing Systems / Installations; and (c) First-Aid Fire Fighting Equipment.

(1) The first two, Fire Detection / Alarm Systems and Fixed Fire Extinguishing Systems, are both fixed installations, and the third, First-Aid Fire Fighting Equipment covers mainly portable fire fighting equipment like Fire Extinguishers, except Hose Reels, which are normally included under First Aid Fire Fighting Equipment, although it is, in fact, a type of fixed installation.
(2) Automatic fixed extinguishing systems have proved to be the most effective means of controlling fires in buildings. For understanding

the capabilities of these systems, knowledge of the main principles involved in their installation, uses and applications are necessary.

(3) Apart from the sound design and installation of the systems, an essential requirement for ensuring fail-free operation of all types of fire protection systems is that all persons who may be expected to inspect, test, maintain or operate fire extinguishing systems shall be thoroughly trained and kept trained in the functions they are expected to perform.

3.4.1 Fire Detection and Alarm Systems

(1) Among the fire protection requirements for a building, fire detection and alarm system has an important role to fulfil. If properly designed, installed and maintained, automatic fire alarm systems can be a substantial help in minimising losses of lives and property from fires in buildings of all types of occupancies.

(2) One of the prime objectives of good fire protection system is to reduce, to the possible extent, the time delays which follow a serious fire outbreak, viz., the alerting time, the reaction time, evacuation time, response time and extinguishment time. This objective can be achieved to any satisfactory level only if the building has been provided with a well designed and reliable automatic fire alarm system.

(3) Automatic fire alarm systems are used primarily for the protection of lives, and secondarily for the protection of property. Building Codes may stipulate, sometimes, partial coverage by detection systems. But, it will be good if the designers and builders keep themselves aware of the fact that recent fire research and analytical studies have come to the conclusion that partial detection does not often, if ever, provide early warning of a fire condition.

(4) Detectors are designed to detect one or more characteristics of the fire (also known as "fire signatures" as per NFPA), viz., heat, smoke, (aerosol particles) and flame (radiant energy-IR, visible, UV). No one type of detector can be considered as the most suitable for all applications and the choice will depend on the type of risk to be protected. Different types of fires can have widely different fire characteristics (fire signatures). For e.g., Some materials burn intensely giving out high levels of thermal energy, but with little or no smoke, whereas

smouldering fires have no visible flame and usually have low heat output. Under the circumstances, proper selection and siting of the fire detectors are essential for achieving the fire protection objectives.

(a) Heat Detectors
There are two types of heat detectors;
(1) Fixed temperature detectors; designed to operate when the detecting mechanism or element reaches a pre-determined temperature. These can again be subdivided into two types: (i) Point detectors, which are small, each protecting a limited area, or (ii) Line detectors, which have a linear sensing device usually protecting a larger area.
(2) Rate-of-rise detectors; designed to operate when the temperature rises abnormally quickly, or when a pre-determined temperature is reached. Note: The temperature range normally adopted for heat sensitive (point) detectors is from 55^0C to 180^0C, inclusive if the rate of rise of the temperature is less than 1^0 C/min.
(3) Methods to detect heat; (i) Fusible metals or metal alloys, which melt when a pre-determined temperature is reached, which operates an electrical circuit, and which in turn activates the fire alarm?; (ii) Heat sensitive covering in cable assembly (thermostatic cables), in this, two conductors are insulated from each other by a heat sensitive covering. At the rated temperature the covering melts and the two conductors come into contact initiating an alarm; & (iii) Expansion of metals, the movements created by expanding metals or bi-metal strips are used to make or break electrical circuits. Figures depicting these principles are given below:
(iv) Expansion of Gases (pneumatic detector), this consists of an air chamber having a flexible diaphragm which can move an electrical contact. Heat causes the air pressure in the chamber to increase, making the diaphragm flexed to close the electrical contact.

(b) Smoke Detectors
These are two types;
(1) Ionisation Detector; Ionisation type of smoke detector responds to the invisible products of combustion, including, small particles of smoke, and are of two types; for, "non-fire condition" as well as, "fire condition".

The basic principles involved in Ionisation detector are: The detector head consists of one (or two) Ionisation chamber(s) connected to form a balanced electrical circuit. The Ionisation chamber contains two electrodes, across which a potential difference is maintained, and a radioactive source (usually an alpha-particle source – usually Americium 236) ionises the air producing positive and negative ions which get attracted to the electrodes of opposite polarity. This flow of ions creates a current flow across the electrodes. When smoke particles enter a chamber, the charged ions attach themselves to some of the particles thereby slowing the movement or flow to the electrodes. This results in a reduction in the current flow in the chamber which actuates an alarm.

(2) Optical Detector; this detector, as its name implies, reacts to the visible products of combustion. An Optical detector has two important components, a light source and a photo-electric cell. The critical factor in the operation of the detector is the amount of light falling on the photo-electric cell. Some optical detectors are designed so that, in a fire situation, more light is thrown onto the photo-electric cell. These are called the light scatter type. Others are designed so that less light is thrown onto the photo-electric cell in a fire situation. These are called the obscuration type.

(c) Flame Detectors

A fire apart from producing hot "gases", releases radiant energy in the form of;
(i) Infra-red radiation; (ii) Visible light; and (iii) Ultra-violet radiation

These forms of energy travel in waves radiating from the point of origin, and radiation detectors (flame detectors) are designed to respond to this radiation. These detectors are designed to respond specifically to-(i) Infra-red radiation, or (ii) Ultra-violet radiation, or (iii) Combination of IR/UV radiation

(1) Infra-red Detector: (a) Conventional Infra-red detector: The basic components of this detector are lens and filter which allows only Infra-red radiation to fall on a photo electric cell. On getting the radiation, the cell will transmit a signal to the filter / amplifier. The flame has a distinctive flicker, normally in the frequency range of 5Hz – 50Hz. The filter / amplifier will amplify signals in this range as well as filter out

signals which are not in this range. The signals in this range are then fed to the integrator / timer which will activate the alarm circuit only if the signal persists for a pre-set period of normally 2–15 secs. Thus, false alarms are avoided or minimised. The detector has a neon flasher to indicate which head has been activated.

(b) Infra-scan radiation detector: The conventional Infra-red detector is designed to protect small areas. For larger areas with a more open plan, infra-scan radiation detectors are provided. In this, the detector monitors 360 degrees in the horizontal plane, and a wide angle on the vertical plane. The moment the photo-electric cell is struck by deflected infra-red radiation, the filter amplifier identifies it and the integrator stops the deflector so that the radiation falls continuously on the photo-electric cell. The timer checks whether the flame flicker persists for more than the 2–15 secs period, and then raises the alarm. This detector is able to provide protection for a large area, even upto a radius of approx. 90m.

(2) Ultra-violet detector: This detector responds only to ultra-violet radiation emitted from flames, and normally operates in the range of wavelengths from 200 nm to 270 nm. Solar radiation in this range is absorbed by the high altitude ozone layer, and hence UV detectors do not normally respond to sunlight.

Principle of operation of this detector is similar to that of ionisation detector. When ultra-violet radiation strikes the gas filled tube it ionises the gas in the tube. A small current is set up between the two electrodes, and the alarm is raised as there is a change in the current flow. The integrator helps reduce false alarms caused by external sources of ultra-violet radiation like lightning or even sunlight. This type of detector is commonly used for specialised applications, as for aircraft engine, fuel storage tanks, oil rigs, warehouses, paint spray booths etc. Infra-red detectors are preferable to ultra-violet types in a smoky fire because the former penetrate smoke better.

(3) Multi-sensor fire detectors: These detectors are designed as point type resettable multisensory fire detectors installed in buildings, incorporating at least one smoke sensor and another sensor which responds to heat, and in which the signal(s) of the smoke sensor(s) is combined with the signal(s) of the heat sensor(s).

(d) Choices / Selection of Fire Detectors

(1) Automatic fire detectors should be suitable to the risks and the environmental conditions of the premises to provide the earliest reliable warning as each type of detector responds at a different rate to different kinds of fire.

(2) The main characteristics of different types of fire detectors are given below. This information will help in the selection of detectors for providing protection for various kinds of fire situations as indicated against them: (a) Smoke detectors give faster responses than heat detectors, but more prone to give false alarms also. (b) Ionization smoke detectors are unsuitable for smouldering /PVC/ polyurethane foam / clearly burning fires like Hydrogen, certain grades of petroleum fires etc. (c) Optical smoke detectors are more sensitive to the larger, optically active, particles found in optically dense smoke, but are less sensitive to the smaller particles found in clean burning fires. (d) Both types of smoke detectors have sufficiently wide range of response for general use. (e) Smoke detectors cannot detect clean burning liquids such as alcohol, which do not produce smoke particles. (f) Optical beam smoke detectors incorporating thermal turbulence detectors are particularly suitable for clean burning fires. Ionization smoke detectors are suitable for detection of rapidly burning fires (g) In a life safety situation it is essential to pay primary attention to early detection of smoke and to protect escape routes, ensure operation of detectors on escape routes before optical density exceeds 0.05 dB / m (visibility falls below 20 mtr) (h) Heat detectors being slow are not suitable for life safety installations and in slow burning fire / or fire in air-conditioned premises. (i) Heat detectors are suitable in compartments / areas where heat producing equipment are used (e.g. kitchen, pantry etc) and in other unsupervised spaces / areas with low value contents. (j) Heat detectors with rate-of-rise elements are more suitable where ambient temperature is low or vary only slowly, while fixed temperature detectors are more suitable where the ambient temperature is likely to fluctuate rapidly over short period. (k) Flame detectors are particularly suited for outside applications, and for general surveillance of large open areas in warehouses etc. or for critical areas where flaming fires may spread very rapidly, e.g. at pumps, valves or pipe work containing flammable liquids etc.

Detailed guidelines for selection, installation, system design and maintenance etc. (for fire detection and alarm systems for buildings, selection/choice of fire detectors etc.) are given in relevant national/international Standards like IS: 2189: 1999; BS: 5839: Part-1: 1988; NFPA-72: 2002 etc.

(e) General requirements for automatic fire detection and alarm systems
(1) It consist of detectors and manual call points connected to panels which, in turn, are connected to Control and Indicating equipments.
(2) The protected area should be divided into zones, each zone covering only one storey of the building or any other prescribed area like stairwell, lift well, other vertical shafts etc.
(3) Individual zones/sectors are necessary if the number of detectors in any area exceeds 20.
(4) One of the chief objectives of zoning is to make it easier to determine the location of fire.
(5) Fire alarm should be an electronic hooter / horn / electric bell having frequency range of 500 Hz to 1000Hz.
(6) The distribution of fire alarm sounders should be such that they have a minimum sound level of either 65dB (A) or 5dB (A) above any other noise source likely to persist for more than 30 secs in the area, whichever is greater, and that the alarm is heard at all designated locations in the building.
(7) A multi-state addressable analogue detector system is designed to reduce the incidence of false alarms.
(8) In large or high-rise or special buildings it may be necessary to have two-stage alarms for facilitating evacuation of areas involving greater life hazard. In this case, while alert signal will be sounding in all areas, the evacuation signal will be restricted only to the floor area as well as other areas immediately affected by the fire.
(9) A Control Centre should be provided especially for high rise and special buildings, preferably in the ground floor, wherein following facilities should be made available:
   (i) The Control Centre should have an area of approx $16 - 20$ m^2.
   (ii) All C&I equipments, and other fire protection ancillary panels should be installed in the Control Centre.

(iii) It should have instant emergency lighting system for at least 6 hours of operation.
(iv) It should have direct telephone & intercom facilities, for a direct hot line to local Fire Brigade Control Room.
(v) It should have attached WC bath, drinking water facilities and other appropriate furniture etc.
(vi) It should have a mimic panel of the premises protected and all the fire protection systems.
(vii) Copy of the Fire Orders with approved plan layout for the premises should be prominently displayed.
(viii) It should have preferably an independent A/C – system, to take care of during shut down of main system.
(ix) All relevant records etc. should be maintained in the Centre & original kept in safe in safe with office head.
(x) The Centre should be manned 24 hours by trained competent fire and / or security staff.

3.4.2 Fixed Fires Extinguishing Systems / Installations

(a) Portable and mobile firefighting equipments: Portable and mobile firefighting equipments like Fire Tenders and other vehicle-mounted fire fighting appliances, can be used for tackling fires inside a building or in the open. But for tackling fires inside buildings, structures or in specific areas, fire extinguishing systems for providing adequate fire protection will be required to be installed permanently within the premises.

(b) Fixed extinguishing systems: Fixed extinguishing systems can be based on various extinguishing media used for protection, as below;
(i) **Systems / Installations based on water:** (1) Hydrant Installations; (2) Automatic Sprinkler Installations; (3) Automatic Water Spray Installations; (4) Automatic Deluge and Drencher Installations.
(ii) **Systems / Installations based on foam:** (1) Automatic foam using low expansion foam; (2) Automatic foam installations using medium expansion foam; (3) Automatic foam installations using high expansion foam.

(iii) **Systems / Installations using CO2:** (1) Automatic CO_2 installations (High Pressure Type); (2) Automatic CO_2 installations (Low Pressure Type)
(iv) **Systems / Installations using dry chemical powder**
(v) **Systems/Installations based on clean gaseous extinguishing agents:** (1) Automatic Halon extinguishing systems; (2) Automatic Halon Alternative extinguishing systems

(A) Systems / Installations based on Water

(A-1) Hydrant Installations

(1) Water being the main extinguishing medium, major fires has to be controlled and extinguished by use of water from fire fighting hoses operated by regular fire services. This water is usually obtained from hydrants installed on public mains or other premises or inside the premises as in Indian case.
(2) Hydrant Systems can be of two types: (a) External Hydrant System, where the hydrants are installed in the open, like the city or town water mains, or hydrant systems installed in the open areas in industrial or such other occupancies; and (b) Internal Hydrant System, installed in buildings or structures to be protected
(3) Basic requirements of any hydrant systems are: (a) Water reservoir or source of supply for firefighting purposes; (b) Pumps of appropriate power & type for pushing water through pipe lines to make water available at the required pressures for firefighting purposes; (c) Pipelines, laid underground or above ground, for delivering water under pressure to the required places; and (d) Hydrants which are the outlets installed on the pipelines at strategic locations on the water mains for drawing water, using delivery hoses, for firefighting purposes.

External Hydrant Systems

(a) These systems are essential & important requirements for fighting fires in high-rises' and individual premises. The guideline for provision, installation, inspection & maintenance of these systems are as in IS: 13039–1991-Code of Practice for Provision, and Maintenance of External Hydrant System.
(b) The guidelines regarding the water reservoirs and such other details for water supply are given in IS: 9668–1991. Code of Practice for Provision and Maintenance of Water Supplies for Fire Fighting.

(c) The capacity of pumps required for these systems have to be worked out based on requirements of output and pressure for the systems. Provision has to be made for standby pumps fed from a different source of power at the rate of 50% of total number of pumps, and subject to a minimum of one. The static fire fighting pumps should conform to the requirements given in IS: 12469–1988.

(d) Pressure requirements of systems; Pressure systems are normally designed on practical considerations and specific needs. A minimum residual pressure 1.5kg /cm2 (20 psi) should usually be maintained at hydrants delivering the required flow. In some foreign countries, a separate system designated high pressure system is maintained under the control of the fire department which is utilized for firefighting purposes only. For instance, in San Francisco US, a high pressure system has been provided with water flow rate of 20,000 GPM at 250 psi has delivered to most of the principal mercantile districts. All the pipes are of heavy cast iron tar coated and lined, and tested to 450 psi. The system was provided primarily because an earthquake can put the regular public water system out of service. A few other cities in US also have provided similar systems.

(e) Indian scenario:
   (1) In other developed countries, well maintained hydrant water mains (which may be either a combined system for domestic as well as for firefighting purposes or separate fire fighting water mains, as in some cities) do exist in all cities and towns. Fire service vehicles, on a fire call, report to the scene, connect to the hydrants, draw water from for firefighting operations, sometimes even for several hours. These fire fighting mains are capable of providing non-stop water flow of even up to 20,000 GPM or more.
   (2) As compared to the above in India, we do not have such reliable hydrant water mains even in our metropolitan cities, not to speak of towns. No doubt, in some major cities, there are hydrants available even now in some roads and streets. These hydrant mains were installed during the pre-independence periods, and many of them are either un-serviceable, or not presently traceable due to constructional changes in between. In practice, they cannot be taken in to account for availability for firefighting

purposes. Consequently, most of our city fire brigades are forced to maintain a large fleet of heavy Water Tenders / Tankers for replenishment of their firefighting vehicles. This arrangement is no substitute for having regular fire fighting water mains, which can only guarantee continuous supplies of water for firefighting purposes.

(3) The existing water supply arrangements for our cities and towns are generally based on the formulae recommended in the Manual on Water Supply and Treatment issued by the Ministry of Works and Housing, Govt. of India, sometime back. According to this, no separate provision is made in city water supply for firefighting purposes while calculating per capita consumption of water. However, the system, in some of the cities, is designed to meet broadly the following requirements:

(a) Minimum size of distribution main is kept as 100 mm (as against 150 mm in many foreign countries).

(b) For firefighting purposes, at least 4 streams, each capable of delivering 450 LPM for about 4 hrs should be available within reasonable distance with pressure of $1 kg/cm2$ to $1.5 kg/cm2$. In major towns this may be increased to 6 to 8 streams (In foreign countries like USA the fire fighting water mains are of much larger diameter and capable of handling bigger fire flow rates for operation of 25 streams or more, each of 900 LPM, and added pressures ranging from $5 kg/cm2$ to $8-10 kg/cm2$, even for a city of population of 1.5 to 2 lakh. The fire fighting operations may continue for several hours also without any interruption).

(c) As per the Manual, the following amounts of water are to be provided in the service reservoirs for firefighting;

S. No	Population of less than	Water Capacity kilo litres	S. No	Population of less than	Water Capacity kilo litres
1	5,000	50	4	30,000	300
2	10,000	100	5	40,000	350
3	20,000	200	6	50,000	400

(f) Pressure and flow in Mains
   (i) The pressure of water flowing in the water mains can be expressed either in kg/cm2 or bars (1atmosphere = 14.7psi or 101.325 k N / m2 or 1.013 bar) (1 bar = 100kN/m2), or as meters head (1meter head = 0.0981 bar)
   (ii) The amount of water a hose or pipe will deliver in a given time depends on its size (cross-sectional area) and its velocity of flow. While flowing through the hose or pipe some loss of pressure will be there due to friction loss. The five principal laws governing the loss of pressure due to friction in hoses or pipes are: (1) Friction loss varies directly with the length of the pipe (for double the length of hose, the friction loss will also be doubled); (2) For the same velocity, friction loss decreases directly with the increase in diameter (If the diameter of the hose is doubled, the friction loss will be reduced to one-half, but the quantity of water is increased to four times); (3) Friction loss increases directly as the square of the velocity (if the velocity of the water is halved, friction loss reduces to one-quarter); (4) Friction loss increases with the roughness of the interior of the pipe/hose; (5) Friction loss, for all practical purposes, is independent of pressure.
(g) Distribution system for Water supplies; It generally consists of; (i) Water mains (trunk, secondary and service mains), (ii) Service mains (used for supply to premises from the streets), (iii) Service reservoirs (including overhead tanks, water towers etc.), and (iv) If the length of the mains is excessive provide Booster pumps at intermediate points.
(h) Water Mains-Broad Features;
   (1) increasing the pipe diameter increases water flow. The relative increases are indicated as under:

S. No	Size of pipe, inches	Relative capacity	S. No	Size of pipe, inches	Relative capacity
1	6	1.0	4	12	6.2
2	8	2.1	5	14	.3
3	10	3.8	5	16	13.2

(2) Pipe systems should be arranged in loops wherever possible. This allows hydrants and other connections to be fed from at least two directions and greatly increases the water flow without excessive friction loss.

(3) In course of time the internal cross section of cast iron pipe may be reduced or its interior surface become coarse, because of tuberculation, incrustation, or sedimentation. The addition of cement lining usually retards or prevents such deterioration.

(4) Plastic pipes (approved unplasticized PVC class 4 pipes are being installed increasingly and are immune from Tuberculation and corrosion problems)

(5) Ductile iron pipes have the corrosion resistance of cast iron and almost the same strength and ductility of steel, and are now being used in place of cast iron.

(6) Cathodic protection methods are widely used for the external protection of iron and steel water mains. This protection is a technique of imposing direct electric current from a galvanic anode to the buried pipe line.

(7) Buried pipe needs a coating to protect against soil corrosion.

(8) Forces acting on pipe laid in the ground are mostly internal static pressure of the water; water hammer; load from the back fill, and load and impact from passing trucks and other vehicles.

(9) All pipe lines should be subjected to hydrostatic pressure tests, at not less than 1.5 times the working pressure for not less than 2 hrs minimum period.

(i) Fire Hydrants

(i) Fire hydrants are source of water from the water mains for fire fighting. The water main is provided with a branch or T-piece to which the hydrant is attached either directly or with a short length of pipe.

(ii) There are two types of hydrants – stand-post type, or underground type (sluice-valve type)

(a) Stand-post type hydrant – General requirements: (1) Shall have one or two sluice-valves; (2) Road surface boxes; (3) Duck foot bend; (4) Flange riser, and (5) Stand post column fitted with one (single headed) or two (double headed)

63mm male couplings (male couplings with blank caps are normally provided for city or street hydrants, and female couplings with blank caps are normally provided for internal private hydrants).

Note: Generally, in the case of private systems, beside each standpost type hydrant, a hose box or cabinet will also be provided which usually contains the following items:

(I) Two lengths of 63 mm fire fighting hose conforming to Type – A of IS: 636–1988 with couplings;

(II) One universal branch pipe conforming to IS: 2871–1983 (jet and spray branch);

(III) Spare rubber washers for the couplings.

Since the water mains are charged, fire hoses can be directly connected to the hydrants and fire fighting operations can be carried out without delay.

(b) Underground type of hydrants (Sluice Valve Type): General Requirements: (1) these hydrants are placed underground alongside the water mains on a short branch, water flowing horizontally past the sluice valve; (2) The hydrant consists of three main castings, the inlet piece which is connected to the pipe, the sluice valve itself and the duck foot bend leading to the outlet; (3) For operating the hydrants certain hydrant fittings are required such as hydrant stand pipe, hydrant cover key, hydrant key, water iron, hydrant bar etc.; (4) For locating the hydrants, prominently marked hydrant plates are provided on the ground surface.

(j) Other factors to be taken into consideration in the provision and use of external hydrant systems:

1). Minimum size of mains should be not less than 150 mm; 2). Underground mains should be laid not less than 1m below ground level; 3) Above ground mains should be adequately supported at regular intervals not exceeding 3.5m; 4) The fire hydrant mains should always be laid in rings or loops; 5) Adequate number of shut-off valves (isolation valves) should be provided at strategic locations in the system for the purpose of isolating any portion for maintenance, repairs etc.

5) Fire fighting mains in industrial premises should not be utilised for any other purpose such as process use etc.,6) Normally in cities and towns, hydrants should be provided at intervals of 100m, but this can be varied according to the risks in the area; 7) In case of industrial premises, the intervals for hydrants can be 30m for high hazard occupancies, 45m for moderate hazard occupancies and 60m for light hazard occupancies; 8) Hydrants should be readily accessible to fire appliances and for fire fighting operations;9) No portion of a protected building should be more than 45 m from an external hydrant;10) For systems in cities and towns hydrant inspection should be carried out at intervals not exceeding one month, and for industrial establishments, once every week; 11) Testing of pressure and output in different areas covered by the system should be carried out at least every quarter. (Details of the tests and maintenance of the systems are given in IS: 13039–1991,. Code of Practice for Provision and Maintenance-External Hydrant Systems. 12) For high hazard occupancies, the hydrant system shall be so designed that when half the aggregate pumping capacity is being discharged at the farthest point, and the other half in the most vulnerable area enroute, a minimum running pressure of 5.25 kg / cm2 is available at the former point and rate of flow of water does not exceed 5 m / sec. anywhere in the system; 13) For ordinary / light hazard occupancy, the pressure requirement at the most remote point can be restricted to 3.5kg/cm2; 14) Minimum output of hydrants is generally accepted as 1125 LPM (250GPM) at a minimum pressure of 5.25 kg/cm2; 15) Water monitors fixed at strategic points in the hydrant system can be of various sizes with various outputs as given below:

Size of monitor inlet	Size of monitor outlet (nozzle dia)	Output (LPM)
63mm	32mm	1750
75mm	38mm	2580
100mm	45mm	3500

16) Pump Capacities of Hydrant Systems (for private systems only): The capacities of pumps for hydrant systems can vary according to the risks to be covered and is as below;

Type of Hydrants	Litres/sec	Litres/min	M3/hr
Internal Hydrants	27	1620	96
	38	2280	137
	47	2850	171
External Hydrants (Industries)	76	4560	273
	114	6840	410

Internal Hydrant Systems

(a) These systems are generally installed for fire protection of multi-storey buildings or special structures. An internal hydrant installation comprises of the following elements:

(i) Static or terrace tank for storing water for fire fighting purposes; (ii) Rising mains, down comer mains or external mains to feed water from the source to the required point under pressure; (iii) Fire fighting pump(s) with all fitments and components; and (iv) other necessary components like internal hydrants (also called as landing valves, external hydrants (also called as yard hydrants), hose reels, hoses and branch pipes, in cabinets.

(b) The main features and requirements for the internal hydrant systems are listed below:

(1) Capacity of underground static / terrace water tanks vary according to the fire risks involved in the occupancy;

(2) Internal hydrants form part of any of the following systems – (i) Dry-riser system, (ii) Wet-riser system, (iii) Wet-riser-cum-down-comer system, and (iv) Down-comer system.

(3) Dry riser system is not normally charged with water but could be charged either through the fire service inlet provided at the bottom, or through an installed pump when required, or directly from a fire appliance;

(4) A wet riser system remains charged throughout so that by connecting delivery hoses, fire fighting operations could be carried out immediately. Generally, hose reels are also connected to this system alongside landing valves. The landing valves are required to be located so as to ensure that no part of the building is more than 30m from a valve. This system is normally charged

by static fire fighting pump installed in the building. However, a fire service inlet is also provided for charging it from fire service appliances. The fire service inlet for 100mm and 150 mm internal dia rising main should have a collecting head with 2 no of 63mm inlets, and with 4 no of 63 mm inlets respectively. The down comer system is connected to a terrace tank through a terrace pump.

(5) In addition to wet riser systems, first aid hose reels are required to be installed on all floors of high rise buildings or special type of buildings. The hose reel is generally taken directly from the rising main by means of a 37mm socket and pipe to which the hose reel (generally of 19mm dia) is attached;

(6) The internal hydrant system should conform to IS: 3844–1989. The hose reel should conform to Type-A of IS: 884–1985,. First-aid hose-reel for fire fighting.

(7) For a wet-riser system, two automatic pumps should be installed to independently feed the wet-riser main, one of which should act as stand-by, each pump being supplied by an independent source of power. However, an interlocking arrangement will ensure that only one of the pumps operates at a time;

(8) For bigger systems, it is desirable to install a small pump of approx. 180–300 LPM capacity, with pressure switches for automatic start and stop, which is known as jockey pump;

(9) The system should be tested before use for a minimum pressure of 7kg/cm2 for at least 30 min, after which a flow test should also be carried out;

(10) Details of periodical tests and maintenance etc. are given in IS: 3844 1989, and also in Fire Protection Manual (12th edition – 1998) issued by Tariff Advisory Committee (TAC).

(A-2) Automatic Sprinkler Systems

(a) Automatic sprinklers are devices for automatically distributing water upon a fire in sufficient quantity to extinguish it completely or to prevent its spread, by keeping the fire under control, by the water discharged from the sprinklers. The water for fire fighting is fed to the sprinklers through a system of piping, normally suspended from the ceiling, with the sprinklers installed at intervals along the pipes.

(b) The orifice of the sprinkler head, incorporating the fusible link or bulb of the automatic sprinkler, is normally kept closed, which is thrown open on the actuation of the temperature-sensitive fusible link or fusible bulb.

(c) Automatic sprinkler systems are quite effective for ensuring life safety, since they give early warning of the existence of fire and simultaneously start application of water on to the fire which will help control and extinguishment of the fire. The downward force of the water spray from the sprinklers also helps minimise the smoke accumulation in the room of fire besides cooling the environment and promoting survival of the occupants.

NBC Part-4, Fire and Life Safety, also recognises the importance of sprinklers for achieving fire and life safety. The provision of the sprinkler system in buildings helps to offset deficiencies in fire protection requirements in existing buildings and the Code provides," trade-offs," in the matter of various fire protection requirements when automatic sprinkler systems are provided. For e.g., longer travel distances to exits, higher fire load density etc. are allowed with the provision of sprinklers. However, it has to be mentioned that partial coverage of the buildings by sprinkler protection is neither advisable from fire protection point of view nor from cost effectiveness. In case a fire originates from an unprotected area and after growing into a well developed fire spreads to the protected area, it would have generally developed sufficient intensity to overpower the sprinklers. *Note:* For more details on sprinkler systems installations and types please refer pp 100–110 of, "Handbook on Building Fire Codes".

(A-3) Automatic Water Spray Installations
(1) General
    (i) Water Spray System is a special fixed pipe system connected to a reliable source of pressurised water supply and equipped with spray nozzles for application on area / equipment to be protected. The system can be operated automatically by connection to an automatic detection and alarm system or manually, or both.

    (ii) These systems can be used for any one or more of the under mentioned purposes:
Extinguishment of fire; Control of fire; Exposure protection (cooling); and Prevention of fire (cooling).

(iii) The suppression or extinguishment of fire is achieved by cooling, dilution of oxygen supplies (smothering), dilution (or removal) of the liquid fuel (starvation or emulsification).

(iv) Water spray systems are generally used for fire protection of flammable liquid and gas storage tanks, piping, pumping equipment, electrical equipment such as transformers, oil switches, rotating electrical machinery etc. and for protection of openings in fire walls and floors.

(v) The type of water spray required will depend on the nature of the hazard and protection required.

(vi) Size of the system: Since most systems perform as deluge systems, large quantities of water are required. Normally, a design discharge rate of about 13600 LPM (3000 GPM) is the limit for one system.

(vii) Strainers are required to be installed in the supply lines of fixed piping spray systems to prevent clogging of the nozzles. Water spray nozzles having very small water passages may have their own internal strainers.

(2) Types of water spray systems: There are two basic types of water spray systems installed as fixed systems. One of these is used to extinguish oil fires and usually referred to as. Water Spray Projector System.; the other is mainly used to provide protection to plant, processes, equipment, and to prevent explosions, and is generally known as a "Water Spray Protector System".

(a) High velocity system: This is generally used for extinction of fires in flammable medium and heavy oils or similar flammable liquids having a flashpoint above 65^0 C. (e.g.: Transformer fires)

The system projects water in the form of a conical spray, with the droplets of water travelling at high velocity. Extinguishment is achieved by the three principles of emulsification, cooling and smothering. Some of the water droplets while passing through the flames get converted into steam, thereby achieving the smothering effect. The high velocity sprays of water are discharged through specially designed projectors.

These systems can be operated either manually or automatically. The high velocity spray system for transformers should be well designed to have adequate coverage of the entire transformer unit including the conservation tanks, the bushings and the bottom area. The positioning of nozzles should be such as to protect all surfaces of the transformer and to give a discharge rate of not less than 10 LPM / m2 of the area to be protected. The system should be of pre-action type.

The water spray systems should have isolation facilities so as to enable periodic testing, maintenance etc. Normally, all cut-off valves should be locked open. The high velocity water spray system for transformer protection operates on the same principle as a deluge system. The detectors mounted on a separate pipeline on detecting the fire releases the compressed air within the pipe, thereby operating the deluge valve. This allows the water to flow out through the projectors in the form of high velocity water spray and extinguishes the fire. Generally, a water motor operated gong (as in the case of sprinkler systems) sounds the fire alarm. Sometimes, additionally an electrical alarm may also be provided.

(b) Medium Velocity Water Spray Systems:
  (i) It applies water in finely divided droplets at medium velocity and is mainly used for fire protection of areas with fire risks from low FP flammable liquids (FP below 65^0 (C) and also for fire extinguishment of water miscible liquids (alcohols etc.). It gives protection to tanks, structures, equipments etc. by cooling, by controlled burning of flammable liquids and also by dilution of explosive gases.
  (ii) These systems are similar in operation and lay out to the high velocity systems.
  (iii) Operation can be done automatically as well as manually.
  (iv) Application rates for water spray systems are as follows: (a) for extinguishment — 8.1 LPM to 20.4 LPM/m2 of protected surface; (b) For fire control — not less than 20.4 LPM / m2 (for protection of pumps, glands other critical areas); and (c) For exposure protection (for cooling) — not less than 10.2 LPM/m2

(A-4) Automatic Deluge Installations

(1) These installations are fitted with open spray nozzles, controlled by a single deluge valve and operated on the actuation of automatic fire

detectors, or sprinkler heads, so that the entire area to be protected is sprayed with water. The installation can be controlled manually also. These systems are provided where there is a concentration of highly flammable liquids like aircraft hangars, tank farms filling gantries etc. and for cooling purposes.

The deluge system primarily caters for special hazards where intensive fires with a very fast rate of fire propagation are expected, and it is desirable to apply water simultaneously over complete area.

(A-5) Automatic Drencher Systems
(1) While sprinkler system provides protection for buildings from internal fires, drencher systems (placed on roofs, windows and external openings) protect buildings from damage by exposure to fire in adjacent premises.
(2) The system comprises of drencher heads, generally similar to those of sprinklers, which may be sealed or open (in the latter case water is turned on manually).
(3) Drenchers are of three main types: (a) Roof drenchers – From the roof edge they throw a curtain of water upwards which then runs down the roof; (b) Wall or curtain drenchers — These operate in the form of a flat curtain over the wall openings or portions of a building most likely to be exposed to fire. The usual practice is to put a line of drenchers just below the eaves of the building, so that they provide a water curtain over the wall; and (c) Window drenchers — These are used to protect window openings and placed on the top level of the windows so as to provide a water curtain over the windows.
(4) The installation should normally be connected to the same supplies which cater to hydrant systems. Besides, a fire brigade inlet should also be provided at the bottom. The maximum horizontal spacing of 2.5m is normally kept between the drencher heads.
(5) Do not fix more than 12 drenchers on any horizontal line of pipe, & 6 on the vertical feed pipe respectively.

(B) Systems / Installations based on foam
(1) (a) General; (i) the general characteristics of foam as an extinguishing agent have already been dealt with under para 3.3.2. (ii) A foam system consists of an adequate water supply that can be pressurised, a supply

of foam liquid concentrate, a proportioning device, pipe work or hose for transportation, and foam applicators / pourers for distribution of foam over the risks. (iii) There are certain criteria in general, to be adhered to for foam to be effective:

a) The burning liquid must be below its boiling point at NTP; b) Care must be taken in application of foam to liquids with a temperature higher than 100^0 C. At these temperatures, foam forms an emulsion of steam, air and fuel. This may produce an increase in volume which may lead to slop over or boil over; c) the liquid must not be unduly destructive to the foam; d) the liquid must not be water-reactive; e) fire must be a horizontal surface fire. Three dimensional fires (of falling fuel) cannot be extinguished by foam unless the liquid has a high flash point.

(iv) It is widely accepted that foam is the only permanent extinguishing agent used for extinguishing fires in flammable / combustible liquids. A foam blanket over a liquid surface is capable of preventing vapour formation. Fuel spills are quickly rendered safe by foam blanketing. Foam may also be applied as a protection against accumulation of toxic and flammable gases in hidden enclosures or cavities. It is an essential extinguishing agent for aircraft as well as flammable liquid storage and handling areas.

(2) Foam extinguishing systems are of three types: (a) Low expansion foam systems; (b) Medium expansion foam system; & (c) High expansion foam systems.

Note; for detailed description of the system please refer pp-117 to 135 of, "Handbook on Building Fire Codes".

(C) CO2 Extinguishing Systems / Installations

(1) General; (i) CO2 is suitable for extinguishing the fires of types: a) Fires involving smouldering carbonaceous solid materials (Class A fires); b) Fires involving flammable and combustible liquids (Class B fires); c) Fires involving combustible gases, except where explosive atmospheres are likely to develop (Class C fires); and d) Fires involving live electrical apparatus and installations.(ii) CO2 is not suitable: a) Chemicals containing their own supply of oxygen, such as Cellulose nitrate, chlorates etc; b) Reactive metals such as sodium, potassium, magnesium, titanium and zirconium, and their halides.

(2) There are two types of CO2 extinguishing systems: (i) High Pressure System: This system consists of a battery of one or more cylinders of CO2 interconnected by a manifold, and feeding into a high pressure distribution pipe work. Special discharge nozzles are fitted in the pipe work and on operation of the installation, the gas is discharged into the protected space with considerable noise. (ii) Low Pressure System: In this system the gas is stored in a refrigerated tank at a temperature of (-) 18^0 C and at 20 bars pressure. The tank is connected by pipe work to the protected spaces with discharge nozzles sited at strategic points on the pipe work.

Note; For detailed description on design, methods etc refer pp-136 to 143 of, "Handbook on Building Fire Codes".

(D) Dry Chemical Extinguishing Systems / Installations

(1) General; (a) General characteristics etc of dry chemical powder as an extinguishing agent have already been covered under para 3.3.2. Dry chemical powder is a highly effective extinguishing agent possessing the unique property of quick knocking down of fires. In addition, it has negligible toxic effects. However, on discharge of dry chemical powder system, there will be visibility problems as well as the need for lot of cleaning up after use.(b) When applied on fire, the flame is immediately put out (knocking down). Smothering, cooling and radiation shielding contribute to the extinguishing efficiency of dry chemical. However, research has proved that its power of chain breaking reaction (inhibition of free radicals) in the flame is the principal cause of extinguishment.

(2) Uses and Limitations of Dry Chemical System: (a) Dry Chemical is primarily used to extinguish flammable liquid fires; (b) Being electrically non-conductive, it can also be used on flammable liquid fires involving live electrical equipment; (c) Due to its quick extinguishing ability it is useful for surface fires involving ordinary combustible materials; (d) The systems are used primarily for flammable liquid fire hazards such as dip tanks, flammable liquid storage rooms and flammable liquid spill areas; (e) It's not recommended for use on delicate electrical equipment such as telephone switch boards and electronic computers since such equipment are liable to damage by

dry chemical deposit; (f) Regular dry chemical will not extinguish fires that penetrate beneath the surfaces; (g) They will not extinguish fires that supply their own oxygen by combustion; (h) Dry chemical though can knock down fires quickly, but extinguishing effect is not permanent. So after dry chemical, another permanent extinguishing agent, like water or foam has to be applied on the fire for achieving permanent extinguishment.

(3) Methods of Application: Two basic types of dry chemical systems are fixed systems and hand hose line systems. Portable extinguishers constitute another method of application of dry chemical. (a) Fixed Systems: It consist of a supply of dry chemical, an expellant gas, an actuating mechanism, fixed piping and nozzles through which the dry chemical can be discharged into the hazard area. Fixed dry chemical systems are of two types: total flooding and local application.

(i) Total flooding: In this system, a pre determined amount of dry chemical is discharged through fixed piping and nozzles into the protected area. Total flooding is applicable only when the risk is totally enclosed, or when all openings can be closed automatically. Only where no re-ignition is anticipated can total flooding be resorted to.

(ii) In local application system, the nozzles are arranged to discharge directly into the fire. The principal use of this system is to protect open tanks of flammable liquids. Here again, re-ignition possibilities have to be considered.

(iii) Hand hose line systems consist of a supply of dry chemical and expellant gas with one or more hand hose lines to apply the dry chemical on to the fire. The hose stations are connected to the agent container directly or through piping. These systems are quite useful for protection of gasoline loading racks, flammable liquid storage areas, diesel and gas turbine locomotives, and aircraft hangars.

Note; for detailed description on design, methods etc refer pp-145 to 147 of, "Handbook on Building Fire Codes".

(E-1) Systems/Installations based on Halon (clean gaseous) extinguishing agents

(1) General:
- (a) Halons (Halogenated Agents) are hydrocarbons with one or more hydrogen atoms replaced by halogen atoms: fluorine, bromine, chlorine or iodine. This combination not only makes these agents non flammable, but imparts flame extinguishment properties also. Halons are used both in portable fire extinguishers and in extinguishing systems.
- (b) Halons were first introduced into commercial use during the 1960s, which possessed exceptional effectiveness in fire extinguishing and explosion prevention and suppression. These agents are clean, electrically non-conductive, and leave no residue (coming under the category of clean agents). This unique combination of highly desirable properties led to these agents being selected for a wide range of fire protection applications.
- (c) Halons which are commonly used as fire extinguishing agents are: (i) Halon 1301 – Bromotrifluoromethane – Chemical formula – CF_3Br, (ii) Halon 1211 – Bromochlorodifluoromethane – Chemical formula – CF_2BrCl, and (iii) Halon 2402 – Dibromotetrafluoroethane – Chemical formula – $C_2F_4Br_2$.
- (d) Halon 1301, which is a gas at room temperature, has been used widely in fixed systems throughout the industrial, commercial, marine, and defence including aviation industries. Halon 1211, a vaporising liquid at room temperatures, was preferred for use in portable fire extinguishers, and to a limited extent in extinguishing systems for unoccupied areas. Halon 2402, a low boiling liquid, has primarily been used in the defence, industrial, marine and aviation sectors in Russia and other former Soviet Union countries, and to a limited extent in Indian Navy.
- (e) In the early 80s, as a result of research studies, it was revealed that certain substances, including Halons, were responsible for catalytic depletion of the stratospheric ozone layer. In fact, it was found that the Ozone Depletion Potential (ODP) of the Halons was much higher than other substances involved. The ODP for Halon 1301 has been found to be 13 times more than CFC (Chloro

Flouro Carbon) who's ODP is reckoned as 1. The ODP for Halon 1211 was found to be 3 times more than CFC.

(f) On account of the coming into force of the international agreement known as "Montreal Protocol" on substances having ODP, production of the Halons ceased in the developing countries by 1 January 1994, and will be phased out in the developing countries also by the year 2010.

(g) Many Halon systems still remain in use throughout the world and, therefore, the specification standards concerning these fire fighting agents and their systems still continue to be relevant.

However, the development as well as utilisation of Halon alternative agents is currently in progress worldwide. Apart from the Halon extinguishing systems which are in existence currently, Halon extinguishing systems continue to be used for critical applications. Recycled Halons, recovered from less critical applications, are now providing the source of supply for specialised applications such as defence equipment, aviation use and explosion prevention / suppression applications that remain dependent on Halons.

(2) (a) Halon extinguishing agents achieve flame extinguishment primarily by inhibiting flame chain reactions (a process known as chain breaking). Among the Halogens, bromine is much more effective in this process than chlorine or fluorine. In total flooding systems, the effectiveness of the Halons on flammable liquids and vapour fires is quite phenomenal. Rapid and complete extinguishment is obtained with low concentrations of agent.

Note; for detailed system design, methods etc refer pp-149 to 153 of, "Handbook on Building Fire Codes".

(E-2) Halon alternative Extinguishing Systems/Installations

(1) General: (a) Phase-out of Halon production had dramatic impact on the fire and explosion protection industry. Since Halons occupied an important place in fire protection, their replacement for various applications has been posing several challenges and problems for the fire protection communities all over the world. The process of developing and application of Halon alternatives has been making rapid progress

during the past few years (b) Clean fire suppression agents are fire extinguishants that vaporise readily and leave no residue. Clean agent Halon replacements fall into two broad categories: (i) Halocarbon agents: These are compounds containing carbon, hydrogen, bromine, chlorine, fluorine and iodine. They are grouped into five categories: a) Hydrobromofluorocarbons (HBFC); b) Hydrofluorocarbons (HFC); c) Hydrochloroflurocarbons (HCFC); d) Perfluorocarbons (FC or PFC); and e) Fluoroiodocarbons (FIC).

Their common characteristics are: a) Electrical non-conductivity; b) Are clean agents which vaporise readily leaving no residue; c) Are liquefied gases; d) Can be stored and discharged from typical Halon 1301 hardware (except HFC 23); e) All use nitrogen super pressurisation for discharge purposes (except HFC 23); f) All are less efficient fire extinguishants than Halon 1301, in terms of storage volume and agent weight; g) All are total flooding gases after discharge; h) All produce more decomposition products (mainly HF) than Halon1301; and i) All are more expensive than 1301 on weight basis.(ii) Inert Gases and Mixtures: These includes: (a) Inergen – IG – 541 (mixture of N2 52%, Argon (A) 40% and CO2 8%; (b) Argonite – IG – 55 (mixture of N2 50% and A 50%; (c) Argon – IG 01 (A) 100%; and (iv) Nitrogen – 100%

These are clean agents stored as pressurised gases and, hence, require substantially greater storage volume. They are electrically non-conductive, form stable mixtures in air, and leave no residue.

Note; for detailed properties, system, methods etc refer pp-155 to 163 of, "Handbook on Building Fire Codes".

3.4.3 First-Aid Fire Fighting Equipment

(1) General; (a) All fires start small, and if immediately tackled with proper type and amount of extinguishing medium, can be easily extinguished; (b) In the earlier days, in the absence of any other present day equipment, portable buckets filled with water and sand were used for tackling incipient fires-water buckets for tackling ordinary fires, and sand buckets for oil fires. Even now, for general fire protection, Railway and other remotely located public premises, water

and sand buckets could be seen displayed for tackling small fires; (c) Portable fire extinguishers are specially designed for tackling fires in their initial stage, and they are now very commonly used for the same purpose. Infact, they are now considered as the first line of defence in fire fighting operations and have assumed the front position among the fire protection measures for all types of occupancies as well as fire risks. (d) The term "portable", when applied to fire extinguishers, implies that they can be carried manually to any desired fire scene and operated by one person. As per standards, maximum weight of a portable extinguisher has been specified as 23 kg. (e) Apart from portable extinguishers, which are of comparatively smaller size, there are bigger sizes, which are trolley-mounted and could be pulled to the desired spot. These bigger sizes of extinguishers also come under the broad term of first-aid fire fighting Eqpts.

(2) Types of Extinguishers:
   (a) Portable fire extinguishers can be divided into 5 categories according to the extinguishing agent they contain: (i) Water type extinguishers; (ii) Foam extinguishers; (iii) Dry powder extinguishers; (iv) CO_2 extinguishers; and (v) Halon / Halon alternative type extinguishers.
   (b) They can also be grouped into categories according to their method of operation. Extinguishers can be operated by the use of air or gas pressure in the upper part of the container, which forces the extinguishing medium out through a nozzle. They can also be operated using a cartridge containing an inert gas (normally CO_2) under pressure. When the cartridge is pierced, the gas which comes out of the cartridge drives out the extinguishing medium. In other types, the pressurising agent (air or inert gas) is stored inside the upper portion of the extinguisher itself, and therefore the body of the extinguisher remains permanently pressurised. The first one is known by the name, gas cartridge type of extinguisher, and the second one is known by the name stored pressure type of extinguisher.
   (c) Water (gas cartridge) type extinguisher: (i) In this pressure is released from a cartridge which is stored inside the body of the extinguisher. The cartridge is pressurised with CO_2 gas (to

a pressure of approx. 35 bars). On puncturing the cartridge, by striking the knob on the top, the gas is released, and on coming out of the cartridge, it expels the water from the body of the extinguisher. The expelled water comes out through the nozzle of the extinguisher in the form of a small jet, which can be projected on to the fire; (ii) The liquid capacity of the extinguisher, when filled to the specified level, is 9 litres; (iii) The gas cartridge is screwed on to a holder which is fitted on to the cap of extinguisher. Maximum size of gas cartridge is 60 g for a 9 litre extinguisher.(iv) On operation, the water jet should give an effective throw of not less than 6m for a minimum period of 60 sec., and at least 95% of water in the extinguisher should be discharged.

(d) Water (Stored Pressure) Extinguisher: (i) the extinguisher is filled with water and dry air pressurised up to 10 bars. The air can be supplied by compressed air cylinders or by certain type of pump; (ii) Operation is performed by withdrawing the safety pin, depressing the valve lever and directing the water jet by means of the hose; (iii) As this type of extinguisher is permanently pressurised, it can only be opened for inspection after discharged; and (iv) Normal capacity of this extinguisher is also 9 litres.

(e) Mechanical Foam Extinguisher (9 L): (i) It is filled with pre-mixed foam solution (AFFF); (ii) Foam extinguisher can either be of the stored pressure type, or gas cartridge type; (iii) The operation of these types is similar to what has been stated under the water type extinguishers; and (iv) The figures of the two types are shown below:

(f) Dry Powder Extinguisher: (i) Out of various types of dry powder extinguishers available in market, some are filled with dry powders suitable for class B & C fires, and some for class A B C fires; (ii) As already stated under Extinguishing Media chapter, dry chemical powders have excellent fire knocking down properties. However, for permanent extinguishment, their use has to be followed with discharge of extinguishing media like foam or water; (iii) Dry Powder Extinguisher (stored pressure type): The construction of this type of extinguisher is similar to that of water (stored pressure type). The pressure maintained inside the extinguisher is about 10 bars. It is normally fitted with a pressure gauge and a fan-shaped nozzle; and (iv) Dry Powder Extinguisher (Gas Cartridge) type: 4 sizes of extinguishers of this type are available in the market – 1kg, 2kg, 5 kg & 10 kg capacities. The sizes of the gas cartridges also vary according to the extinguisher size. This type of extinguisher is quite common as a requirement for various types of occupancies.

(g) CO2 Extinguisher: The main features of this extinguisher are: (i) It consists basically of a high pressure cylinder; (ii) The CO2 is retained mostly in a liquid condition at about 51 bars pressure (at a temperature of 15oC); (iii) Different capacities are available, viz., 2kg, 3kg and 4.5 kg (portable types); (iv) Trolley mounted types are of capacities 6.5kg, 9kg and 22.5kg.

(h) Halon Extinguishers: In India Halon 1211 extinguishers are still available although they are – getting phased out; The standard capacities of these extinguishers are 1.25 kg, 2.5 kg, 4 kg, 5 kg & 6.5 kg; They are quite effective on fires in electrical / electronic equipment; They are getting replaced gradually by other extinguishers containing Halon alternatives. (Soda Acid & Chemical Foam Extinguishers have already been phased out, and there IS withdrawn).

(3) Selection and Installation of Extinguishers:
  (a) the most important considerations while selecting extinguishers are the nature of the area to be protected, and the nature of the hazard involved. Another factor to be considered is the human element involved. An individual's reaction to a fire will be largely influenced by his familiarity with the extinguishers, his training and experience in its operation and his self-confidence. Training, therefore, is vital.
  (b) Detailed instructions regarding selection, installation & maintenance of first-aid fire extinguishers are given in IS: 2190 – 1992. It is essential that all users (at least organisations) should be familiar with these instructions, so that maximum advantage can be gained for promotion of fire safety standards for their own benefit.

(4) Inspection and Maintenance of Fire Extinguishers; (a) An inspection is a quick check that visually determines that the fire extinguisher

is properly placed & will operate; (b) Maintenance; a complete and thorough examination of each extinguisher; and (c) A maintenance check involves opening the extinguisher, examining all its parts, cleaning, replacing defective parts, recharging and repressurising the extinguisher, wherever necessary.

3.5 Fire Protection and Safety Management for Various Types of Occupancies

(a) General

Experience has proved that it will be too ambitious and impractical to expect that prevention of fires can be achieved 100% in all types of occupancies and situations, when several unpredictable factors, including vagaries of nature and acts of human commission and omission are bound to occur.

Nevertheless, all those concerned and responsible for enhancement of building fire safety standards continue their untiring effects to mitigate losses of lives and property due to fires. The best possible way to achieve this laudable objective is to develop an integrated system of balanced fire protection that combines the best of different design features of both active and passive fire protection systems for the buildings. This is what all framers and implementing agencies of national and local level Building Codes and Regulations, as well as the entire building construction community should aspire for.

(b) Types of Occupancies

As per code all buildings, whether existing or hereafter created shall be classified according to the use or the character of occupancy in one of the following groups: Group A: Residential; Group B: Educational; Group C: Institutional; Group D: Assembly; Group E: Business; Group F: Mercantile; Group G: Industrial; Group H: Storage; and Group J: Hazardous

(c) Business Buildings (Group E)

Business buildings shall be further sub-classified as follows:

Subdivision E-1: Offices, banks, professional establishments, like offices of architects, engineers, doctors, lawyers and police stations.
Subdivision E-2: Laboratories, research establishments, libraries and test houses.

Subdivision E-3:   Computer installations. Subdivision E-4: Telephone exchanges

Subdivision E-5:   Broadcasting stations and T.V. Stations.

Note: Broadcasting Radio Transmitter station may require more measures as some activities of this will fall into Group G: Industrial and Group J: Hazardous, so it may require additional fire protection and safety measures.

(d) Graphic Symbols for Fire Protection Plans

While making out Fire Protection Plan Drawings certain standard graphic symbols are required to be used for identifications of various fire protection equipment and systems recommended for incorporation in the Fire Protection Plans of the premises concerned.

All those concerned with the design, construction of buildings and installation inspection and maintenance of the fire protection systems and equipment and the implementation of the Plan in its entirety are required to be well conversant with these symbols. Refer IS: 12407–1988, which details Graphic Symbols for Fire Protection Plans.

(e) Fire Protection-Safety Signs

Safety signs pertaining to fire protection/means of exit have come to be universally accepted and adopted as international standards, particularly in the Tourism leisure (Hotels) and Aviation Industries. These have come to be adopted for Assembly buildings like Cinema Theatres, Auditoriums etc., and Refer IS: 12349–1988 of BIS for Fire Protection-Safety Signs for information and guidance.

(f) Legislation for Fire Safety and protection in India

A comprehensive list of Rules, Acts and other regulations for reference, are; 1) Factories Act 1948 (as amended 1987), 2) State Factories Rules, 3) Petroleum Act, 1934, 4) Petroleum Rules 1976, 5) Indian Explosives Act 1884, 6) Explosive Substances Act 1908, 7) Explosive Rules, 1983, 8) Gas Cylinder Rules 1981,9) Carbide of Calcium Rules1937, 10) Indian Mines Act 1952, 11) Oil Mines Regulations 1984, 12) Environment (Protection) Act 1986, 13) Environment (Protection) Rules 1986, 14) Air (Prevention and Control of Pollution) Act 1981, 15) Merchant Shipping Act 1958, 16) Merchant Shipping (Fire Fighting Appliances) Rules1990, 17) Merchant Shipping (Fire Fighting Appliances) Rules 1969, 18) Hazardous Wastes (Management and Handling) Rules 1989, 19) Chemical Accidents (Emergency Planning, Preparedness and Response) Rules1996, 20) Manufacture, Storage and Import of Hazardous Chemicals Rules 1989, 21) Cinematograph Act 1952 (With State Acts and Rules), 22) Public Liability Insurance Act 1991, 23) Motor Vehicles Act 1988 (As amended in 1994), 24) Central Motor Vehicles Rules 1989, 25) Model Fire Service Bill (As circulated to State Governments in 1958), 26) State Fire Service Acts and Rules (In various States), 27) Delhi Fire Prevention & Fire Safety Act, 1986, and 28) Delhi Fire Prevention & Fire Safety Rules, 1987.

3.5.1 Fire Protection for business buildings

Fire protection & means of escape requirements are: (a) Business Buildings have no restrictions in height; (b) They should be of fire resistant construction; (c) Materials used for interior finish, furnishings etc should be of Class-1 flame spread; (d) Occupant load for business building should be $10m^2$/person; (d) Occupants per unit exit width should be 50 for stairways, 60 for ramps and 75 for doors; (e) Exits shall be so located that the travel distance on the floor shall not exceed 30m; (f) No exit doorway shall be less than 1m wide; (g) Exit doors shall not open immediately upon a flight of stairs; (h) All means of exit including staircases, lifts, lobbies and corridors shall be adequately ventilated and lighted; (i) Internal staircases shall be constructed of non-combustible material and should be protected staircases; (j) No gas piping shall be laid in the stairway; (k) No part of the building should be utilised for storage of flammable liquids or hazardous materials; (l) Individual floors shall be prominently indicated on the wall facing the staircases as well as lift landings; (m) In case of high rise buildings, all necessary fire protection as well as means of exit requirements as per NBC Part IV should be incorporated; (n) In case of high rise buildings, fire lifts should be provided as per Code; and (o) All requirements of Emergency and Escape Lighting as per Code should be implemented.

3.5.2 Life Safety requirements for business buildings

Due to following reasons, the life safety requirements in business buildings amount to early alerting the occupants in case of fire or any other emergency, and providing adequate facilities for evacuation to ensure life safety of the occupants as per the type and nature of the building: (a) Building remains occupied generally only during day time; (b) Occupants are generally expected to be alert; (c) Occupants are expected to be familiar with the lay out and exits of the building; (d) Most of the occupants are expected to be physically capable of self-evacuation in case of fire or other emergency. Assistance or special evacuation help will have to be rendered to physically-challenged (disabled) people; and (e) Total evacuation may be difficult in case of a high rise building fire, in which case, phased evacuation may have to be adopted.

3.5.3 Other factors

(a) The fire load in Business buildings is lower than Storage, or hazardous occupancies. But, the occupant load, and thus the life hazard potential in these buildings, is quite high and, therefore, it is necessary that the Business buildings, especially those with larger areas and more floors, must have the stipulated scales of fire protection requirements as per Codes;

(b) The designer and the Code officials will have to undertake a pre-evaluation of life safety as well as fire protection systems required for Business buildings taking into consideration factors like the need for support services for such premises like cafeterias, kitchens, auditoriums, storage areas, car parking facilities, small retail outlets etc. Similarly, the Code as well as mandatory special design and construction requirement for access, movement, accommodation as well as for the means of exit requirements for physically challenged people also has to be adequately incorporated in the buildings.

3.6 Materials used in Radio stations

Almost all inorganic material used in radio engineering, like porcelain, glass & mica, are non-combustible. However most of the organic materials, especially those used as conductor insulations, are combustible to some degree but whether or not they produce a self propagating fire depends on the environmental conditions like air supply to the equipment, temperature, rate at which heat is lost, shape of the wiring form & several other factors. The use of modern insulating materials, such as epoxide and polyester resins and polyvinyl chloride (PVC), has enabled many components, fitting, and supporting devices to be made more compact, thus saving valuable space and increasing the reliability and safety of the equipments. However these insulators are to some extent flammable and their decomposition temperature are not very high.

The components and materials used in radio engineering under normal working conditions should not emit toxic or corrosive fumes, liquids or other deleterious materials which could be detrimental to operating personnel or the operation of other plants. Further any component when subjected to overload current, due to circuit fault conditions, should neither interfere with the operation of, nor cause damage to, adjacent components or materials. The construction of components used in radio engineering should be of such design that they are fail-safe by not causing excessive heat or flaming.

Oxygen index test have been introduced in recent times in connection with the quality control of the insulating materials. Flammability test apparatus which can measure the critical oxygen index of materials to 0.1% level is in use by several organizations in this field. Some specifications for Radio equipments restrict the use of materials whose oxygen indices are below a certain specified percentage. A critical oxygen index is defined as the volume in% of oxygen in a mixture of oxygen and nitrogen in which a given substance will continue to burn.

There are many flammable materials in use in radio stations which even burn in gas mixtures with less oxygen than normally found in air. Other materials like Polytetrafluoroethylene (PTFE) have high oxygen indices. Some experiences of engineers with various materials used in radio transmitter and studios being the cause of devastating fire are stated below for view and to consider it while designing of facilities;

(a) In high power radio stations the voltage at feed-out and feed-in insulators at transmitter hall outlet and feeder hut inlet can be very high, sometimes of the order of 200 KV, and this is serious fire hazard if material used is combustible. So no wood or other combustible materials should be used in feed-in or out bowl plates or its frame or the wall opening notwithstanding it is covered with copper sheet and grounded.

(b) A, TV interference filter installed in a HF transmitter caught fire due to heating of insulating materials in the unit as it was installed in an enclosed section of a 20 cm diameter transmission line. The insulating materials used in the filter were polypropylene and fiberglass.

(c) Much concern has been expressed with the use of PVC in radio equipment due to experience of fire which can spread through overhead and under floor chases to other equipments and areas via PVC cabling. The combined effects of hydrochloric acid and smoke constitute a serious health hazards when fire occurs in PVC covered cables. The PVC cable decomposes into two stages. In 1st stage gaseous hydrogen chloride is released when the temperature rises above 200⁰C and this in turn reacts with the moisture in the air or sprayed water to form a hydrochloric acid mist. This mist can be mistaken for steam by fire fighters or staff and if inhaled it acts as a severe irritant and corrosive agent to the respiratory system. 2ndly at higher temperature carbonaceous degradation of the PVC material takes place forming dense black smoke, which can burst into flames causing devastating fire with severe losses.

(d) A fire occurred in the battery room of one transmitter gutting 20 cells of a 50 volt battery installation. The cells in polystyrene cases ignited and burnt out during the night after shut down of station, when no staff was on duty. This happened as adequate separation was not provided between the cases and maintenance procedure was not followed properly as the need was to keep outsides of cases, the battery stands and cabinets free of acid, and dry at all the times to avoid external short circuits, as well as corrosion problems.

(e) Materials used for recording tapes, and polystyrene cassettes or spools or vinyl LP or EP discs and now-a day's CDs are highly inflammable and many cases of fire have taken place. The extinguishing of fire in these materials is not an easy matter because of the problem of re-ignition. Tests with several fires extinguishing media have led to the following conclusions;

   (1) Both CO_2 and bcf (bromo-chlorodifluoromethane) are capable of extinguishing a fire involving spools and tapes in very early stages of fire. However a considerable amount of CO_2 is required on account of re-ignite of the fire.

   (2) When the tape burns it leaves a residue of red hot iron oxide. The polystyrene spool decomposes to give off flammable vapors which are readily re-ignited by the hot ash of the burned tape.

(3) When carbon tetrachloride (CCL4) is used, heavy brown vapors are given off. Chloride is liberated and persists for a long time. This type of extinguisher therefore should not be used for tapes fires.

(4) Water spray is effective in extinguishing the tape and spool fires. It may be interesting to state and note that when the spool and cassette material is involved in fire, it is sometimes possible, by speedy extinguishing of the fire, to salvage the tape. The reason is that the polystyrene material absorbs the heat and frequently burns with very little damage to the tape inside.

(5) High expansion foam is a very efficient type of extinguishing media for this type of fire. It causes very little damage to the tape, the residue is quickly cooled and re-ignition seldom occurs. As the expansion rate is very high, water damage is negligible.

3.6.1 Heat of Combustion for Insulating Materials

Heat of combustion values for insulating materials commonly used in radio equipments for some materials are very high. From the table 3.1 below it can be seen that wide range of heat would be generated in case of fire in these materials, if they are used in manufacturing. It is important that equipment designers and installers should be aware of the fire characteristics of the insulating materials being used in the equipment. The operator also should appreciate this factor and be familiar with the flammability, flash points, vapor density and other fire characteristics of solvents and materials associated with running and operation of the station facilities.

Table 3.3: Typical Heat of Combustion (BTU/lb) for various Insulating Materials

Sr. No	Material	Heat of Combustion	Sr. No	Material	Heat of Combustion
1	Cotton fiber	7200	7	PTFE	2200
2	Natural rubber	10000	8	Polyvinylchloride (PVC)	9500
3	Nylon	8750	9	Printed Circuit Board	18000
4	Pitch	15000	10	Silk fiber	9300
5	Polyethylene	20500	11	Transformer oil	19350
6	Polystyrene	18000	12	Wooden former	8000

3.6.2 Furnishings Property

As part of the general fire problems of materials at a station, the furnishings also should not pose a fire hazard. Cup-boards for any type of purpose fabricated out of combustible material are normally not advisable. The retiring room, dining room as well as kitchen furniture, camp beds, storage racks and lockers for the shift duty staff should be of steel or other non-combustible materials to reduce the fire hazard.

Fabrics of multipurpose or temporary radio or TV studios are a potential fire hazards and these should not come in contact with hot surfaces like hot lighting systems or reading lamps. To take care of this hazard the lamps should be screened with a wire grill and kept well clear of flammable materials. In case of TV studios where performance calls for the use of smoke bombs, fireworks, flares and the like, special precautions are necessary to protect staff, all other personnel, and the furnishings or the electronic effects should be used.

3.7 Fire Prevention, Protection Requirements and Plan

A very important aspect as described below which can save lot of resources and property including man-hour if given due attention by designer's of a Radio station.

3.7.1 Fire Prevention

As the saying goes, "prevention is better than cure". So prevention must be considered before the protection. Apart from the elimination of combustibles from the immediate area, smoking should be prohibited within certain specified areas/equipment rooms. Flammable liquids should not be allowed in rooms with operational equipments. Some organizations have rules which prohibit the use of any floor cleaning substances which contain flammable materials. Many fires in the installations have been caused by the workman igniting the combustible materials with acetylene welding equipment, spark from the arc welding equipment, grinders and careless disposal of cigarettes & match sticks. A continuous watch for potential fire should be maintained during all such potential works which can ignite fire. Even after the day's work is over all the places should be checked by supervisor for the potential fire possibility before closing and locking the work area for the night time.

Given certain condition of faulty wiring, flammable materials etc, potential equipment fires can be anticipated. However, real danger lies in the apathetic attitude of many workmen towards what can be predicted about fires and what can be done once fire has started.

3.7.2 Fire Protection Requirements

Design engineers must assess fire risks and specify & act within means to minimize it. General requirements for fire protection applicable to most of large radio stations are summarized as;

(a) Plan and provide non-combustible type construction for equipment buildings.
(b) Properly seal leftover portion of all type of openings, whether for cable tray or duct feeder or race ways, etc between equipment rooms, floors, ceilings and walls through which interconnecting wires, cables and ducts pass.
(c) Provide adequate fire barriers between transmitter and transformer rooms or any other areas containing large quantity of flammable liquids like transformer oil.
(d) Steam pipe of transmitter's vapour cooling system where they pass through near to combustible materials should be properly isolated and covered with insulation to retain heat inside and not ignite the material.
(e) Store all combustible materials, including paper and the entire one associated with office services, in enclosed metal cabinets in rooms separated from the station equipments. If this is not practicable than should be kept at a safe distance from equipments and all possible source of ignition.
(f) Provide steel shelving in bin store rooms and the storage of combustible parts should be kept to a minimum.
(g) Install Power plants including fuel supplies, as per standard practices laid down for such installations.
(h) Design and provide a properly engineered fire protection system to suit the needs of the complete station. This should include monitored,

supervised automatic fire detection and alarm system to ensure prompt response.
(i) Install automatic fire detecting or other appropriate devices in ceilings, under floors, in chases and other locations where bulk cabling is isolated from the protected areas.
(j) Use, where practicable, fire retardant cables or cables with fire retardant outer jacket. Also use fire retardant paint on other type of cables.
(k) Protect equipment from lightning damage with suitable discharge devices. This should include protection from high voltage induced due to lightning on towers, feeder lines, power supply lines etc.
(l) Protect equipments from power line surges due to switching, unbalance loads and welding machines.
(m) Provide automatic controls to shut down air-conditioning and ventilation system and to close dampers, in the event of operation of fire detection device.
(n) All general purpose extension boards for portable tools and soldering stations should have pilot lamps.
(o) Use non-combustible filters, preferably dry fiber glass type, for all air inlet system, including transmitter cooling, building ventilation, and air-conditioning systems.
(p) Provide liberally very large quantity of portable extinguishers throughout the station, keeping in view the particular type of material in various areas.
(q) Keep ready to use tarpaulins or water proof covers at strategic locations to protect equipments from water damage, in case of spraying to extinguish the fire.
(r) Ensure that all the operating staff including helpers, cleaners is trained in proper use of all type of fire fighting as per material on fire as well as other emergency fire procedures.
(s) Plan and provide safe exit to all station staff form the building in case of emergency quickly within few minutes.
(t) Ensure that all dry grass and other combustible materials are removed to a safe distance from antenna field, fences, building and other important transmitter system like D/G room, sub-station etc.
(u) Use insulation materials having low oxygen indices. The oxygen ratio of air is 0.21 and more the oxygen index exceeds this figure the more difficult is the combustion in air of the particular material.

3.7.3 Fire Protection Plan

Many Radio installations have suffered heavy losses because management has not developed fire protection arrangements sufficiently enough to match technical advances that have been taking place in Radio Equipments. With ever increasing range of hazardous materials continually being introduced into the radio industry by manufactures with varying degree of flammability and toxicity, management must make a positive approach to deal with the problem. It may not be possible to set down a standard fire protection plan to which every administration can work. However guidelines should be developed which, if followed, could bring the organization into a much stronger position in the event of fire threat.

A typical fire protection plan may have following action points to be followed by the staff;

(a) There must be a direct line of responsibility for fire prevention from top station management to technician level.

(b) The organization structure should cater for a fire executive (from existing engineers), whose job should be to report on planning, installation, and construction, service, or operation activities from the fire point of view?

(c) A well designed and engineered fire protection system should be installed as part of engineering services during construction of building itself to ensure the availability of full-fledged fire protection system before the arrival of equipments and staff. Special fire protection system for radio equipments, such as smoke detector in ducts, built-in CO_2 system etc, should be installed with main radio equipments.

(d) The installed fire protection system should provide training to alert management and staff on everyday potential hazards and a better general understanding of fire.

(e) A well thought and designed programme should be implemented to ensure regular checking of not only physical protection facilities but also of such matters as housekeeping, because it is well known that human carelessness is one of the major cause of fire.

(f) A complete list of items to be checked and inspected at regular / periodic interval should be available in the control of the person-in-charge for fire protection responsibility in control rooms.

(g) Mock fire drills of management and staff, must be held at regular intervals to keep the staff aware of the hazard and the need to control it as and when the need arises.

(h) There should be regular scheduled training and refresher courses for the old and new staff from state fire departments and local fire officers.

3.8 Fire Fighting Functions in a Radio Station

Almost every fire begins at a small scale which can be generally extinguished readily if the station staff on the spot are trained and equipped to handle it promptly. Because of the remoteness of transmitting stations from town, the early arrival of fire brigade is often not possible and as such full fire fighting facilities should be provided at the site to cater for an outbreak. The fire brigade should be considered only as back-up service. In a large manned radio station complex, it is necessary to develop first line of defence against fire, consisting of selected group of staff in each shift with a pre-assigned leader, trained in the use of fire-fighting equipments provided at the station. As soon as a fire breaks, these men will execute immediately a predefined plan procedure to attack the fire. Main considerations in a fire fighting function in a station should be;

(a) Immediate indication and location of a fire outbreak.
(b) Prompt initiation of action to deal with the outbreak by the rapid concentration of the fire team on the scene.
(c) Effective and efficient application of appropriate measures.
(d) Evacuation from the transmitter or affected building of those staff who are not required to assist.
(e) Prompt restoration of service after the fire outbreak has been subdued.

Fire extinguishing apparatus installed in the station should be suitable, adequate and kept maintained in ready to use condition for the potential fire hazards which may take place in a radio station. To fail to use the provided fire devices to control the fire because of improper maintenance or ignorance of a workman in their application is a serious negligence and amounts to disciplinary action on the team leader.

The problem of the unattended radio stations is altogether different as planning for providing highly reliable fire fighting system and equipment has to be made. In the event of any malfunction of equipment which may result in fire, no one is present there to see what is happening. So a system of accurate fire monitoring therefore should be provided. Another point is that even though the fault may be detected, but no staff is available to implement initial fire fighting operations, pending the arrival of fire brigade. Therefore more attention must be paid not only to the automatic detection of fire or smoke, but to its automatic control and extinguishing.

3.8.1 Staff Training

As part of normal Radio Station safety training program, attention should be given to fire safety, protection and extinguishment. It is highly desirable that all the station staff must be trained in the practical use of fire-fighting equipments. However it may not be always feasible as it depends on the type and number of staff engaged, like in a large studio complex, there may be hundred of staffs with many being casual, freelance or on part time employment. So all these part time, freelancer or casual employees cannot be imparted training for fire safety or protection or in use of fire fighting extinguishing equipments. However, even such type of employees must be sensitized by the regular employees in the fire protection subject and use of fire-fighting equipments, if need arises, by regular instructions and advises as and when new part time staff is engaged.

Under such situations it is advisable to fully train selected permanent staff personnel in the use of all type of fire fighting equipments and gadgets available to them by installing in the station and in the most efficient and expedient way to control the fire on the premises. In smaller stations / studios and all transmitting centers, regular training for fire protection and safety should be implemented.

Points to be considered in developing a training program for transmitting stations should include followings;

(a) Every staff / workman irrespective of status on the station should know the location of fire extinguishers and all other associated fire fighting devices as well as equipments.

(b) All staff should know how to operate properly each of the common as well special type of extinguishers and gadget installed at the station as per the requirement of type of fire and recommendation of fire office of area.

(c) Every workman should know the proper extinguisher to use on each class of fire and limitation of each type.
(d) Every one deployed in the transmitter station should be thoroughly familiar with the procedure to be followed in the event of fire and of which team, one is part of & what are his duties in such time.
(e) Every official / staff /workman deployed in the station in a particular area, whether, a member of technical, mast technicians, logistics, store keepers, helpers, cleaning, Guards, Drivers, Diesel technicians, labor or office staff, should be made conversant and as such posses knowledge in the emergency methods to shut down the plant or disconnect the power to the area where the fire have occurred.
(f) All the personnel, who are to be deployed at the transmitter station should be got trained from a fire office. The annual recharge and inspection time should be used for imparting training, advice and instruction to all staff for use of each type of fire fighting equipments including the procedure to be followed by demonstrating practically.
(g) A list of the personnel, trained from fire officer should be displayed at prominent location and on the notice board, which always should remain updated after transfer if so.
(h) Every official / staff /workman deployed at station should assume his duties of fire protection and control as soon as it is required without any further instruction from supervisor.

3.8.2 Type and Characteristics of Fire Fighting Equipments

There are two type of equipments used to extinguish and control the fire. First is fixed type which includes; automatic sprinklers, hydrants, hoses, and special pipe system for carbon dioxide, foam or dry chemical and second type is the portable equipments which supplements the fixed type. Fire fighting facilities should meet the following requirements;

(a) They should be reliable and efficient in operation and to use. (b) The operation of the fire-fighting equipments should not be a danger to human life. (c) The extinguishing media should be suitable for the type of fire most likely to occur in the area protected taking into account the nature of the RF or power equipment. (d) There should be minimum damage to the equipment being protected following the operation of the fire extinguishing device.(e) The fire-fighting equipment should require and occupy minimum space in the equipment area. (f) There should be facility for display of local and remote warning signals that the protection system has operated.

The selection of the fire fighting equipments for adequate fire protection solely depends on the local conditions. For a small radio station, hand held extinguisher may be sufficient. Large stations however may require a standpipe or sprinklers system in addition to many portable extinguishers. Hazardous locations such as transformer rooms or enclosures in high power transmitter stations may require the installation of elaborate and specialized extinguishing systems. Even with such systems, however portable hand extinguishers also should be provided. Hand held extinguisher being easy to carry and handy are helpful and important to extinguish the fire in its initial stage itself without having to use the resource of large apparatus.

Large stations should be provided with sufficient quantity of water as reserved in the overhead tank of the station. As power supply may have to be shut down in case fire breaks, a pump specially designed for operation during fire conditions operating from captive power (diesel generator) should also be provided. All large transmitting and complex radio studios center must have this facility.

As you are aware super power MF and high power multi aerial HF broadcasting stations are established on a site covering an area of up to 500 acres. In such complexes controlling grass fire is often a serious problem during hot dry weather conditions. FM radio stations are established on mountain tops surrounded by forests all around. In such cases when fire breaks out it may be a difficult problem to control it. Such sites should be cleared of uncontrolled vegetation for a distance of at least 30 meter from building, masts, radiator or SW towers or feeder huts. Any vegetation planted for beautification purposes should be kept under control with all shrubs well cleared of building, towers and diesel storage tanks etc. Trees on the property should be kept to a minimum and those left standing should be in such locations that should they fall towards the building, there will be clear space of at least 5 meter between the tree and building. In case of high power transmitter installations in addition to fire, often the smoke laden atmosphere also can cause flashovers in open wire transmission lines, antennas and other high voltage equipments. Motorized fire extinguishing apparatus installed on a standard truck chassis are provided at such stations. The water tank, pump, hoses, ladders, hand extinguishers knaps, hand beaters and other implements such as shovels, rakes, and axes are provided on the truck. All such facilities must be checked for their foolproof operations and functions once a year to ensure availability of full-fledged fire fighting facilities ready for use as and when required.

3.9 Design Considerations for fire measures in a Radio Station

Design considerations to be kept in view by a design engineer of a radio station for fire protection measures are;

(a) In Radio Equipments procurement

Modern radio equipments are generally well engineered with excellent in-built protective circuits with few chances of fire outbreak. But still fire do breakout and cost of replacement of equipments and loss of revenue producing time can be considerable. The danger of fire to a transmitter installation unit can come from within the transmitter cabinet or cubicle itself, from within the room housing the transmitter, from the immediate area around the transmitter like the transformer room, and from outside the transmitter building itself. All these risks have to be diligently evaluated and controlled. Almost all transmitter installation requires forced air cooling for proper operation of the equipments. Circulating fans, filters and air washing units are provided for this purpose. High temperature and air flow failure detection devices are provided in such installations to detect the fault and set to operate the control circuit to shut down the equipment and sound an alarm in the event of a high temperature arising from the failure of air supply or other cause.

It is not possible to construct radio component and equipments from non-combustible parts. The fire risk components and parts found in radio equipments are, insulation materials of wiring, insulation boards, fan and pump motors, oil filled components, resistors, capacitors, and printed circuit boards. Although components may have varying degree of combustibility but fire can spread quickly once it starts, particularly inside a unit or a cubicle where working temperature normally is higher.

Of late, lot of progress has been made in developing insulation material with a low surface spread of flame. A typical example is glass reinforced plastic material used as cladding panels to prevent ice accumulation on FM antenna panels. Lightning strokes and sparking from poor bonded metal parts or joints has resulted in fire seriously damaging these panels. Contrary to the popular belief, very few fires appear to have been caused by the overloading of the conductors. In practice it is not easy to cause ignition by overloading a conductor. Tests have shown that a sustained current of more than five times the rated capacity of the cable is necessary to create conditions where ignition will occur. Ignition caused by overloading of circuit is thus avoidable if the cable conductor is protected by a suitable device, such as fuse or circuit breaker.

Many fire experts have concluded that damage to the installation either by moisture, heat, wax from components, brittle or perished insulation, corrosion or mechanical fault was the real cause of the majority of the cable and wire fires. In case where there is a deterioration of the insulation material, it could be expected that an appreciable amount of earth leakage current would be produced before conditions leading to ignition state would arrive. An examination of some power circuit of transmitters which caught fire showed that many of the fire attributed to this could have been prevented, if the earth leakage circuit breakers had been installed in the equipment.

Therefore in view of above the most important thing is to design the radio engineering equipment and systems so that the possibility of the fire is reduced to a minimum. The next in importance is to be able to extinguish the fire in their initial stage itself. Luckily now there is a wide range of sophisticated electrical and electronic equipments available for fire detection including not only fixed temperature and rate of thermal rise devices but also based on atomic radiation, infra-red, ultra-violet and light scattering principles. So in design in addition to including the fire detecting and preventing devices or systems in transmitter and other radio equipments, the consideration for provision of detecting, preventing and extinguishing of fire should be kept in the system design for the plant.

(b) In Buildings construction
Buildings provided for housing radio studios or transmitter equipments and their auxiliaries and sub-systems ranges from a small hut, shelter type structure or Porta cabin or a huge complex of buildings and structures for large radio studios and transmitting centers. In case of super power

transmitters it is normally two storeys and in case of studios in metro cities it may be a multi storey with even basements.

Irrespective of type of building, experiences have shown that fire can start in so many ways and spread under such widely different circumstances that complete immunity from fire is almost impossible even with a well engineered installation. Although buildings constructed of non-combustible materials do not constitute a fire hazard, and many may even successfully withstand internal fire to prevent spread to other buildings or areas.

It would be extremely difficult to express fire hazard for all type of radio installations and buildings by a single index. There appears to be no alternative but to assess the fire hazards for each set of conditions encountered and then to minimize the risks. The decision for alternatives should be made with regard to experience and best accepted practices. The design of the building should be influenced considerably by the type of equipment to be installed, also the location and the availability of local material. Certain parameter of constraints which should be considered while planning and constructing a building is as;

(1) In case of an unattended transmitting station situated in an area considered to be of low fire risk from environmental hazards, the main equipment building should generally be of brick or metal construction with steel or metal roofing, and materials should be rated not less than 'fire retardant'.

(2) In case of an unattended transmitting station situated in an area considered to be of high fire risk from environmental hazards, the main equipment building should generally be constructed exclusively from non-combustible materials throughout and generally to be windowless.

(3) Attended transmitting stations where fire protection facilities are provided should be constructed according to local building regulations applicable to the area but using the materials rated not less than 'fire retardant'.

(4) In case of a station where an emergency power generating plant is to be installed under the same roof as the main building, access to and from the generating room should be via fire doors of rating not less than one hour. Similarly wall separating them should be of material fire rated not less than that of external wall of main building.

(5) In case of generator set located in a separate building, materials used for constructions should be not less than 'fire retardant' and the building should be at least 3 meter away from the main building.

Some other general consideration in building constructions should be as below;

(1) Materials used for interior finish, furnishings etc. should be of Class-1 flame spread.
(2) Exits shall be so located that the travel distance on the floor shall not exceed 30 m.
(3) No exit doorway shall be less than 1m in width and no exit door shall open immediately upon a flight of stairs.
(4) All means of exit including staircases, lifts, lobbies and corridors shall be adequately ventilated and lighted.
(5) Internal staircases shall be constructed of non-combustible material and should be protected staircases.
(6) No gas piping shall be laid in the stairways as well as No part of the building should be utilized for storage of flammable liquids or hazardous materials.

(c) In provision of Trenches / floor Ducts and Chases

Electrical power cabling either under the floor trenches or above the false ceiling is one of the major fire hazards in a radio station. In number of cases, it was found that the temperature rise of cables within the ducts or chases was greater than that was assumed, probably due to the fact that some of the modern electrical insulating materials also have rather better heat insulating properties than those used previously, so that a heat build-up takes place in the cable duct or chase. As a result of this some designers are providing smoke detectors in ducts and chases.

Trenches and chases run room to room and floor to floor without any fire stopping facilities being provided in between. So in such situation fire can spread from room to room and also to other floor. One method is the installation of fire stopping facilities to retard the spread of fire and smoke is a wise decision in such cases. Some transmitter and studios installations opt for permanent sealing of ducts and chases openings by using cement mortar to preserve the integrity of wall or floor. Second method is using a flexible pillow type mass consisting of fire retardant bag containing non-combustible mineral wool. The bag is pressure fitted into the opening to tightly pack the space around the cables and has been found effective in the retardation of fire, smoke and fumes. Loose mineral wool is also sometimes used but this is subject to dislodge from the opening in time. Approved bags have same fire rating as asbestos. Tests have shown that they will not allow the fire to pass through an opening for 60 minutes at about 1000ºC. This meets the requirement to retard the fire during the critical and initial stages and minimize the fire loss and associated problems.

Third and another method which enables easy extension of trenches cables as required, and at the same time prevents fire or smoke from spreading, is to use about 40 mm thick asbestos dicalcium silicate plates which have 2 hour fire rating. These plates can be easily worked for drilling and cutting. The gap between cables can be filled with a fire resistive caulking material of incombustible inorganic fiber to prevent smoke spreading.

Experience has shown that cost of service restoration following a fire in trenches is influenced considerably by the manpower efforts required in removing the water, drying out and cleaning up. In some case it was noticed that the even tough damage by fire was relatively minor, the cost of restoration was nearly equal to capital cost of the equipment. At one combined studio transmitter installation, a studio helper seeing a minor fire in studio carpet, attacked it with a high pressure fire hydrant installed outside building, instead of beating it or extinguishing through hand held CO_2. The consequences were disastrous as the water flooded the studios, filled the under floor ducts and chase and flowed down the ducts into the transmitter room which was at lower level. This water entered the blower intake and was thrown onto the transmitter, fracturing the glass tubes, valves and causing fire in the MT supply contactor. This tripping made the transmitter to shut down. The cost of restoration in time and money was exorbitant. This is a glaring example of carelessness on the part of a workman and shows the lack of basic training by the management in fire prevention, fighting and initial actions in case of a fire.

These types of situations are avoided by grading equipment room floors to suitable waste water outlets to minimize the water flow in the trenches and chases. Such type of arrangement is highly recommended and desirable in case of transmitter rooms where water or vapor cooled tubes are used because of the danger of the leaks in pipes and pumps. There is possibility of failure of container of oil filled component and a resultant spread of oil on floor, in trenches and chases. For such type of case, the component having less than 20 litres of oil should be mounted in a receptacle of suitable size and shape to contain the oil. However for the components like smoothing or modulation choke or modulation /plate transformer or power transformers having more than 20 litres of oil should be provided with a drainage system to take the spilled oil out of building suitably.

Where under floor trenches and chases are essential, it is common practice to install all wiring, junction boxes etc. on the side walls and leave the bottom free and clear of any wiring or device. In addition the trench floor should have 1 in 30 slopes grading to take away waste or spilled water/ oil. In such trenches the cables can be drawn over an iron grid suspended above the floor.

(d) In Under floor Cabling

In transmitter and studios installation where huge number of consoles, equipments and other accessories are operational requires hundreds of interconnecting cable bunches. To meet specific requirements and to give maximum flexibility, many of these equipments are fixed over raised floors or mechanical racks / structures. When considering the protection of cabling installed in the space beneath the floor, generally 35 to 45 cm, consideration should be given to the grouping of cables, the in-built factor of safety, care in handling during installation, ventilation to remove heat due to I^2R losses, the combustibility of cables and termination / junction boxes, and the means of access to cables for firefighting purposes.

Where practicable, the flooring should be constructed of metal decking type materials but where the use of combustible material like wood cannot be avoided; the timber should be impregnated or coated with an approved fire retardant treatment. As the under floor spaces are high hazard areas, they should be properly protected by providing following protection measures;

(1) Raise an alarm immediately a fire starts if no automatic alarm system is installed or it fails to work.
(2) Take immediate action to extinguish the fire as per plan of the station for such eventualities.

(3) Allow the fire to be fought and extinguished with normal means.
(4) Shut-down the affected equipment immediately by disconnecting the power supply to it.

Installation of ionization type of detector is the most common method of detecting the fire in such situations. In large installations, the area should be divided in many zones, separated by non-combustible bulkheads and the fire alarm panels so wired that it will indicate the specific zone where fire has occurred. The decision to incorporate a fixed automatic carbon dioxide or other type of system depends on the importance of the installation, the hazard risk of the equipment and the particular area where the equipment is installed.

(e) In Storage of Flammable Liquids
Stock of flammable liquids in a radio transmitting station should be kept to a minimum, preferably in accordance with the normal day to day requirements. However for very large transmitter set up in an isolated and remote location, it may be necessary to keep large supplies of transformer oil, lubricating oil etc on hand to meet either emergency situations or normal maintenance requirements. As you know individual transformers associated with high power transmitters may have an oil capacity of up to 4500 liters. There is at least couple of such transformers in addition to power transformers of sub-station at a transmitter station. So it may not be normal to hold this much quantity in station stock, but it would be normal to hold about six drums of 200 liters capacity each for an isolated site of transmitter complex for top up requirements in case of overload faults. This is a small quantity of total oil in transmitter equipments.

In stations where the quantity of flammable liquid is comparatively small, say less than 200 liters total, and is stored in drums or other portable closed containers, these should be stored in a specially constructed steel storage cabinet. Each container in the cabinet should be limited to 20 liters capacity. Where large quantity as in case of above, it should be stored in an isolated fire resistant store away from potential fire sources with full-fledged fire fighting arrangements. Particular care is necessary for such storage area in bonding and earthing all the metal parts of the building to prevent sparking from the induced RF voltages. Even the doors of such shelters should have an earth straps or braded wire bonded to it at few places as a safe guard to prevent sparking while closing or opening it due to induced RF.

At many radio transmitter stations captive diesel generator sets of up to 2 x 1200 KVA capacities to sustain the uninterrupted broadcast service due to unreliable utility power supply or being tapped from a long rural feeder having average regulations, frequency stability and reliability, are provided. Such capacity diesel generator set consumes lot of diesel fuel and for meeting day-to-day fuel consumption requirement of the station as per the outage of utility power supply, huge amount of diesel is required to be stored in underground tanks with fuel lifting pumps as well in drums also for meeting emergency requirements if station is far off from the main city and diesel supply from the main bunk of supplier by diesel tanker is irregular. The storage of this fuel, including for the station vehicles if so, should be in accordance with well documented standard procedure as per the regulations of local authorities. In high power RF fields' special care is to be exercised to bonding and earthing arrangements of diesel tanks, piping, engine sets and switchboards. Although the fire break out record of diesel engine sets at radio stations is reasonably good, even at those stations where continuous local generation of power, due to severe outage of utility power supply is involved, there were lots of fire breakouts to indicate that the fire potential should not be under-rated. Even though some fire were caused by failure of electrical equipments but the majority were associated with diesel fuel and lubricating oil. The use of daily service tank for fuel oil minimizes the risk of large quantity of oil being dumped in to engine room in the event of a major pipe fracture. Experiences however have shown that in some cases there is sufficient oil to cause complete destruction of D/G plant and room.

The storage of gasoline associated with vehicles, station cranes needs if any, requires great care, particularly at high power transmitter stations, to ensure that the system is thoroughly bonded and connected to a low impedance earth system. When vehicles are serviced they should be earthed with firm connection through lugs, and as a general rule, insulated filing hoses should be used to minimize the risk of sparking. In such cases all conducting materials in the immediate vicinity which are above ground level should be bonded and connected to a low impedance earth at intervals not exceeding 10 m, in case of high power HF broadcast stations. Where the storage is located beyond a distance of 500 m from the antenna of a MF or HF transmitter operating with a maximum effective radiated power up to 500 KW, the hazard to flammable material is substantially reduced. The design and configurations of the earthing requirements would be a matter of engineering judgments.

(f) Domestic type of Appliances and misc equipments

There are lots of domestic type appliances, general gadgets for day-to-day use and radio equipments which are required in a radio stations for associated use and other office as well watch and ward services. Statistics of fire in domestic type appliances, considered over a period of 10 years, indicate that the combined total of radio and television set fires is 3rd highest in a list of 13 electrical appliances. They contribute to about 12% of the fire. The cooking and space heating appliances have higher figures. From reports of several hundred known outbreaks of fire, about 60% were confined to equipments, 15% spread beyond equipment and damaged the contents of room wherein they were located, and 24% resulted in damage to both room contents and building.

In a break-up of the figures, the number of fires in mains operated television sets is nearly twice that of in radio sets. Some of the possible cause for hazards of domestic type of equipments is as below;

(1) Un-insulated wiring and bus bars between high voltage components,
(2) Paper based phenolic circuit boards,
(3) Inadequate separation between power transformers and components made of combustible materials,
(4) Faulty ac switches, relays, other components, and breakers,
(5) Faulty deflection yokes surrounding cathode ray tubes, and
(6) Faulty automatic tuning devices.

However of late in modern days the risk of fire in television and radio sets have reduced considerably due to improvement made in components materials as the devices have become solid state, which generates less heat, which earlier used to be the main reason for fire problems.

3.9.1 Fire detection and Protection Facilities

Fire detection and protection facilities, such as smoke, fire & UV detectors, and automatic alarms, CO_2, water sprinklers, fire hydrants, and other such facilities required for a radio station, should be planned as part of main engineering design and installed along with construction of building. Portable appliances, signage, instructions, first-aid instructions, fire exit plan map should be placed immediately after the access to occupy the building and permission to start operation is given.

The location of fire hydrants and portable appliances is very important and every care is to be taken that they are placed in prominent positions and are readily accessible. The recommendation of local fire authorities must be obtained for numbers, type, capacity and placement of the fire fighting equipments inside the building and layout of fire hydrants around the building along with storage quantity of water for the requirement of fire men.

Due to advancement of technology, fire hazards in modern radio equipment are not usually large. But in high power transmitters where high current and power are very common, working temperature of the cubicle may often be very high and breakdown of insulation may occur very easily. Solution is to use only those materials which are non-combustible, but for various reasons such as inferior performance, difficulty of fabrication or high cost of non-combustible materials, this is not a practical possibility. In case of oil filled components used in high power transmitters, emphasis is on preventing excessive damage rather than on protecting the building against the spread of fire from within.

In case of transmitters with fully enclosed steel cubicles, a fire within cubicle would generally be restricted to within but early indication of approaching breakdown by means of a suitable detector may result in reduced damage by allowing early action to be taken and to facilitate quick restoration of service.

For rack mounted equipments, fire hazard is restricted to possible ignition of transformer insulation, capacitor dielectric, resistors or fabric of insulated cables. Oil filled components are seldom used and section of the equipments are segregated by the use of metal covers. The apparatus is well protected electrically as a general rule, and fires are infrequent. Within the equipment heat damage rarely extends from one section of enclosed equipment to next. The risk of the building of fire taking place in such type of equipments is very uncommon.

3.9.2 Fire Detection Devices

Now-a-days market is flooded with wide range of devices for detecting the fire, and the type of unit to be installed in a particular location depends upon the nature of fire which is likely to occur in that area. Smoke detector is foremost mean of fire and detection is capable of providing earliest possible indication and its application has been widely recommended for the protection of radio studios and transmitter equipments. Modern smoke detectors are very sensitive and the radio technicians should be careful while working with soldering irons in their vicinity. Care should be taken to install smoke detectors clear of obstructions to air flow and in an area of maximum effectiveness. Now-a-days these are installed in each cubicle of transmitter including matching, tank coils and combining cubicle.

A new type of smoke detector called Vesda (very early smoke detector apparatus) is suitable for use in duct application. It is based on the light scattering from minute dust or smoke particles. Two inlets are provided for the measuring point of this apparatus, one to monitor air flow through duct and other to monitor air outside the building. The air samples are compared and variation from pre-set levels is used for triggering alarm of smoke.

Heat actuating devices, fixed temperature or rate-of-rise detectors are desirable for fire detection in engine plant rooms, since the products of combustion in engines or high velocity air currents may cause operation of smoke detectors, if used.

The detection of smoke in equipment air cooling or exhaust system is often a problem. The purpose of having smoke detection in duct system is primarily to shut down the equipment, particularly the blower motors. In some installations fire and smoke dampers are operated. The presence of smoke in the duct can be detected in two ways; either detector may be fitted within the duct or a sample of air flow can be bypassed to a detector chamber located on the outside of the duct. Because of the stratification of smoke in the duct, sampling of the entire cross section is the only consistently method of detection. The positioning of detectors is very important for their proper functioning and prompt detection of the smoke.

Presence of smoke has a very interesting behavior on the voltage withstand and current carrying capacity of conductors in RF fields inside a transmitter cubicle. Tests have shown that in the presence of smoke, an arc or discharge will develop at a very much lower potential. Depending on the concentration of smoke, the distribution and its source, the voltage rating of a conductor can drop to a level as low as $1/3^{rd}$ of the normal rating. As many components & assemblies are designed with a safety factor of 2 to 3, chances of flashover in a smoke filled cubicle is very high. Therefore an earliest detection of smoke in a cubicle is essential & must to enable removal of power quickly to shut down the cause of fire or smoke and protect the transmitter from severe damage.

Various devices normally used for the detection of arcs inside transmitter cubicles are; ultra-violet, infra-red, and magnetic sensing devices. Tests have shown that commercially available ultra-violet detectors are suitable for detection of power supply and RF arcs, but the infra-red detectors are generally unsatisfactory because of their insensitivity to radiation from RF arcs. However as ultra-violet detectors can be triggered by ambient radiation such as sunlight and artificial lighting, care in installation is necessary to ensure that they are properly protected in their installed position. The ultra-violet sensors detects an arc by the flash; the magnetic arc sensor detects an arc by the noise pulse. The magnetic arc sensor is sort of an antenna which couples a sample of RF energy and any arc induced noise to a low pass filter which passes only the noise pulse to trigger a circuit. The infra-red detectors are used to detect flash inside ceramic valves of high power transmitters.

3.10 Planning for fire Protection facilities

The extent of fire protection facilities to be provided at transmitter sites will vary from organization to organization, and also with the power of the transmitter. Another important factor is whether transmitter is manned or unmanned. The most important, however is to design the transmitter equipment so that the possibility of fire will be reduced to a minimum. It also need to be ensured that fire is self extinguishing, by the choice of appropriate materials or that the rate of spread of fire is so slow that there is good chance of extinguishing the fire in early stage. Next and another very important item are to be able to get access to the seat of fire with extinguisher equipment before much damage s done to the equipment.

Factors which should be taken into consideration in planning the installation and operation of unattended stations should include followings;

(a) Unmanned transmitter operational sites should be provided with adequate and reliable control and protective circuitry engineered specially for unattended operation.
A highly reliable transmitter when operated by skilled staff may not necessarily be reliable if left unattended and operated under automatic conditions.
(b) If the entire installation is to be operated unattended, than it has to be planned that way only as it need special engineering and design work. Special precautions which usually are not needed must be taken for an unattended operation, to ensure adequate protection of the station, taking into consideration all aspects of the environment.
(c) In an unattended transmitter site, time switches up to one hour delay, to switch off all the lights and non-essential power outlets, after visit of maintenance staff, should be provided.
(d) Unattended stations must be inspected regularly and properly maintained by skilled technical staff under the supervision of experienced radio engineer.
(e) When remote equipment is installed in cubicle type housing, designer should install an automatic extinguishing system in each cubicle to take care of fire in that cubicle.
(f) As an additional control measure, many remote stations have smoke detectors fitted into air exhaust ducts. When smoke is detected, power to the equipment is automatically disconnected, and an alarm signal is sent to the control station.
(g) If automatic CO_2 flooding system is fitted than arrangement must be made to introduce a manual safety switch to disconnect the control of the automatic discharge of the CO_2 system from fire alarm circuit. The switch should be fitted with visual and audible off-normal indications. There is a possibility of accidents of maintenance staff in these automatic types of installations, and the operation of manual safety device during maintenance should be part of the normal safety precaution routines.

These are must as CO_2 has a very high asphyxiate risk at extinguishing concentration. So a warning must be prominently displayed at all places where such type of automatic devices is used. Extinguishing concentration of CO_2 in% volume to air is 28 whereas dangerous concentration level for asphyxiation is 9 in% of volume to air.

(h) Plan for bromo-chlorodifluoromethane (b.c.f.) for automatic extinguishing, another type of popular fire extinguishing gas, which is much more safer than CO_2 as its extinguishing concentration is 5.2–10% of volume to air and dangerous concentration value in% volume of air of 24, is much higher than extinguishing rate. his means even after extinguishing the fire the concentration is at a very safe level.

(i) At manned high power transmitting stations, suitable extinguishers should be located at strategic points throughout the equipment room areas. Theses units should be located in halls, equipments rooms, workshop room, test rooms, retiring room, kitchenette, armed guard rooms and other shelter areas.

For such requirements CO_2 type 3 kg capacity units are the most appropriate choice supplemented with 22 kg trolley mounted units at in alcoves close to each transmitter hall or one in each corner and in front of HT or transformer rooms.

(j) When a station has large technical staffing complement, there would be considerable building requirements for administrative support staff and facilities. At high and super power transmitter complexes catering to national and international broadcasting functions, the officers also are located in the transmitter building, while at others, a separate building is provided. Fire protection for such office type buildings do not require the same facilities as the equipment buildings but foam type units, water gas units or hose reel s may be provided.

(k) Plan for fixed automatic sprinklers protection system as a means of reducing fire damage in large multi-transmitter installation and in studio buildings. If properly installed and maintained, a sprinkler system provides very effective control and extinguishment with very little hazard to personnel and with no measureable increase in damage to radio or electrical equipment as compared with damage caused by heat, flame, and smoke.

(l) Water is main requirement for sprinklers, with supply being from main supply or overhead gravity tanks. As most of the gravity over head tanks are not sufficiently high to provide high pressure of water, an automatic self-starting electrical diesel or petrol booster pump is installed for this requirement. Overhead tanks in a radio station are normally constructed to have reserve capacity always for meeting the fire requirement.

(m) In the initial design of a radio transmitter building and equipment works, plan for installation of a public address system to warn and guide the staff and other personnel to emergency exits in case of fire breakouts.

In addition any other requirement as per the direction / recommendation of local fire authorities must be implemented before transmitter complex facilities are commissioned.

3.10.1 Sprinkler Systems

An automatic sprinkler system performs three important functions;

(a) It detects an outbreak of fire in the area it is installed to serve.
(b) It automatically releases water or other liquid as per design under pressure at the spot of the fire.
(c) It actuates audible and visual warning system to occupants of the building as well can send a signal to a fire station or control point as per the design criteria.

There are normally two types of sprinklers as explained below;

(a) Wet type of Sprinkler
The most common type of wet sprinklers used in transmitter installations is the soldered cantilever link and the bulb types. The soldered cantilever link uses an alloy of several metals built into a solder acting as a bonding material for a fusing element. The link is held in position at constant tension by two cantilever arms fulcrum center wise. As the solder alloy melts due to heat from a fire, the cantilever arms parts in opposite direction allowing water to impinge on a deflector plate and covering a large area with spray. In the bulb head type a volatile liquid within a glass bulb expands and fractures the bulb allowing water to impinge on deflector plate.

Such types of sprinklers have been found to be more than 96% efficient in broadcasting stations. From the records available it is established that over 70% of fire in radio installations which were extinguished by automatic water sprinklers were controlled by the operation of only one sprinkler head in a multi head installations. This means that water damage was restricted to a salvageable level for minimum loss of equipment as well as machines and that the system used only the amount of water necessary to control the fire.

(b) Dry type of sprinkler system
Some broadcasters are reluctant to install and use wet type sprinkler system in their studios due to the risk of damage to equipments, acoustics, carpets, & furnishings in the event of accidental operation of sprinkler system. An alternate most popular & widely used protection is the dry pipe type sprinkler system which features a piping system containing air or carbon dioxide under pressure, instead of water. Only if a fire is detected by a separate ionization or thermal detector system water is allowed in the pipes. The sprinkler head must then fuse at 100°C, before permitting the water to discharge on the fire. The water is automatically stopped when separate sensing device cools, even though the sprinkler head remains fused. This dual control greatly reduces the risk of accidental operation and consequent water damage, yet provides the added safety of the wet pipe system. Standards set down for automatic system require that, in most cases, power to the radio equipment is cut-off and it is shut down before the application of water. Also the sprayed water must have positive drainage to minimize the damage to the equipment and associated wiring due to flooding.

A main disadvantage of the dry pipe system is the time delay between the opening of sprinkler and the discharge of water. For a rapidly spreading fire this could be disastrous. However this problem could be partly overcome by the provision of devices which are quick opening and by ensuring the whole system, particularly the dry pipe valves are properly maintained. Some causes of failures of the dry pipe system either corrosion or deposits on seats & moving parts and use of grease, paper or paint to make the seats air or water tight. The remedy lies in resurfacing the seats or replaces the rubber linings or gaskets.

3.10.2 Automatic Protection facilities for switchboards

A question, whether to provide automatic sprinklers or not, in the high voltage equipment rooms such as switchgears and switchboards, always arises among planners due to problems arising because of spraying of water on live electric equipments. The practical evidence from the reports of much fire indicates that in comparison with the risk of allowing a fire started by a high voltage unit to go out of control, the consequences of discharge of water on the unit is of minor nature. It is obvious that, if a fire originating in high voltage equipment is unchecked, it may within minutes turn into inferno and may require the help of fire brigade with high speed, high pressure hose reels to control it resulting into water all around. So it is clear that of the two evils, the question of water damage is minor compared with the risk of serious fire. So each organization must have a policy whether to use the water in case of fire in high voltage components /equipment or not as without using water it may not be possible to extinguish the fire with dry chemical type of powder normally used in fire fighting in such cases.

3.11 Plan for Reduction of Fire protection Needs

In order to have optimized fire fighting needs for a radio station, the factors which will reduce the amount of fire protection requirements needs to considered at the time of designing phase of equipment and building. Such factors are summarized below;

3.11.1 Factors to be considered in General Design

(a) HT components and transformers requiring more than 900 liters of oil should be fitted with protection devices like Buch-holz relay which should be interlocked with transmitter control circuits to shut down in case of over current / overloads, sparking or short circuits.
(b) The transformers, smoothing and modulation chokes / transformers as well as oil filled capacitors should have adequate safety margins and protection facilities.
(c) Oil should be regularly checked for quality and breakdown voltage rating.
(d) Large transformers and other oil filled components should be installed in a separate secured HT component room with adequate fire barriers protections for adjacent rooms. Cable or other floor/wall openings

should be sealed or effectively boxed with metal to prevent the oil draining from the opening.

(e) The HT component room floor should be so constructed that in the event of catastrophic failure of the largest transformer, the oil will be rapidly drained or sump properly constructed for this purpose. The relative level of the floor of the HT component room and transmitter hall as well as other room of transmitter accessories should be such that in the event of spillage, oil will not flow into the transmitter or other rooms of the building.

(f) The transmitter proper if of component floor mounted type; wherever practicable should be fully enclosed in sheet metal cubicles to ensure that fire which may start within it do not endanger the surroundings external to it.

(g) Equipments should contain maximum amount of non-combustible materials. Inorganic materials such as glass & porcelain are non-combustible, but organic materials used widely in radio equipments vary in combustibility. Most organic materials particularly those used as conductor insulation are combustible to some degree. But whether or not they produce a self propagating fire depends on many external conditions such as air supply to the equipment, the rate at which heat is lost & other factors. Polytetrafluoroethylene is excellent insulation for conductors, being non-combustible under normal conditions, but due to high cost, its use is mainly restricted to RF type cables and critical components. But in almost pure oxygen environment it is also combustible.

Both PVC and polythene are combustible as flexible conductor insulation materials. Rigid PVC is relatively non-combustible, but to obtain flexibility for wiring, mixing of plasticizers, which themselves are flammable, seriously impairs the fireproof properties. Polythene is not easy to ignite but if construction of the wiring form is such that molten material is held in the combustible zone, polythene will burn very rapidly.

Fiber glass which is widely used in radio equipments and components has been the cause of extension of many fires. One of the most commonly used resins polyester in it is not readily ignited but the very presence of the glass fiber of the reinforcement material has a wick effect which tends to increase the flammability.

Also the use of plastic in the construction of radio equipments since world war-II is enormous and considerable. This is extensively used in active and passive components of radio transmitter and equipments. Fire can extensively damage the plastic materials as it is highly combustible.

(h) Battery installation if provided, need to be properly installed and maintained as it is prone to fire from acid fumes and other short circuits.

(i) Good housekeeping is most essential and a sure method to reduce the chances of fire outbreaks. A neat and clean transmitter HT components room can be had by providing drip trays under the outlet/drain plugs or taps of oil filled components, by removing flammable material and banning smoking by staff during maintenance. A transmitter area should be strictly NO SMOKING ZONE. Supplies of papers including manuals, log books and other combustible materials which is in excess of the minimum required for efficient operation should be removed from the equipment areas and transmitter hall as well control room.

(j) All air-conditioning, ventilation, air-circulation or transmitter cooling system ducts, including the cable ducts and chases, must be made out of non-combustible materials and should be fitted with smoke detectors.

(k) Filter used in inlet-air ducts to transmitter hall and in transmitter cubicle or in air-handling units must of non-combustible type or at least of type that will not burn freely or emit excessive smoke during fire outbreaks. (l) Activity or occupancy not related with installation should not be permitted in transmitter hall or equipment area.

(m) Workshop type activities should be restricted to workshop, test room or other room earmarked for this purpose. Some specific works applicable to installed equipment, which cannot be removed from equipment room, can be done in equipment room, but such works should be restricted to bare minimum.

3.11.2 Factors to be considered for accessibility

During the initial design and installation stages, it is necessary to provide for the contingency of occurrence of fire, and make sure that all parts of the equipment are accessible to fight the fire. Cabinets and cubicles should have doors or easily removable panels so that hand extinguishers can be discharged into them quickly. Similarly cable ducts, trays and air ducts should have removable sections or panels for this purpose.

The most satisfactory, efficient and useful hand held fire fighting appliances in a radio station are CO_2 and b.c.f extinguishers. These types should be available in such a quantity that wherever and whenever a fire in the transmitter building occurs, one can be brought in operation by the station staff without loss of time.

3.11.3 Isolation of Power supply

While fighting a fire in a radio station, the criteria should be to minimize the loss to equipment and personnel working in the station. The safest practice and first priority is to isolate the power supply, and if the fire continues then use the extinguishing means most suitable for the prevailing conditions.

Where power has to be isolated during fire, it should be done only for the affected part of the system; otherwise there may be danger of cutting off the lights needed for the fire fighting crew or power to operate the pumps etc. If power is not isolated, additional electrical damage may result and there is danger of electrical shock to fire fighter.

It is not advisable to use soda-acid, foam or loaded stream type extinguishers on fire on radio equipments unless water is not available, since these liquids leave salts on transformer windings, on insulators, terminals and other conducting parts prone to subsequent failure. If it is essential to use these, due to lack of approved type extinguishing agents, then power supply must be disconnected before use. As these liquids are conducting media, they may cause additional electrical damage and involve danger of electric shock to the fire fighter.

3.11.4 Oil filled Components

All oil filled components like transformers, circuit breakers, electrolytic capacitors, reactors, modulation transformers, filter and modulation chokes and other similar equipments containing oil involves the additional hazard of oil fire. It should be planned to use oil having relatively high flash point, but it may be heated and ignited by excessive current or an electric arc, if the oils is unclean and have impurities in it. After power supply has been isolated, such fires can be extinguished by various methods of extinguishing oil-fires. If the power supply cannot be immediately isolated, then use vaporizing liquid, carbon dioxide or dry chemical type extinguisher without danger of electric shock. But these hand extinguishers are not ordinarily effective for extinguishing large fires.

Improperly sealed wax filled condensers have been found to be cause of much concerns, particularly where they are mounted in horizontal position, as it can cause fire due to dripping of wax on other components say on a standoff insulator or a cable bunch etc. Similarly fire can breakout due to improper earthing of oil filled capacitors in matching network of feeder hut where RF field is very high. Even the improper or loose grouting of copper sheet shielding in a feeder hut can be a cause of fire in the matching network.

3.12 Typical Firefighting Equipments and facilities for a Radio Station

Provision of Fire fighting equipments, detection and alarm facilities at a radio station will depend on many factors including the size of the station, whether manned or unmanned, proximity to public fire brigade services and importance of the station from public service point of view and the security point of view.

In case of a typical unmanned transmitting station, facilities will include a combined thermal and early warning detection system covering all rooms of the station building. The equipment rooms should be fitted with smoke detectors while the other rooms fitted with heat activated detectors. At stations where captive or emergency power generating plant is provided, it also should be fitted with a heat detector system. A central control panel need to be located at a convenient point enables all detection devices to be brought to a common point. The panel should have facilities for activating a visible and audible alarm, as well as putting into operation an automatic fire extinguishing system if provided, and also extending an alarm to a local fire station.

For a typical unmanned station where the value of the equipment warrants a smoke detection system associated directly with the equipment cubicles should be provided. The detectors should be wired to a control unit, which can shut off equipment and blower of the unit from where alarm had originated. It should be capable of sending an alarm signal to local fire office also.

Except in very large stations, automatic extinguishing systems covering large area are not normally installed. Where systems have been provide they protect HT component rooms having huge oil filled components. Portable hand operated extinguisher of various types and internal hose reel as per fire brigade specifications need to be provided throughout the building and in location recommended by local fire officer.

Water storage facilities with Overhead tank of capacity up to 100KLs should be provided at super power transmitting stations with minimum 20KL as reserve for firefighting use. A diesel powered generator is also generally recommended for feeding water at required high pressure into a reticulated hydrant service.

For bush fires or paddock fire, a fully equipped fire tender truck is recommended. As a minimum facility, fire beaters and knapsack spray units should be provided. For an unmanned station, an automatic extinguishing flooding system is installed where the value of the equipment and its importance warrants such protection. Portable hand equipments are also provided for use when staff is on site. Although it is not usual to provide water storage solely for firefighting purposes at very remote stations, visiting staff requires water for drinking purposes and single water tank supply usually serves the needs of both requirements.

Typical facilities for fire detection, protection, fighting, and monitoring as well as for sending remote alarm, in case of a large transmitting manned station in a remote location are as below;

(a) A properly planned and installed combined smoke and thermal automatic fire alarm system throughout the station technical rooms and building. The system should cover; (1) Transmitter hall and its associated rooms like HT room, Blower rooms, pump rooms, and heat exchanger rooms,(2) Speech, emergency play back and program downlink as well STL facilities rooms,(3) Air-conditioning plant rooms,(4) Power supply sub-station and all switchgear room including substation transformer rooms or enclosures,(5) All test rooms, workshop and repair rooms,(6) Diesel Generator and oil storage room,(7) Antenna matrix, switching and matching hut,(8) Administrative and amenities building,(9) Godowns and Garages,(10) Armed Guard dormitories, watch and ward buildings.

(b) Main fire alarm control panel should be located in the control room of transmitter building or in transmitter hall where engineers remains on duty with an extended indicator panel at the entrance of the administrative building.

(c) Smoke detection facilities should be provided in; (1) Each transmitter enclosure,(2) Transmitter air exhaust ducts,(3) Transmitter HT component rooms,(4) Control room,(5) Program input room,(6) Studio transmitter link and Radio Networking Down link, (7) Test, repair and workshop rooms, and (8) Air-conditioning ducts.

(d) Manual fire alarm points should be provided; (1) adjacent to main control panel in control room,(2) one at each end of transmitter hall,(3) one near to repeater indicator panel in administrative building,(4) One each on external wall of Diesel generator room and power House,(5) one on external wall of A/C plant room,(6) one on external wall of antenna switching & other such rooms, (7) one on external wall of garage, godowns, (8) one on 1st floor housing heat exchanger room etc., and (9) one each on external wall of antenna tuning and other field buildings.

(e) Overhead tank Water storage facilities of at least 50K liters with 10 to 20K liters of reserves capacity for firefighting purposes should be provided. A fire hydrant grid system running around the entire building connected to OHT with 10 cm dia pipe should be provided to cater for interior and external protection of all building and the immediate area around it. Each hydrant point should be provided with a cluster of six 20 mm outlets in addition to the fire hose reel connection for use by the fire brigade. Hydrants boxes should be so placed that all exterior portions of the building and the immediate site area may be protected by using 1x40 m and 5x20 m lengths of 20mm (4ply ratings) rubber hose fitted with a variable 5 mm nozzle. A manual diesel engine booster fire pump coupled to a 10 cm dia hydrant pipe system should be provide to be located away from building in a clear area, say near

OHT. This pump should be controlled from the control panel kept in control room or transmitter hall. The entire hydrant system should be earthed in accordance with the building earthing and bonding specifications.

(f) Portable which can be easily held in hands and carried swiftly by staff should be provided at following places;

    (1) Transmitter building; all rooms housing working equipments including the store rooms with one 3 kg CO_2 unit near each door on wall. Two numbers of 22 kg trolley mounted CO_2 units each in the transmitter hall on each corner and two numbers one each outside HT equipments room should be provided.

    (2) Antenna matrix switching and ATU building; two numbers of 3 kg CO_2 one inside and another outside each building mounted on wall near the entry to should be provided.

    (3) Workshops, repair and testing rooms; one 3kg CO_2 unit and another 9 liter foam type should be provided and these should be inside on wall and another mounted on wall near exit of room.

    (4) Filter cleaning and service rooms; one 3 kg CO_2 should be mounted on wall near door.

    (5) Power supply switchgear and Diesel Generator room; Two numbers of 9 kg foam type units should be provide adjacent to each diesel generator set. One 22 kg trolley mounted CO_2 type should be provided inside power supply switch gear room near entry rolling shutter. Additionally, one 22 kg trolley mounted CO_2 type should be provided inside covered enclosures at entry point if erected. Two 3 kg CO_2 type should be provided at the main switchgear and one 3 kg CO_2 type should be provided on outside wall of this area.

    (6) Retiring rooms and kitchenette facilities room; one 3kgCO_2 type in each room mounted on wall. One 9 liter foam unit and one 9 liters water gas container type should be provided in the corridor by the side of these rooms.

    (7) Garage and Godowns building; two 9 liters foam type in each vehicle parking slots and one 9 kg co_2 type in the godowns. One number each of 3 kg co_2 wall mounted type also should be provided in each room.

(8) Unloading Area; one 9 liter capacity water gas container type should be provided near the gantry of each unloading area.

(9) External plant depot & shed; two 9 liter foam type units should be provided at the entrance to both areas.

(g) If the station is located in remote area and the time taken by the regular fire fighting department is appreciable as well it has a vast antenna field which normally remains full of grass and shrubs, then it must have a motorized fire unit mounted on a standard truck chassis as part of station fire fighting facilities.

The unit should be fitted with facilities to deal with grass fires in the field and outbreaks in the out huts /buildings. It should be fitted with an integral water tank of 9000 liters capacity with pump and 50 m long hose reel with heat resistance properties. This truck also should have 4 numbers of 3 kg CO_2 type fire extinguishers and two 9 liter foam type extinguishers, 4 knapsacks, 6 hand beaters for fire and a tool box compartment containing 2 axes, 4 shovels, and 2 rakes.

In addition this vehicle should have facility to tow a mobile tanker in rear side and fitted with a winch on front side having 50 m steel rope with at least two tonnes breaking strength. This vehicle should have a walkie-talkie set and complete range of first-aid as well as a satellite phone (for remote locations only in case no other communication facility is available).

References

1. Handbook on Building Fire Codes (pp 40 to 174); (Document No.: IITK-GSDMA-Fire05-V3.0, Final Report: C-Fire Codes IITK-GSDMA Project on Building Codes); by – G.B. Menon Fire Adviser, Govt. of India {Retd} Cochin – Ex-Chairman CED-22, FireFighting Sectional Committee Bureau of Indian Standards; J.N. Vakil, Asst General Manager{Retd}, TAC/GIC, Ahmadabad – Ex-Chairman CED-36 Fire Safety Sectional Committee Bureau of Indian Standards.
2. Handbook for Radio Engineering Managers – J.F. Ross – Fire Hazards – Chapter – 31, (PP 500–50 7)
3. Bureau of Indian Standards IS on the subject like IS: 12407:1988 & IS: 2189: 1999 etc.

CHAPTER 4
LIGHTNING HAZARDS

4.1 History

Benjamin Franklin (1706–1790) endeavored to test the theory that sparks shared some similarity with lightning using a spire which was being erected in Philadelphia. While waiting for completion of the spire, he got the idea of instead using a flying object, such as a kite. During the next thunder storm, in June 1752, it was reported that he raised a kite, accompanied by his son as an assistant. On end of the string he attached a key, and he tied it to a post with a silk thread. As time passed, Franklin noticed the loose fibers on the string stretching out; he then brought his hand close to the key and a spark jumped the gap. The rain which had fallen during the storm had soaked the line and made it conductive.

Although experiments from the past time of Benjamin Franklin showed that lightning was a discharge of static electricity, there was little improvement in theoretical understanding of lightning (in particular how it was generated) for more than 150 years. The impetus for new research came from the field of power engineering: as power transmission lines came into service, engineers needed to know much more about lightning in order to adequately protect lines and allied equipments.

How lightning initially forms is still a matter of debate. Scientists have studied root causes ranging from atmospheric perturbations (wind, humidity, friction, and atmospheric pressure) to the impact of solar wind and accumulation of charged solar particles. Ice inside a cloud is thought to be a key element in lightning development, and may cause a forcible separation of positive and negative charges within the cloud, thus assisting in the formation of lightning.

4.1.1 General

Lightning is not only one of the most breathtakingly beautiful natural phenomena but also one of the most dangerous. Lightning can cause serious risk to life as well as ruinous damage to broadcast infrastructure including equipments. Lightning strikes the highest object on ground and take the path of least resistance on its way to ground. Thus a broadcast tower is most attractive target for lightning discharges due to the fact that it is an excellent electrical conductor also and offers many low resistance connections to earth. The average intensity of a lightning flash is estimated to be 20,000 amps; however lightning intensities of up to 2000,000 amps have been registered. A lightning strike no thicker than a few inches can illuminate one million 100 watt light bulbs.

A lightning flash is composed of a series of strokes with an average of about four. The length and duration of each lightning stroke vary, the typical average being about 30 microseconds. The average peak power per stroke is about 10^{12} watts. Lightning frequency increases as you get closer to the equator and/or higher the altitude.

Damages to the equipment results mainly when these discharge current pulses occur close to the technical installation. The real world of radio transmitters unfortunately is one, where periodic lightning storms occur and cause some finite incidence of antenna and power line strikes. The actual strike incidence will vary widely through general geographical locations, topography, height of the transmitter mast, routing of feeder and incoming power lines. Unless definite precautions are taken, such strikes can damage the transmitter, particularly final RF stage amplifier, HT rectifier and other solid state devices used in transmitter. A super reliable transmitter in laboratory environment could become very unreliable in the field except in the regions where lightning is non-existent.

Because of the great risk that lightning presents, it is important to comprehend its nature, and it's key to understand the methods of protecting your equipments and yourself from lightning-induced damage. This chapter presents a simplified summary of lightning data from several sources in a systematic approach to the problem. Starting with protection principles the chapter concludes with definite system recommendations.

4.2 Formation Process, Type and Triggering of Lightning

Lightning is created by weather conditions, most prevalent in the spring and summer and in tropical climates. When a warm air mass meets a cold air mass, tremendous up surging convection air currents are created. Friction between these air masses creates static electricity in the cloud formations. A negatively charged cloud layer will positively charge the ground beneath it, resulting in a condition much like the two opposing plates of a capacitor. When this charge overcomes the dielectric resistance of air, a current will flow between them to neutralize the charge, in the form of lightning. These discharge currents can pass either from cloud-to-ground or cloud-to-cloud. The latter can be just as destructive as a direct ground strike, because they create sympathetic currents in the ground beneath the discharge.

A general view of lightning is shown in fig 4.1 and it is considered as an atmospheric discharge of electricity, which typically occurs during thunderstorms, and sometimes during volcanic eruptions or dust storms. In the atmospheric electrical discharge, a leader of a bolt of lightning can travel at speeds of 60,000 m/s (220,000 km/h), and can reach a temperatures approaching 30,000 °C (54,000 °F), hot enough to fuse silica sand into petrified lightning, known scientifically as glass channels or fulgurites which are normally hollow and can extend some distance into the ground. There are some16 million lightning storms in the world every year.

Followings are some of the processes which generate lightning;

(a) Charge separation
Charge separation is the first process in the generation of lightning. There are two Hypotheses describing this process;

(I) Polarization Mechanism Hypothesis
The mechanism by which charge separation happens is still the subject of research, but one hypothesis is the polarization mechanism, which has two components: (a) Falling droplets of ice and rain become electrically polarized as they fall through atmosphere's natural electric field; (b) Colliding ice particles become charged by electrostatic induction. Ice and super-cooled water are the keys to the process. Violent winds buffet tiny hailstones as they form, causing them to collide. When the hailstones hit ice crystals, some negative ions transfer from one particle to another.

The smaller and lighter particles lose negative ions and become positive, the larger and more massive particles gain negative ions and become negative.

(II) Electrostatic Induction Hypothesis

According to the electrostatic induction hypothesis charges are driven apart by as-yet uncertain processes. Charge separation appears to require strong updrafts which carry water droplets upward, super-cooling them to between -10 and -20 °C. These collide with ice crystals to form a soft ice-water mixture called graupel. The collisions result in a slight positive charge being transferred to ice crystals and a slight negative charge to the graupel. Updrafts drive lighter ice crystals upwards, causing the cloud top to accumulate increasing positive charge. The heavier negatively charged graupel falls towards the middle and lower portions of the cloud, building up an increasing negative charge. Charge separation and accumulation continue until the electrical potential becomes sufficient to initiate lightning discharges, which occurs when the gathering of positive and negative charges forms a sufficiently strong electric field.

Fig 4.1: A general view of Lightning's

There are several additional hypotheses for the origin of charge separation. According to one such hypothesis; charge separation is initiated by the ionization of an air molecule by an incoming cosmic ray.

(b) Leader Formation

Leader Formation as a thundercloud moves over the Earth's surface, an equal but opposite charge is induced in the Earth below, and the induced ground charge follows the movement of the cloud. An initial bipolar discharge or path of ionized air starts from a negatively charged mixed water and ice region in the thundercloud. The discharge ionized channels are called leaders. The negative charged leaders, called a "stepped leader", proceed generally downward in a number of quick jumps, each up to 50 meters long. Along the way, the stepped leader may branch into a number of paths as it continues to descend. The progression of stepped leaders takes a comparatively long time (hundreds of milliseconds) to approach the ground. This initial phase involves a relatively small electric current (tens or hundreds of amperes), and the leader is almost invisible compared to the subsequent lightning channel.

When a stepped leader approaches the ground, the presence of opposite charges on the ground enhances the electric field. The electric field is highest on trees and tall buildings. If the electric field is strong enough, a conductive discharge (called a positive streamer) can develop from these points. This was first theorized by Heinz Kasemir. As the field increases, positive streamer may evolve into a hotter, higher current leader which eventually connects to the descending stepped leader from the cloud. It is also possible for many streamers to develop from many different objects simultaneously, with only one connecting with the leader and forming the main discharge path. Photographs have been taken on which non-connected streamers are clearly visible. When the two leaders meet, the electric current greatly increases. The region of high current propagates back up the positive stepped leader into the cloud with a "return stroke" that is the most luminous part of the lightning discharge.

(c) Discharge

Discharge; when the electric field becomes strong enough, an electrical discharge (the bolt of lightning) occurs within clouds or between clouds and the ground. During the strike, successive portions of air become a conductive discharge channel as the electrons and positive ions of air molecules are pulled away from each other and forced to flow in opposite directions.

The electrical discharge rapidly superheats the discharge channel, causing the air to expand rapidly and produce a shockwave heard as thunder. The rolling and gradually dissipating rumble of thunder is caused by the time delay of sound coming from different portions of a long stroke.

(1) **Gurevich's runaway breakdown theory;** A theory, known as the "runaway breakdown theory", proposed by Aleksandr Gurevich of the Lebedev Physical Institute in 1992 suggests that lightning strikes are triggered by cosmic rays which ionize atoms, releasing electrons that are accelerated by the electric fields, ionizing other air molecules and making the air conductive by a runaway breakdown, then "seeding" a lightning strike.

(2) **Gamma rays and the runaway breakdown theory;** It has been discovered in the past 15 years that among the processes of lightning is some mechanism capable of generating gamma rays, which escape the atmosphere and are observed by orbiting spacecraft. Brought to light by NASA's Gerald Fishman in 1994 in an article in Science, these so-called Terrestrial Gamma-Ray Flashes (TGFs) were observed by accident, while he was documenting instances of extraterrestrial gamma ray bursts observed by the Compton Gamma Ray Observatory (CGRO). TGFs are much shorter in duration, however, lasting only about 1ms. Professor Umran Inan of Stanford University linked a TGF to an individual lightning stroke occurring within 1.5 ms of the TGF event, proving for the first time that the TGF was of atmospheric origin and associated with lightning strikes. CGRO recorded only about 77 events in 10 years; however, more recently the RHESSI spacecraft, as reported by David Smith of UC Santa Cruz, has been observing TGFs at a much higher rate, indicating that these occur about 50 times per day globally (still a very small fraction of the total lightning on the planet). The voltage levels recorded exceed 20 MeV.

Scientists from Duke University have also been studying the link between certain lightning events and the mysterious gamma ray emissions that emanate from the Earth's own atmosphere, in light of newer observations of TGFs made by RHESSI. Their study suggests that this gamma radiation fountains upward from starting points at surprisingly low altitudes in thunder clouds. Steven Cummer, from Duke University's Pratt School of Engineering, said, 'These are higher energy gamma rays than come from the sun. And they come from the kind of terrestrial thunderstorm that we see here all the time'.

Early hypotheses of this pointed to lightning generating high electric fields at altitudes well above the cloud, where the thin atmosphere allows gamma rays to easily escape into space, known as 'relativistic runaway breakdown', similar to the way sprites are generated. Subsequent evidence has cast doubt, though, and suggested instead that TGFs may be produced at the tops of high thunderclouds. Though hindered by atmospheric absorption of the escaping gamma rays, these theories do not require the exceptionally high electric fields that high altitude theories of TGF generation rely on. The role of TGFs and their relationship to lightning remains a subject of ongoing scientific study.

(d) Re-strike
High speed videos have shown that most lightning strikes are made up of multiple individual strokes. A typical strike is made of 3 to 4 strokes. There may be many more. Each re-strike is separated by a large amount (typically 40 to 50 milliseconds) of time. Re-strikes can cause a noticeable "strobe light" effect. Each successive stroke is preceded by intermediate dart leader strokes again to, but weaker than, the initial stepped leader. The stroke usually re-uses the discharge channel taken by the previous stroke. The variations in successive discharges are the result of smaller regions of charge within the cloud being depleted by successive strokes. The sound of thunder from a lightning strike is prolonged by successive strokes.

4.2.1 Types of Lightning

Some lightning strikes take on particular characteristics; scientists and the public have given names to these various types of lightning. Most lightning is streak lightning. This is nothing more than the return stroke, the visible part of the lightning stroke. Because most of these strokes occur inside a cloud, we do not see many of the individual return strokes in a thunderstorm. The return stroke of a lightning bolt, which is the visible bolt itself, follows a charge channel only about a half-inch (1.3 cm) wide. Most lightning bolts are about a mile (1.6 km) long.

(a) Cloud-to-Cloud Lightning
Lightning discharges, as shown in Fig 4.2, may occur between areas of cloud having different potentials without contacting the ground. These are most

common between the anvil and lower reaches of a given thunder storm. These lightning can sometimes be observed at great distances at night as so-called "heat lightning". In such instances, the observer may see only a flash of light without thunder. The "heat" portion of the term is a folk association between locally experienced warmth and the distant lightning flashes. Another terminology used for cloud-cloud or cloud-cloud ground lightning is "Anvil Crawler", due to the habit of the charge typically originating from beneath or within the anvil and scrambling through the upper cloud layers of a thunderstorm, normally generating multiple branch strokes, as shown in Fig 4.3, which are dramatic to witness. These are usually seen as a thunder storm passes over you or begins to decay. The most vivid crawler behavior occurs in well developed thunder storms that feature extensive rear anvil shearing.

(b) Dry Lightning

Dry lightning is a term in the United States for lightning that occurs with no precipitation at the surface. This type of lightning is the most common natural cause of wildfires.

(c) Rocket Lightning

It is a cloud discharge, generally horizontal and at cloud base, with a luminous channel appearing to advance through the air with visually resolvable speed, often intermittently. It is also one of the rarest of cloud discharges.

Fig 4.2: Cloud-to-cloud lighting

Fig 4.3: multiple paths cloud-to-cloud lighting

(d) Cloud-to-Ground Lightning

Cloud-to-ground lightning is a great lightning discharge between a cumulonimbus cloud and the ground initiated by the downward-moving leader stroke. This is the second most common type of lightning, and poses the greatest threat to life and property of all known types.

(e) Bead Lightning
Bead lightning is a type of cloud-to-ground lightning which appears to break up into a string of short, bright sections, which last longer than the usual discharge channel. It is fairly rare. One of theories is the observer sees portions of the lightning channel end on, and that these portions appear especially bright.

(f) Ribbon Lightning
Ribbon lightning occurs in thunder storms with high cross winds and multiple return strokes. The wind will blow each successive return stroke slightly to one side of the previous return stroke, causing a ribbon effect.

(g) Ground-to-Cloud Lightning
Ground-to-cloud lightning is a lightning discharge between the ground and a cumulonimbus cloud from an upward-moving leader stroke.

(h) Ball Lightning
Ball lightning is described as a floating, illuminated ball that occurs during thunderstorms. They can be fast moving, slow moving or nearly stationary. Some make hissing or crackling noises or no noise at all. Some have been known to pass through windows and even dissipate with a bang. Ball lightning has been described by eyewitnesses but rarely recorded by meteorologists.

(i) Upper-Atmospheric Lightning
Reports by scientists of strange lightning phenomena above storms date back to at least 1886. However, it is only in recent years that fuller investigations have been made. This has sometimes been called mega-lightning.

4.2.2 Triggering of Lightning

Lightning has been triggered directly by human activity several times.

(a) Rocket-Triggered
Lightning struck Apollo-12 soon after takeoff. It has been triggered by launching rockets carrying wire spools into thunderstorms. As the rocket

ascends; wire unwinds providing a path for lightning. Such lightning bolts are straight due to the path created by the wire.

(b) Volcanically-Triggered
Extremely large volcanic eruptions, which eject gases and material high into the atmosphere, can trigger lightning. This phenomenon was documented by Pliny the Elder during the AD 79 eruption of Vesuvius, in which he perished.

(c) Laser-Triggered
Since the 1970s, researchers have attempted to trigger lightning strikes on demand. Such triggered lightning is intended to protect rocket launching pads, electric power facilities, and other sensitive targets. Researchers' generated filaments that lived too short a period to trigger a real lightning strike. Nevertheless, a boost in electrical activity within the clouds was registered channel of ionized molecules

4.3 Lightning and a broadcast Transmitter System

A transmitter site and its electronic equipments alogwith radiating tower antenna must be considered as a system for the purpose of lightning protection. A broadcast tower is the most attractive target for lightning discharges due to the fact that it is an excellent electrical conductor and offers many connections to earth ground. Also it's the highest tall structure in the open surroundings, even sometimes on elevated ground plane.

Broadcast transmitters historically have been manufactured using high power vacuum tube technology. In the past few decades, however, the broadcast industry has been progressively converting to all solid state transmitter designs, because of their greater reliability and efficiency compared with the vacuum tube transmitters. But solid state transmitter's internal protection circuits/devices, unlike tube type transmitter, are not able to protect the transmitter from damage as the result of a direct lightning hit to its antenna, by themselves. So they must be supplemented by additional protection devices, at the antenna and at the point wiring entering the transmitter cabinet, which will divert lightning stroke energy away from the transmitter.

The power vacuum tubes, in old transmitters had the disadvantage that their performance slowly deteriorates and fails with use, and also they consume much more input power for a given output than its solid state counterparts. However, it has a major advantage over solid state in respect of that it can withstand a great deal of abuse by way of over-current, power surges, transients and lightning strikes and still keeps on working. Tube transmitters can operate into substantial load mismatches and have been known to absorb a large number of power line and antenna surges and also transients with little or no damage. Transistors and other solid state devices, on the other hand, will operate almost indefinitely without degradation if their maximum operating values are never exceeded, but can be destroyed in just a few milliseconds of over-voltage or over-current conditions.

All solid-state products share these risks, but broadcast transmitters face much greater risk because of their location and class of service. A broadcast transmitter usually is installed at a rural, remote location, often at the end of a long power line, where it receives all voltage instabilities, surges and transients caused by users farther up the line. Further, it is often installed at high altitude locations and directly connected to a tall radiating guyed tower, which performs as an excellent lightning rod. At the same time, unlike many other types of commercial and consumer electronics, high reliability and continuous uninterrupted operation without outage is required for its class of service, a failure means economic & revenue loss and possible loss of image of the broadcaster as critical public service in the time of emergency.

Therefore it becomes imperative, that much greater care and planning of the installation is required when replacing a tube type transmitter with a new solid state model. It is incumbent on the broadcast engineer to understand methods of surge protection and grounding necessary for a proper installation. If installation procedures recommended by manufacturer and planning engineer are followed correctly, the solid state transmitter will perform with reliability and trouble free.

Lightning can cause greatest damage to broadcast equipments out of all or any externally generated events. Lightning will strike the highest object and will take the path of least resistance to ground. So a broadcast tower becomes an attractive target, by nature of the fact that it is an excellent electrical conductor and offers many connections to earth ground.

A broadcast tower antenna functions as an interface between nature (environment) and transmitter. Fig 4.4 below illustrates the principle of coupling of electro-magnetic energy of lightning to a broadcast antenna structure. The amount of energy coupled depends on the strength of the lightning pulse in KA (kilo ampere) and the speed of lightning going to the ground in KA / ms, the distance to, and height of the antenna. Essentially an electrical discharge between positive and negative regions of a thunderstorm, lightning is highly visible form of energy transfer. With lightning strike, huge discharges of current pulses flow from the clouds to earth within a fraction of micro-seconds. These pulses consist of a heavy DC component and have a frequency spectrum up to megacycles.

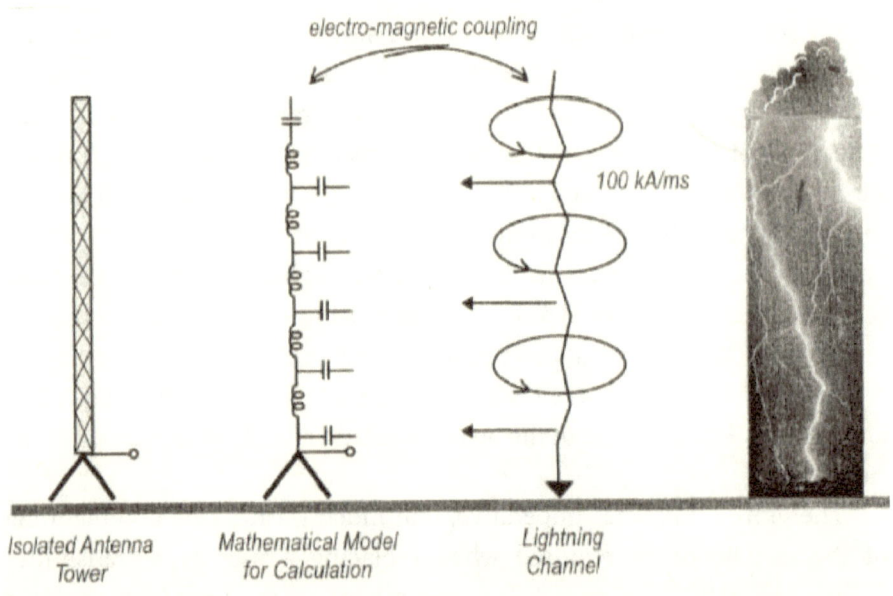

Fig 4.4: Principle of Coupling of Electromagnetic energy of Lightning to a broadcast tower

A broadcast tower antenna standing high and often at isolated places, and connected equipment are highly vulnerable to lightning. During lightning events, each piece of metal with a certain length induces voltages. This is due to the electro-magnetic coupling between the lightning channel and the metal structure.

As explained lightning strikes the tallest conducting objects in the vicinity. Thus a tall tower will shield the area surrounding it just like an umbrella, and this area has the shape of a cone with the tower at its apex. The taller tower will shield larger area from the effect of lightning.

There are many ways to deal with the possibility of lightning strikes on a radio tower. One is to take steps to prevent lightning from striking the tower. The most effective means of doing this is to install an ionization dissipater at the top of the tower. This is an array of multiple sharp steel points, connected to ground by a separate substantial conducting cable. It has long been known that a sharp point in the air will attract the ionized air particles, drawing the charge out of the air in the area surrounding the point. There is a limit to how much current each point can draw, and so to make it more effective, many points are installed. These take varied forms, from the relatively inexpensive "bottle brush" style to large, umbrella-like dissipation arrays which can be expensive and add considerable weight and wind-load to a tower structure. The use of all but the largest ionization dissipaters can be a double edged sword: there are limits to how much charge each point can attract, so the current drain is limited by the number of points. When the charge buildup exceeds the ability of the device to drain it, the device can act as a lightning rod and becomes a lightning COLLECTOR, actually attracting the largest lightning strikes. Also, its effectiveness is only as good as its ground connection.

There is another, simpler approach to lightning damage mitigation: to simply accept the fact that lightning will strike the tower from time to time, and to take steps to minimize the damage by providing an attractive and direct path to ground. Concurrently, any paths that pass to the ground through the station equipments are made as unattractive as possible. To successfully achieve this requires some understanding of how lightning currents behave in a conductor.

The average lightning strike has maximum current of 20,000 Amps, and there are occasional strikes that exceed 100 thousand amps! The typical lightning surge has a rise time of 5 microseconds, and then takes another 40 microseconds to decay to half amplitude. Because the earth has a measurable resistance, it will be impossible to attract all of this current into the ground. A typical radio tower site will have a ground resistance of from 5 to a few hundred ohms, consisting of the resistance of the soil itself plus the resistance of the electrode's connection with the soil. Additionally, there are many paths to ground at a typical site, each having its own impedance.

Thus, the path to ground will resemble a parallel resistor network, with different levels of current flowing in each leg. Some of these paths to ground will be direct, but others will invariably pass through the station equipment. The installation objective should be to improve the effectiveness of the direct earth connections, and reduce the currents in those paths through station equipment to the extent possible. If we assume a ground resistance of 50 ohms, then as per Ohm's law, with a current of 20,000 Amperes there would be one million Volts peak voltage drop between the tower ground system and true earth. This voltage will appear between the shield of the coaxial cable and ground, and some of if it will try to find a path to ground through the transmitter's cabinet. Further, any current flowing in the coax shield will be inductively coupled to the center conductor, which will appear at the transmitter's output network. This is a frequent cause of lightning-induced arcing inside a transmitter.

4.4 Lightning Characteristics and Incidences

The para will detail the discharge of energy caused by the lightning strike from an electrically charged cloud to ground. Most of the electrical storms of this type are localized, short in extent, and are caused by localized air heating and convection. A less common but more troublesome type of storm is the frontal type, extending up to several hundred kilometers, of the meeting point of warm-moist and cold air masses. The incidence of electric storms and lightning with regard to their type are prepared by weather bureaus of each country and maps are available for such purpose, which must be used for knowing the average, peak numbers of strikes per square miles per annum including the contour factor of the ground.

As you know that any structure or for that matter an antenna of height (h) shields an area of approximately $9\pi h^2$ square feet, a radius of three times the height, then the actual strike incidence at a particular antenna site, where frontal storms are predominant, will be multiplied by a contour number $0.375 \, h^2 \times 10^{-6}$. The places where convection storms are predominant, the factor approx reduces to 75% of the frontal storm value. Let us consider two examples as below;

(a) Frontal storms; data; contour factor – 100, Antenna height – 500 ft
Antenna strikes per annum; $100 \times 0.375 \times 500^2 \times 10^{-6} = 9.4$ strikes

(b) Convection storms; data; contour factor – 10, Antenna height-500 ft
Antenna strikes per annum; $10 \times 0.375 \times 500^2 \times 10^{-6} \times 0.75 = 0.7$ strikes

Note; if the antenna is not situated on flat area but on a small local hill, then antenna height should be increased by hill height for this calculation.

The probability of lightning currents exceeding the peak crest currents for tower structures is shown in fig 4.5.

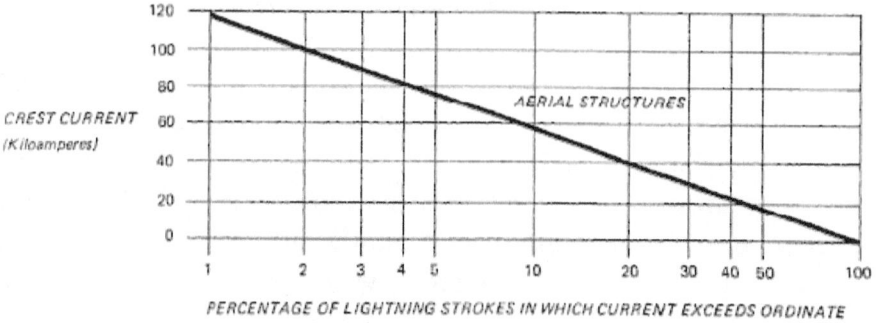

Fig 4.5: Magnitude distributions of currents in lightning strokes to structures

The next important considerations are actual electrical parameters of the stroke. Some of the assumptions are;

(a) Main stroke of a lightning is characterized by a rapid rise and near exponential decay of current from a high impedance source comprised of a long column of ionized air. Presumably the inductance of the air path determines the rate of rise of the current and the resistance determines the current peak value and decay rate.

(b) Hence a median main strike pulse may be considered as a unidirectional near exponential pulse of 20,000 amperes peak amplitude lasting 40 microseconds to half amplitude. There is 5% probability that the pulse amplitude is 4 times greater than the median value.

(c) The rise time of a typical lightning strike pulse is of the order of 5 microseconds to peak amplitude.

4.5 Lightning Protection Principles

A lightning strike is a discharge from a charged cloud to semi-infinite reservoir, which is normally referred to as "ground". Unfortunately, at the surface of earth an ideal terminal connecting to the ideal ground, which should have impedance ranging from few ohms to several hundred ohms, is rarely available. If a lightning strikes a radio tower with local grounding either directly (grounded tower or a shunt fed tower) or via a spark gap (in case of insulated tower) then the large current pulse flowing through the local ground impedance would develop a very high potential with respect to ideal ground e.g., with a medium current pulse of 20,000 Amperes and impedance to ground say 50 ohms, the potential would be 10^6 peak volts.

If the tower antenna local ground is connected to remote grounds through surface cabling, then a substantial part of discharge current would flow through these connections into the remote grounds. The real connection to ideal ground becomes a parallel combination of all possible paths to the ground. This is the actual situation, because the local tower antenna ground is one terminal of the tower antenna for transmission purposes and requires a drive connection, which usually is the outer shield of the coaxial cable or outer ground wires of the open wire feeder line cage in case of overhead feeder lines.

From this description it may be inferred that the first and foremost important principle is to provide the lowest impedance (best possible) local ground at the base of the tower. The tower antenna ground mat, normally being used, cannot be assumed to have low impedance to ideal ground. In poor soil conductivity or in frozen soil it may function as a good counterpoise type ground mat, yet may have high resistance to ground. As such these ground mats must be supplemented by a ring of copper rods driven into ground to a depth of 450 mm on outer periphery of the mat to connect to ideal ground.

Even after providing best practical ground to a tower antenna, it still will have a finite impedance to ideal ground. Therefore some component of the lightning stroke current will flow through the outer cage ground wires of a open wire feeder line or outer shield in case of a coaxial cable to remote grounds. This portion of current will induce a high voltage between the inner and outer of the coaxial cable or the feeder line at the transmitter end of the connection. The transmitter needs protection against this induced current. But the 1st concern should be the actual path which the discharge current takes in reaching the remote grounds. The 2nd concern should be that this current should not pass through trandsmitter components. To ensure this principle, a clear understanding of all possible remote grounds paths like; local ground connection at the base of tower, incoming ac power supply line, audio & antenna current remote monitoring and control cables, and also Phaser and other cables coming to ATU hut, is required. This also requires a careful arrngement of ground connections within the transmitter building. Fig 4.7 which is RF earth schematic for a transmitter set up, illustratres this principle for a typical site. As shown in this figure, the current resulting from a direct strike on the tower antenna will flow to the ground via tower lightning ground system provided at the base of the tower as well as via remote grounds composed of the remote control / monitoring cables, ac power supply line cables coming from the transmittter hall to antenna tower.

The desired paths for these currents to remote ground should be achieved by ensuring that all ground paths which interconnect the tower antenna and the transmiter building, including the ground straps and the shield of any cables in the system are connected directly and solidly to a single point within the transmitter building reffered to as the stations reference ground point. This point in turn should be connected through shortest thick and wide heavy duty conductors to the ground terminals of surge arrestors connected directly across all incoming cables such as ac power suply line and control /monitoring cables. In addition, all connections to the transmitter should be done through the center of ferrite cylinders, which acts as chokes, blocking the flow of lightning discharge current through th etransmitter cabinet.

The importance of protecting transmitters from damages coming from lightning strikes has greatly increased with the introduction of solid-state design in broadcast industry. In the old times there was no such problem as the power vacuum tubes were able to withstand unfavorable conditions, like high transients, lightning strikes and power surges without interrupting operation, even though they had the disadvantage of limited lifetime.

Thus, even though it is theoretically possible to take steps to prevent lightning striking a tower, experience shows that it is much simpler and more effective to accept the fact that lightening will strike and one need to take steps to minimize the damage to the highly sensitive solid-state transmitter broadcast systems.

Protection devices inside transmitters themselves and most important of all an appropriate systems configuration to include transmitter building and antenna grounding systems can reduce lightning related damages significantly.

4.6 Static Charge and a Broadcast tower System

Before understanding static electricity, we first need to understand the basics of atoms and magnetism. All physical objects are made up of atoms. Inside an atom are protons, electrons and neutrons. The protons are positively charged, the electrons are negatively charged, and the neutrons are neutral.

Everything is made up of atoms, and atoms are made of tiny particles, some of which are electrically charged. Most atoms are electrically neutral; the positive charges (protons in the nucleus or center of the atom) cancel out the negative charges (electrons that surround the nucleus in clouds). Opposite charges attract one another and similar charges repel each other. Sometimes the outer layer (the negatively-charged electrons) of atoms is rubbed off, producing atoms that have a slight positive charge.

Therefore, all things are made up of charges. Opposite charges attract each other (negative to positive). Like charges repel each other (positive to positive or negative to negative). Most of the time positive and negative charges are balanced in an object, which makes that object neutral.

4.6.1 Static Charge

Static charge is stationary electricity that builds up on a material. A common example of static electricity is the slight electrical shock that we get when we touch a doorknob during dry weather. The static electricity is formed when we accumulate extra electrons (negatively-charged particles) on our body which we get when we rub the carpet with our shoes. So an imbalance between negative and positive charges in objects is static electricity.

Another case, when one walk across a room to pat one's dog, but got a shock instead? Also, you took your hat off on a dry winter's day and had a "hair raising" experience! Had it been in humid weather it would not have happened. During dry weather, the excess charges do not dissipate very easily, and you get static electricity. But during humid weather, these electrons flow through the damp air and the object become electrically neutral. The object that did the rubbing will accumulate a slight negative charge as it gets extra electrons. Or, may be you have made a balloon stick on the wall after rubbing it against your clothes? Static electricity results due to imbalance of negative & positive charges in an object. These charges build up on the surface of object until they find a way to be released or discharged. One way to discharge is through a circuit.

The rubbing of certain materials against one another can transfer negative charges, or electrons. For example, if you rub your shoe on the carpet, your body collects extra electrons. The electrons cling to your body until they can be released. As you reach and touch your pet, you get a shock.

Explaining "hair raising" experience, as you remove your hat, electrons are transferred from hat to hair, creating that interesting hair do! Remember, objects with the same charge repel each other. Because they have the same charge, your hair will stand on end. Your hairs are simply trying to get as far away from each other as possible!

When you rub a balloon against your clothes and it sticks to the wall, you are adding a surplus of electrons (negative charges) to the surface of the balloon. The wall is now more positively charged than the balloon. As the two come in contact, the balloon will stick because of the rule that opposites attract (positive to negative).

4.6.2 Static charge on a broadcast tower

Friction between cold moist and warm air masses creates static electricity in a cloud. A negatively charged cloud layer will induce positive charge on the metal parts, buildings or the ground beneath it, resulting in a condition much like the two opposing plates of a capacitor. Fig 4.6 below illustrates the principle of static charging of broadcast towers. Assuming linear voltage distribution between clouds to ground, isolated tower at a height of 70 to 100 mtr will be charged with 70 to 100 KV. So when a cloud comes near a broadcast tower, it induces opposite charge in it. This charge accumulates across the guy break-up insulators. Once the charge reaches break-down voltage rating of guy break-up insulators, it sparks over and spark is sustained by the RF power of the transmitter till some arrangement is not made for by passing this built-up charge with a low impedance path to the ground.

Fig 4.7 shows an effective and prevalent static discharge protection system of an AM tower. As you know the most likely object for a lightning hit in a broadcast transmitting radio station is the AM tower standing 1/4 to 5/8th wavelengths tall in the sky and sometimes reaching to a height of more than 350 meter in MF band and up to 600 meter in LF band. Protecting the tower is one of the best ways to protect all other RF equipment from lightning.

Think of tower as a large lightning rod protruding in the sky line. It is always highest structure in the area by design and requirement. This makes it, the most likely object to take the strike of a thunderstorm. Lightning likes all other electricity prefers and takes the shortest (lowest impedance as per Ohms law) path to ground. The shortest physical distance is not always the lowest impedance. However, when air is the only thing between highly charged clouds and ground, the tallest conductor in the area is most likely the lowest impedance path.

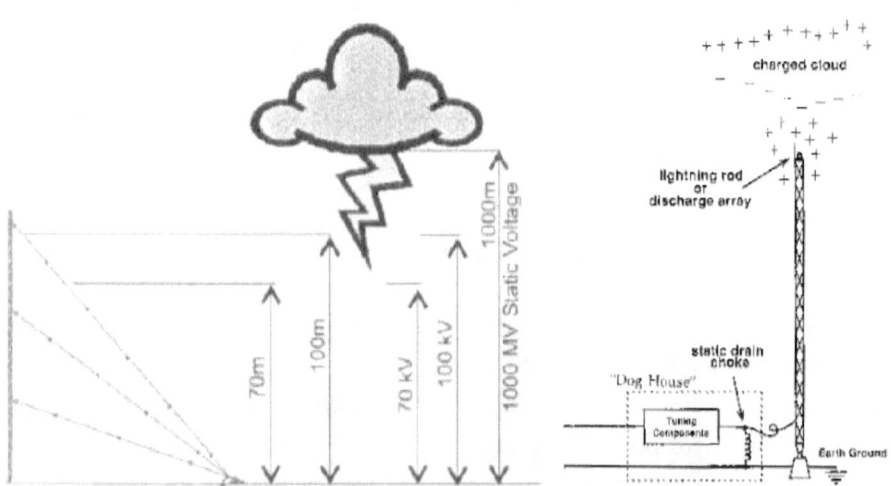

Fig 4.6: Principle of Static charge on a tower

Fig 4.7: Static Discharge Protection of AM Tower

You may be surprised to know that a lightning rod works best by not getting hit by the lightning. The whole purpose, in fact of this pointed rod over the tall metallic structures is to drain off the induced charge before it can build up high enough to cause an actual lightning. The idea is that sharply pointed objects discharge static charge much better than rounded objects. Rounded objects tend to build up a charge whereas pointed objects drain a charge. Prior to a strike, lightning is nothing more than a big static charge. The clouds builds up one polarity and the ground builds up opposite. Anything connected to ground shall have same charge as the ground itself.

Now if the tower has the smallest gap to the clouds and same charge as everything else, it is surely going to be hit by lightning. Better to reduce the potential between the clouds overhead and the tower and let some other charged object take the hit. You can do this by having one or more sharp points on the tower and having the tower grounded for DC.

Most towers have a pointed metal, "lightning rod" attached to the top of the tower and sticking up above the beacon. This rod needs to have a low resistance DC connection to ground through the tower members. It is best if the rod is welded to the top section and the sections are welded together. If the connections are corroded, the DC resistance will be high. A corroded tower is also a poor radiator if it is being used as an AM antenna. The bottom of the antenna needs a good earth ground. For FM, TV, and Communication tower, this involves simply connecting the tower frame itself to a solid earth ground. AM tower antennas are trickier because connecting the tower to ground will short circuit the signal. The requirement is of a device providing DC path to ground for static discharge but which simultaneously provides very high impedance or almost open circuit to RF signal. Such a device is called a, "Static Drain Choke".

The Static Drain Choke is a coil of wires mounted on an insulating form and connected between RF feed line and tower ground. As such this is in parallel to or across the tower base insulator but kept inside the Feeder Hut. The dia of coil need not be big because the static discharge currents will be less than an ampere. The coil has many turns for catering to entire MF band and it has large inductive reactance at MF frequencies for high impedance to RF signal. But for DC signal it is almost short to ground as such is lowest impedance path to static discharge. This coil is part of standard items supply list in ATU and as per manual depending on frequency of operation number of turns can be selected at site to offer high impedance to RF and lowest to static charge.

After installation of ATU components & static drain choke or in a working station after shut down of transmitter, open the feed line, it may be worth to measure the DC resistance from tower feed to ground. It should be of the order of few ohms or even less. Otherwise choke is defective or there is some loose connection.

Sometimes one pointed rod is not sufficient to remove the charge fast enough. If one has this problem either due to location of site or a tall tower, you can still do something to prevent strikes. In such cases install an array of discharge points. A dissipation array consists of sharp barbed wires spiked stick or like a broom. This material is formed into a dome over the tower, a ring around the tower top or an arm that protrudes above the tower top. The principle is that many points will drain the static charge faster than one point. This is similar to idea that many parallel wires will offer lower DC resistance than a single strand. You may be aware that ships and aircraft use multiple discharge points to safely drain static charges.

4.6.3 Static Charge estimation

The most common type of transmitting antenna for use in the long and medium wave bands comprises of a vertical steel latticed triangular or cylindrical mast, insulated and driven at its base and kept in an upright position by means of steel guy wire ropes. The height of the mast is normally between 0.1 and 0.625 of the transmission wavelength, depending on the desired coverage, vertical radiation pattern and the radiated power. Towards the lower end of the MF band, antennas may therefore be up to 300 mtr high if they are to have vertical radiation patterns having so-called "anti-fading" properties. In the LF band, anti-fading mast antennas are not feasible and the tallest LF masts built so far is 500 m high.

As you know that in order to suppress induced RF currents in the guy wires, and to prevent unwanted radiation from these, the guy wire ropes are normally divided into short sections by inserting guy insulators. This guy break-up insulator is subjected to two kinds of voltages:

(a) Induced RF voltages due to the RF currents flowing in the mast; and
(b) Electrostatic voltages caused by atmospheric or clouds electrostatic fields in the vicinity of the antenna.

Whereas the induced RF voltages are practically constant in time and depend on the radiated power of the transmitter but the electrostatic voltages are variable and can reach very high values. Immediately before, and during a thunderstorm, electrostatic voltages as high as 200 kV to 300 kV may occur across an insulator and values even as high as 400 kV are possible. These voltages are the result of very strong electrostatic fields between the clouds and the earth (maximum field-strengths of 5 kV/m to10 kV/m have been observed). Despite the use of very expensive and bulky guy break-up insulators, such high voltages can cause flashovers between the metal fittings at either end of the insulator which can lead to severe damage.

This problem of static discharge across guy insulators plagues many high-power LF and MF transmitting stations with tall guyed mast antennas. Although the problem was observed and described as early as 1939 by Mr. G.H. Brown, but it gained more importance at the time of steady increase in the powers of LF/ MF band transmitters while upgrading their powers (often in range of several hundred or even few thousand kilowatts) to restore the service area and to counter high noise floor due to industrialization. Therefore greater attention has been paid to the problem in these years and a number of articles about the influence of electrostatic and RF fields on guy insulators have been published in the specialized radio journals like PIRE (Proceeding of Institute of Radio Engineers) etc. An accurate method of computing the electrostatic field in the vicinity of a guyed mast, and the electrostatic voltages across the guy insulators is presented by, ""J. V. Surutka and D. M. Velickovic in an article entitled; Static voltages on the guy insulators of MF and LF broadcast tower antennas in the journal, "The Radio and Electronic Engineer", Vol. 43, No. 12, pp. 744–50, December 1973.

The static voltages across guy insulators mentioned in the previous paragraph were calculated using the method described in the referred article. The static voltages and the flashovers they cause are not, in themselves, particularly dangerous when the transmitter power is low. Problems do arise, however, in the case of high-power transmitters. Once the static voltage has caused a breakdown of the insulator spark gap, the induced RF voltage of the transmitter can maintain the arc even if the voltage is much less that needed to trigger a flashover.

If the arc occurs at the insulator (which is without RF choke or some other bypass arrangement), which connects the guy to the mast structure, there will be an appreciable change in the antenna feed point impedance and the transmitter's reflectometer protection system will momentarily interrupt the transmission thus extinguishing the arc. The only fault in this case will be a very short break in transmission.

In contrast, a flashover across the insulator other than the first one, which is not in contact with the mast, will not activate the reflectometer and RF energy from the transmitter will maintain the arc; in high-power transmitters the power in the RF arc can be considerable and the resulting thermal stresses may damage the insulator. In the absence of a system or a device serving to extinguish such arcs, the safety of the mast can be jeopardized. In particular, prolonged arcing can cause catastrophic damage to strain insulators in which the fittings are not interlinked. In any event, the replacement of a broken insulator is time-consuming and costly.

Contrary to what might be expected, a lightning strike which touches the mast directly or which falls in the immediate vicinity represents an incomparably smaller risk as regards antenna safety than a flashover triggered by the general level of the electrostatic field. A direct strike will cause flashovers on all the insulators on a guy and this, in turn, will trigger the reflectometer, interrupting the transmission.

4.7 Methods for protecting a mast antenna from static discharges

Large static potentials are developed across break-up insulators due to approching clouds or during the conditions of blowing snow, dust, high dry winds and lightning discharges close to the antenna system. In the absence of static drain devices like RF Choke across the break-up insulators, the combination of peak RF voltage and the static charge can exceed the insulator breakdown voltage. Although highly dependent on antenna configuration and local RF field, a power arc condition initiated by static charge can be sustained by the transmitter power. Such conditions heats up the supporting guy wire loops causing failure of guy wire and /or the insulator. In order to avoid annoying or even dangerous consequences caused by the static atmospheric electricity, several devices and methods have been suggested and developed. Some of the devices used for protecting the broadcast towers from damage due to static discharge are described below;

(a) Static Drain Resistors

The oldest and most commonly used method for avoiding static discharges and accompanying RF arcs is the use of high-resistance static leak resistors connected in parallel to each insulator, except to those which connect the guys to the mast structure. Their purpose is similar to that of the static choke. They bleed off any static charge, on the antenna side of the insulators, caused by blowing snow, dust or other environmental conditions, before the static potential buildup is sufficient to cause arcing across the insulators.

These resistors leak the electric charge from the guys to the earth, thus preventing the accumulation of static voltage. Although this may seem to be an ideal solution of the problem, it is far from being perfect in practice: a direct or near stroke may destroy the resistor, so that the insulators remain unprotected with no visible indication.

(b) RF chokes

Although static drain devices like RF chokes, offer a solution to potential power arc problems a better solution yet is the use of an oil filled safety core insulator instead of an egg insulator. This will eliminate the need for a static drain device due to the high stand-off voltage values of these insulators. Oil filled safety core insulators can be a cost effective solution in place of static drain devices for low tensile guy loads up to 10,000 lbs.

Normally this component is part of tower supply and is designed and manufactured by the manufacturer. But there are situations when it may be required to be fabricated in the field on account of change in the operating frequency of the tower or relocation of a tower to some other station for reutilization at different carrier frequency.

As choke coil is connected in parallel with the insulator, a parallel tuning circuit principle is used as an equivalent circuit for calculating the design parameters. The tuning frequency of this parallel circuit is generally 80 to 85% of the carrier frequency of the station. This parallel tuning circuit becomes capacitance at the operating frequency.

(c) Arc Detectors

An interesting form of a detector measures the RF currents in the lowest section of the guys by means of current transforms and compares them. When an arc occurs, these detectors activate protection devices, which interrupt the transmission for a short interval to extinguish the arc. If the system operates correctly, the number of these interruptions may be so high, during stormy weather and immediately before it, that listening to the program becomes unpleasant. Since the arc detectors remove the consequences, but not the cause of arcs, some completely new technical solutions, which avoid the use of guy insulators, have been proposed and realized. Among these the most notable are:

(1) The replacing the steel guys by plastic ones;
(2) The use of self-supporting towers; and
(3) The use of certain antenna types with steel guys without insulators.

(d) Static Drain Choke

A static drain choke, also called RF choke, is installed at the base of the antenna across the air spark gap but inside feeder hut. Its purpose is to provide a DC path to ground and bleed off any static charge accumulated on the tower antenna. Sometimes ATU utilizes matching network inductor to perform this function altogeather elimimnating the need for a separate choke. Depending on its proximity to the antenna, this inductor may or may not be adequate. ATUs often have an internal drain choke connected to their output termination point. The purpose of this choke is to bleed any static charge thru RF feed to ground, before it reaches the RF output filter. If there is a series capacitors in the RF feed between transmitter and antenna, any internal choke will not be able to assist in bleeding off static charges on the antenna.

(e) Static or spark ball gaps

Fixed at tower base across base insulator, these ball gaps are the first line of defense, and perform two valuable functions. First, they limit the static charge build up on the tower by creating an easily ionized path for accumulated static discharge, thus preventing strike potential buildup in most cases. Second, they tend to bypass most of the lightning current flow directly to the ground at the tower base reducing the risk that a lightning strike will overwhelm other mitigation and protective devices in the transmitting plant.

Why balls are Round Shape?; A century ago Frank William Peek determined a spherical shape had the most repeatable arcing characteristics. Thus a ball gap could be set reliably as a static discharge lightning arrester, and the tower base static ball gap was born based on Peek's formulas.

(f) Use of non-metallic Guys
Non-metallic guys are made of modern synthetic fibers. Although these have remarkable electrical and mechanical properties, there is still insufficient experience of their behavior over a long period of time for static discharge function. Also, the material is not resistant to high temperature and fire.

(g) Self-supporting tower
Self-supporting tower such type of antennas have neither guys, nor insulators, so there are no problems of static, but they have two serious shortcomings: high price, and a wide & varying tower cross-section, which makes it impossible to obtain a proper anti-fading vertical radiation pattern, if such pattern is required. Attempts have been made in recent years to construct guyed antennas without insulators. The main disadvantage of this approach is that they cannot provide an anti-fading radiation pattern owing to the strong RF currents in the guys.

The shortcomings of the various technical solutions described above reaffirm (at the present state of technology and particularly in the case of tall anti-fading antennas) the preference for slender mast with insulated steel guys, provided that an effective system can be devised for eliminating static discharges.

An efficient and simple solution for static problem as stated para 4.5.3 was proposed and realized by professor Surutka in 1977, during the reconstruction of the antenna of the main MF transmitter of Radio Belgrade and the article for solution, "The elimination of static discharges on the stays of high-power MF antennas," is available in the EBU Review-Technical, No. 208, December 1984, by Professor J. V. Surutka and D. M. Velickovic.

4.8 Recommended measures for protection of an AM Transmittter set up

One common misconception equates lightning with a huge flow of direct current (DC). But a lightning strike is a dampening, alternating current, generally oscillating between ground and cloud at a rate in the LF and VLF radio frequency spectrum with harmonics well into the MF range. The result is that good transmitting antennas are inherently good lightning receiving antennas. A typical complete chain of an AM transmitter set up alongwith various devices and configurations used for protection from lightning is shown in fig 4.8. The description of each device used for protections is as below;

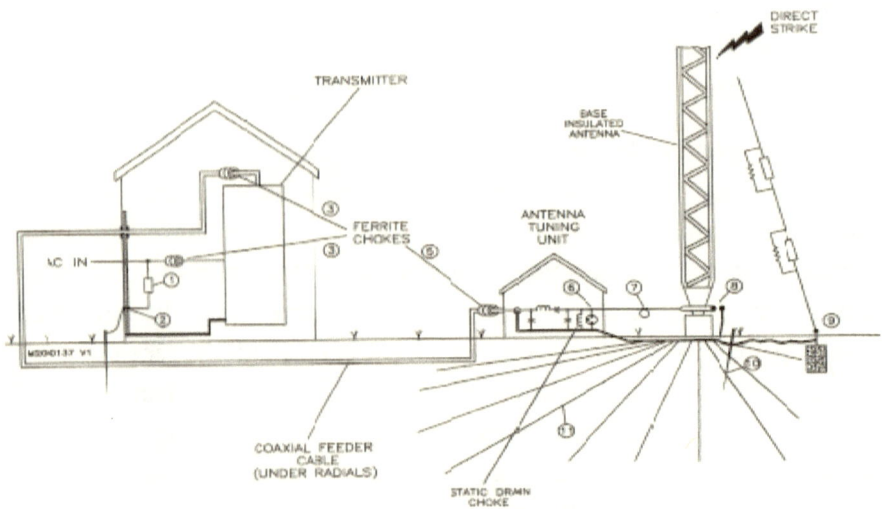

Fig 4.8: Typical devices used in an AM transmitter chain for Lightning Protection

4.8.1 Surge Protectors

Each conductor of all type of incoming cables which interconnects a transmitter plant to remote locations should be bypassed with suitable rating surge protector devices to the station reference ground point as shown below in figure 4.9 (a, b & c). The most important surge protectors are those connected to the incoming audio, control & monitoring cabling as shown in fig 4.9 (a) and incoming 3 phase ac power supply line coming from electricity departments grid substation entering the transmitter building and then single phase lines from the LT switchgear of the transmitter to its power cubicle for various distribution as shown in 4.9 (b) & (c).

The surge protectors used on these lines serves dual purpose of bypassing the surges which originate at sources which are remote from the transmitter site such as lightning strokes on ac power distribution lines and more importantly to provide safe path to currents resulting from direct lightning strikes to statios antenna tower. As stated earlier the current may be of the order of 20 kAmp and as such the surge protector should be rated accodingly. The total surge current, however, is shared between antenna tower grounding circuit/rods and various interconnecting cables that connects the transmitter to various remote grounds depending upon their relative path impedance. Generally the ac supply line will present the lowest impedance path due to its thicker wire size and its wide distribution to other ground connections. Therefore it is recommended that ac supply power line surge arrestors should be rated fro 20 KA without any deterioration and ample margin. The suppressor must also maintain the surge voltage developed across it within the safe limits and must return to a non-conducting state without removal of the normal working voltage following the surge condition. Lower surge ratings may be used for other remote cabling due to their relative sizes and corresponding higher impedance.

Varistors type element is considered the most suitable and satisfactory for all of the above purposes due to their ruggedness and fast self-restoring characteristics. Two back-to-back Zener diodes can also be used to form a passive shunt clipping element for functioning in the same manner as a Varistors. They exhibit lower ratios of clipping to working voltage but are not presently available with comparable surge current ratings.

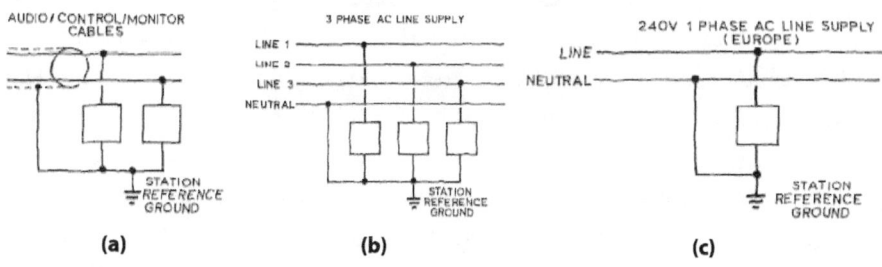

Fig 4.9: Typical Surge protectors' for (a) Audio, control, & misc cables, (b) 3 and (c) 1 phase power supply line

4.8.2 Station reference Ground

The configuration of stations reference ground is as shown as item # 2 of fig 4.8 where all ground connections join together at single point is of utmost importance from all technical points of view. This point must be located in the close proximity of the surge protector of the ac line remote cabling and should be connected to them by shortest length thick straps.

It would not be sufficient to connect interconnecting grounds from the antenna system to a safety ground bus inside the transmitter building which is at an appreciable distance from the reference ground point. Likewise all equipment, safety ground connections should be connected in a radial manner for having shortest path / distance from reference station ground. The station reference ground should be connected with a thick copper or GI strap to at least two earth pit ground rods separated by 2 to 3 times their length and driven up to the water table of the station site.

4.8.3 Ferrite Cores

Shown as item # 3 of figure 4.8, the ferrite cores or also called toroids, are threaded over the cable (i.e. cables passed through them) that connects the transmitter from various sources and called interconnecting cables, prevents the pickup currents from various radiations entering the transmitter cubicle. This device blocks them by increasing the impedance to such stray currents developed in the cable due to pickup. It should be ensured that all conductors of the ac power lines and the inner as well as the shield of all coaxial cables should pass through ferrite cylinders to present high impedance to undesired unbalance currents without affecting the desired signal.

In case of open wire feeder line connecting RF power of transmitter to ATU, provision of horn gaps at both ends (transmitter hall side feed outlet and ATU feed inlet) are must. In case of coaxial cables diameters of size larger than 2 inches ferrite may not be available, so for such cases the use of ferrite can be dispensed with however increase the number of ferrite on remaining interconnections to transmitter and ensure that even the ground wire of ac line cable passes through the toroid. In addition the RF copper feed pipe connection to the radiator tower should be provided with a loop of at least 12 to 18 inch dia to retard the lightning discharge on the tower.

4.8.4 Surge Protectors at transmitter output

To prevent a direct lightning strike discharge on the tower reaching the transmitter output circuit via feeder cable shown as item # 4 in fig 4.8, a ferrite core should also be used at the output just before connecting feeder line or coaxial cable after impedance transform network if the cable size permits. In addition to have another protection, an air ball gap and a gas tube protector should be connected in parallel across the transmitter output circuit. The ball gap provides a high current shunt element for protection. Gas tube has a much more closely defined breakdown voltage but is less rugged and has a life limitation of a finite numbers of surge discharges. The gas tube device capacity rated in ampere-second (coulomb) may therefore require periodic replacement depending on the lightning strike frequency and intensity of discharge. These gas tube devices are a must for solid state modular amplifiers as solid state elements are more sensitive to lightning as compared to old generation transmitters using electronic valves in their amplifiers.

4.8.5 Ground Strap between Antenna and Transmitter

As discussed in previous Para about remote grounds for the lightning path, a ground strap between tower antenna and transmitter can be used to advantage by providing additional path for lightning discharge. Unlike the lightning currents flowing in the screen of the coaxial cable or open wire feeder line, the currents in this ground strap will not induce transients into the inner conductor or cage of the coaxial feeder or line connecting them to final RF amplifier. If the ground strap is used it must be directly connected to station reference ground as shown in fig 4.8. In such case an additional ferrite choke may also be used between transmitter and ATU.

4.8.6 Spark ball Gap at Antenna Tuning Unit end

As shown the item # 6 in figure 4.8, a spark ball air gap at input as well as at output of an ATU in parallel to RF signal is highly recommended and in fact it is must as part of standard design in a radio transmitter station and these two spark ball air gaps are part of standard supply. The size and ruggedness of these spark ball gaps in the ATU is considerably smaller than the one being used at the base of the tower where voltage is much higher. The breakdown voltage of spark gaps is dependent on shape of balls and the spacing of air gap, precise gap setting is never possible. Standard charts are available for adjusting and setting the air gap for breakdown with the manual of each ATU. The air gap, however be set so that breakdown never occur under normal RF working conditions. In case a tuning circuit configuration does not provide a DC path from the antenna feed to ground, a static discharge or leak choke (also called RF choke) must be connected in parallel with spark gap inside the feeder hut.

4.8.7 Loop in Antenna Feed line

Further as stated earlier also and shown as item #7 in figure 4.8, a common, very simple, inexpensive and in-genuine way of reducing the lightning currents in the antenna feed is to provide a loop of 18 to 24 inch dia by bending the copper feed pipe itself. This loop adds a low but finite series inductance to retard the lightning. With lightning strike, huge discharges of current pulses flow from the clouds to earth within a fraction of micro-seconds. These pulses consist of a heavy DC component and have a frequency spectrum up to megacycles. So this loop offers very high impedance to higher spectrum lightning current and enables the protective devices sufficient response time for safety action. This device requires almost no care and maintenance except periodical cleaning and clearing of loop of all webs and other trapped items. A clear view of this loop is shown in fig 4.10.

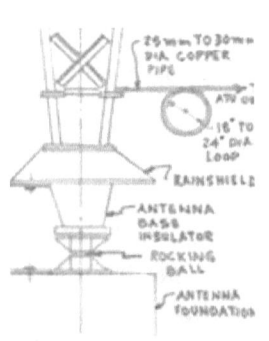

Fig 4.10: A typical lightning retarding loop in RF feed

4.8.8 Spark ball Gap across base Insulator

A very simple, most effective and important protective device against lightning for a base insulator of tower radiator shown as item # 8 in figure 4.8 is a standard fitment for any guyed self radiating tower. It consists of a pair of tungsten or carbon spark balls with an air gap between them. The device is very rugged, easy to repair or inspect or install and is widely used at the base of self radiating towers where it serves as a crude but an effective first line of defense as lightning protector for the transmitter chain. As the spark ball gap is exposed to harsh outside environmental conditions, its mounting arrangements should be rugged and rigid enough to maintain the gap separation during severe weather conditions in rains and storms. The balls should be aligned in the horizontal plane, rather than vertical to prevent rain water or dew drops sticking to it and reducing the effective gap separation. It is recommended to mount it below base insulator rain shield to avoid water drops dripping on it. It should be checked periodically as per maintenance schedule and cleaned of any material clinging to balls or any spider's web to keep the air gap as per design. Clouds of flying insects have been known to initiate premature breakdown. The area around the tower base should be kept free of high grass and other vegetation.

So the most important lightning protection device is a properly adjusted air spark gap, that is located at the base of the antenna and is connected to the lowest possible ground impedance. Its principle function is to shunt the majority of lightning currents to a low impedance ground at the base of the antenna. When properly adjusted, it will minimize the transients flowing towards the transmitter through the coaxial RF feeder.

Air spark gap at the antenna should be a ball gap in preference to some other small radius devices. A spark ball type air gap is known to offer faster breakdown at repeatable voltages. Spark balls that are between 25 mm (1 inch) to 50 mm (2 inch) in dia are best for lightning protection. They create an almost uniform field situation, which improves the predictability of the gap setting needed. It is recommended the balls be manufactured from very hard, low impedance materials such as carbon or tungsten. This will minimize damage to balls in their function. Calculation of spark ball gap and the effect of dia of spark ball is described in the later parts.

4.8.9 Series Capacitor

A series DC blocking capacitor is recommended to be installed in the RF feed between transmitter and antenna. It will greatly help in reducing current flow to the transmitter's RF final PA stage during the "continuing currents phase" of a lightning strike. This phase, which occurs immediately after the very fast rise of the lightning impulse, usually has a duration of 40 to 500 milliseconds and transfers the bulk of the lightning charge (50 to 500A). A representative lightning strike delivers about 25 coulombs, with 75% of this delivered in the 'continuing currents phase'. A series DC blocking capacitor will present a high impedance to low frequency component of lightning and significantly restrict the flow of unwanted transients, from antenna to transmitter, in RF feed coaxial cable.

4.8.10 Grounding of Mast Guy Wire ropes

In the earth ground path schematic figure 4.8 items # 9 is the grounding strap of guy wire rope of a self radiating guyed tower. The lower end of a guy wire rope should be electrically connected to one of the ground plane radials by a thick at least 100mm (4 inches) copper strap or to a separate earth pit system made specifically for this purpose near the anchor block. These connections will allow proper functioning of guy insulators. The Spark ball gaps, guy insulators, RF chokes across each guy break-up insulators and the connecting straps should be checked for sign of electric or mechanical breakdown, after each lightning strike as well as periodically under maintenance schedule for the damages from static or lightning or theft /vandalizing or the loose connection.

4.8.11 Tower Antenna lightning Ground System

As shown in the earth ground schematic fig 4.8 for the protection from lightning item no #10, the tower antenna's lightning ground system must provide lowest possible impedance to ideal ground. An earth pit system on each of the 4 corners of the ATU hut or a ring of four driven ground rods long enough to penetrate up to the level of available water table, will provide a very satisfactory connection. If there is no ATU hut adjoining the tower then the rods is best alternative and they must be separated by 2 to 3 times their length and each of these should be separately connected with a thick (at least 100 mm) copper strap to tower base ground terminal.

4.8.12 Radial Ground System

Item #11 of the fig 4.8, the system of copper earth radials of the tower antenna must be connected in a ring or common point at base end and terminated to copper sheet covering fixed over tower base foundation block. 4 no of 100 mm wide copper straps must connect this common ring to lower end of base insulator with a good electrical connection capable of carrying total RF current. All these wire interconnections must be silver brazed /soldered to ensure perfect electrical continuity even in a corrosive environment. These should never be bolted or screwed as it will not make perfect electrical connection in RF environment.

4.9 Transmitter Building Equipment Layout

The geometry of the interconnections in and around the transmitter building are most significant for the effectiveness of the lightning protection system of a radio tranmitter. The aim is to provide a path for destructive lightning currents flowing from the tower to the ac power line, which does not include the interior of the building. So this section will describe the layout design of the building in a transmitter complex from this point of view.

4.9.1 Ideal Building Equuipment Layout

An ideal transmitter building geometry is normally as shown in Figure 4.11, in which feeder line / coaxial cable and ac line service enter the bulding in close proximity to one another. This facilitates to have station reference ground at a single point. The outer of open wire line, shield of coaxial cable in case of stations having omni radiators, is directly connected to the station ground with the common terminal of the surge protectors device. The line terrminals of the surge protector connects via short low inductance cables to the line of the ac power.

In case of a station having directional antennas with phasor installed within the transmitter building, all open wire feeder line or coaxial cables, should enter at this same point and must be connected to the station reference ground point. In case where a building safety ground ring is installed, it should be connected directly to the station reference ground point. With this arrangement, most of the lightning current will tend to bypass the interior of the building due to relatively low impedance of the path through the surge protector compared to that of the long loop in and out of the building which passes through the transmitter. At power level up to 10 KW, ferrite core toroids may be threaded over the ac power and coaxial feeder cables inside the building which acts as RF choke to the lightning currents, but offfers no resistance to the normal RF currents. This technique may not be possible for high power transmitters as the maximum internal dia of commonly available ferrite cores is limited to about 3 inches.

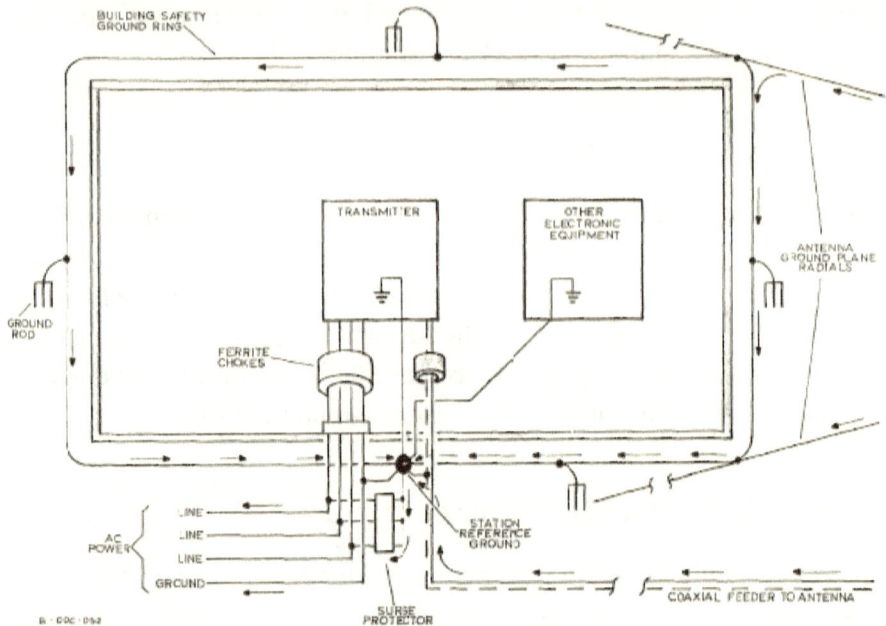

Fig 4.11: Ideal Transmitter Building Layouts

4.9.2 Poor Building equipment Layout

Fig 4.12 shows a very poor transmitter building equipment layout which contains all the ideal elements of previous arrangement but configured and placed so poorly that little or no benefit is obtained from it. This figure establishes the importance of using correct configuration. Followings are the fundamental error in this layout;

(a) The ac power supply line is fed from left hand side while the R F feeder is from right hand side instead of standard enrty /exit from same side of the building as in case of ideal layout.

(b) The earth ground points are taken from the safety ground ring around the building at various locations and no station reference ground point have been provided.

(c) Ferrite core toroids have been provided independently over each line of the ac power source causing them to be completely saturated by the normal operating currnents in these lines.

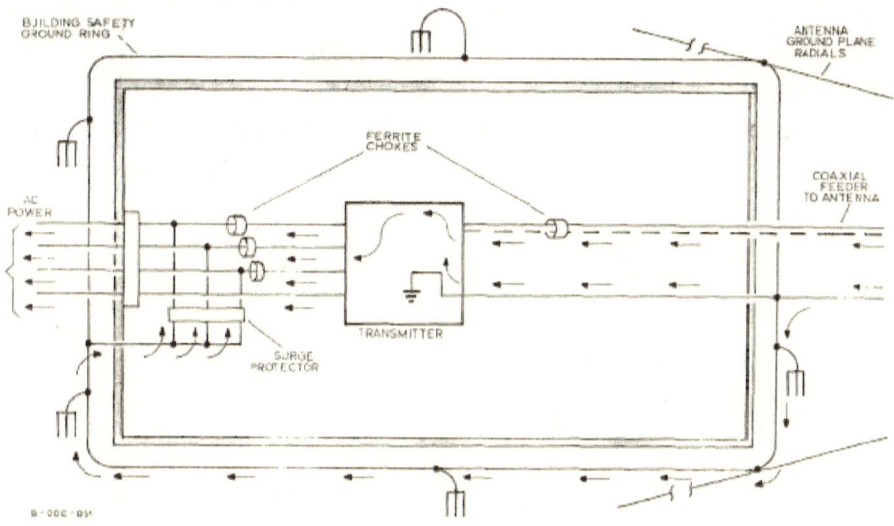

Fig 4.12: Poor Transmitter Building layout

(d) The shield of the RF coaxial feeder cable is directly connected to the transmitter output, permitting lightning currents directly to pass through componenets of final stage RF amplifier.

It is worth mentioning here for this layout that even if the shield of coaxial cable was connected to the safety ground ring running around transmitter building, most of the lightning current would still tend to flow in the direct path thru the trnsmitter due to its relatively low impedance and lesser path length compared to the alternative longer path through the safety ground ring and the surge protector.

(e) Safety ground connection of the transmitter being connected at left hand side of the figure provides another undesired path for lightning currents flowing in other ground interconnect between antenna & transmitter building.

4.9.3 Correcting a Poor building equipment layout

In existing installations it is impractical and absolutly of no use to re-configure the layout to conform to exactly ideal layout arrngement. Fig 4.13 illustrates a method of corrections for a non–ideal building equipment layout, where ac line service is connected at a location which is widely separated from the entry point of the coaxial feeder and the antenna ground strap. It should be noted that the coaxial cable should not contact any grounds within the building prior to being connected to station reference ground with a shortest strap. It should also be well out of reach of personnel working in the building to ensure their safety during lightning storms.

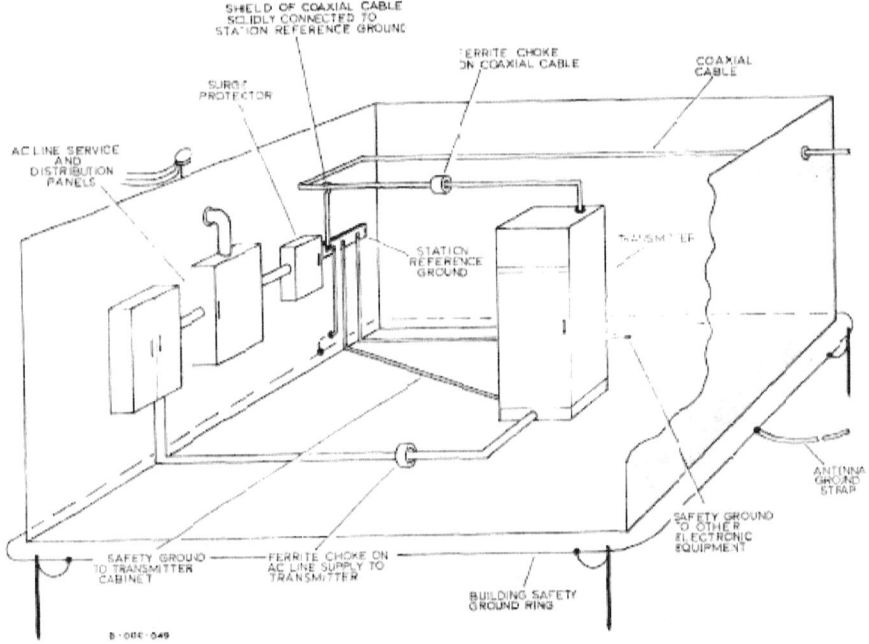

Fig 4.13: Correctioning Poor Transmitter building layout

Following facts however should be carefully considered while attempting to improve the poor layout of transmitter equipments in the building;

(a) The ac power line supply, the coaxial cable or feeder line and all other cables including ground connections which connect to the equipment to be protected, must be brought in close proximity to each other at the station refence ground point before feeding to this equipment.

(b) The equipment in above clause means in the entire transmitter building. With this arrangement, both personnel and all equipment within the building are protected. The principle may in some cases be applied only to an area in the building or to the radio transmitter alone due to logistical difficulties.

(c) All incoming ground conductors should be connected directly to the station reference ground point, which inturn should be connected radially to the ground terminal of all equipments in the building.

(d) A set of varistors or similar devices capable of carying the lightning current should be connected via shortest cables between the station reference ground point and the conductors of the ac line supplly.

4.10 AC line Surge Protectors

Utility AC supply lines to a transmitter building usually represents the lowest impedance path to remote grounds. Therefore it will carry most of the lightning currents flowing away from the transmitter site. The surge protectors connected between station reference ground and ac supply line cables must therefore be rated to carry most of the expected lightning current. It is also important that the potential developed across the surge protectors by the lightning current is balanced with respect to all of the lines, so that no net lightning potential appears between any ac power supply line to the transmitter. In case of three phase supply it is possible, however in case of single phase it is not possible or even in case of three phase supplies that are not balanced with respect to ground. Voltage rating of surge protectors should be choosen such that the prevailing off load steady state voltage is safely below the minimum turn-on voltage with ample safety margin.

4.10.1 Type of Surge Protectors

Surge protection devices protect against surges generated by electromagnetic effects, such as lightning or electrostatic discharge caused by a variety of effects. As such, surge protection may be applied at the mains input to eliminate disturbances on the mains supply external to the operating equipment or internally generated overvoltages usually caused by high inductive load switching. A surge protector may either attenuate a transient by filtering or divert it to prevent damage to the load. Those that divert the transient fall into two broad categories;

(a) **Crowbar devices**; Crowbar is an active device that switch into a very low impedance mode to short circuit the transient until the current is brought to a low level.

(b) **Clamping devices;** An active device that limit the voltage to a defined level as per the capacity of the device.

(c) **Functioning of Crowbar devices;** The crowbar group includes devices triggered by the breakdown of a gas or insulating layer, such as air gap protectors, carbon block detectors, gas discharge tubes (GDTs), or break over diodes (BODs), or by the turn-on of a thyristor; these include overvoltage triggered SCRs. and surgectors.

One advantage of the crowbar-type device is that its very low impedance allows a high current to pass without dissipating a considerable amount of energy within the protector.

On the other hand, there is a finite volt-time response as the device switches or transitions to its breakdown mode, during which the load may be exposed to damaging overvoltage. Another limitation is power-follow, where a power current from the voltage source follows the surge discharge. This current may not be cleared in an ac circuit and clearing is even more uncertain in dc applications.

(d) **Functioning of Clamping devices;** Zener or avalanche diodes and voltage-dependent resistors (varistors) display a variable impedance, depending on the current flowing through the device or the voltage across its terminals. They use this property to clamp the overvoltage to a level dependent on the design and construction of the device. The impedance characteristic, although nonlinear, is continuous and displays no time delay such as that associated with the spark-over of a gap or the triggering of a thyristor. The clamping device itself is transparent to the supply and to the load at a steady state voltage below the clamping level.

4.10.2 Low-cost - High Performance Varistors

Main function of clamp is to absorb overvoltage surge by lowering its impedance to such a level that the voltage drop on an always-present series impedance is significant enough to limit the overvoltage on "critical parts" to an acceptable level. Modern Zener diodes are very effective and come closest to the ideal constant voltage clamp.

However, the avalanche voltage is maintained across a thin junction area, leading to substantial heat generation. Therefore, the energy dissipation capability of a Zener diode is quite limited.

A varistor by contrast, displays a nonlinear variable impedance. The varistor designer can control the degree of nonlinearity over a wide range by exploiting new materials and construction techniques that extend the range of applications for varistors. For example, varistors now offer a cost-effective solution for low-voltage logic requiring a low protection level and low standby current, as well as for ac power line and high capacity utility-type applications.

Compar ed with transient suppressor diodes, varistors can absorb much higher transient energies and can sup-press positive and negative transients.

Furthermore, against crowbar-type devices, varistor response time is typically less than a nanosecond, and devices can be built to withstand surges of up to a 70,000A. They have a long lifetime compared with diodes, and the varistor failure mode is a short circuit. This prevents damage to the load that may result if failure of the protection circuit is undetected. Varistors typically offer cost savings over crowbar-type devices.

4.10.3 Varistor Operation

Metal Oxide Varistors, or MOVs, are typically constructed from sintered zinc oxide plus a suitable additive. Each intergranular boundary displays a rectifying action and presents a specific voltage barrier. When these conduct, they form a low ohmic path to absorb surge energy. During manufacture, the zinc oxide granules are pressed before being fired for a controlled period and temperature until the desired electrical characteristics are achieved. A Varistor's behavior is defined by the relation: $I = KV^{\alpha}$, where K and α are device constants.

K is dependent on the device geometry. On the other hand, α defines the degree of nonlinearity in the resistance characteristic and can be controlled by selection of materials and the application of manufacturing processes. A high α implies a better clamp; zinc oxide technology has enabled varistors with α in the range 15 to 30 significantly higher than earlier generation devices such as silicon carbide varistors. The V-I behavior of a varistor is shown in Fig 4.14 highlighting the distinct operating zones of the varistor.

The slope of the protect region is determined by the device parameter β, which bears an inverse relation to α.

In fact, varistor behavior can also be described by the relation: $V = CI^{\beta}$ (the inverse of $I = KV^{\alpha}$)

where C is also a geometry-dependent device constant.

Fig 4.14 also compares the varistor characteristic with that of the ideal voltage clamping device, which would display a slope of zero, as well as a Zener diode characteristic. The Zener diode comparison highlights the extended protect region the varistor also offers for a comparable current and power capability.

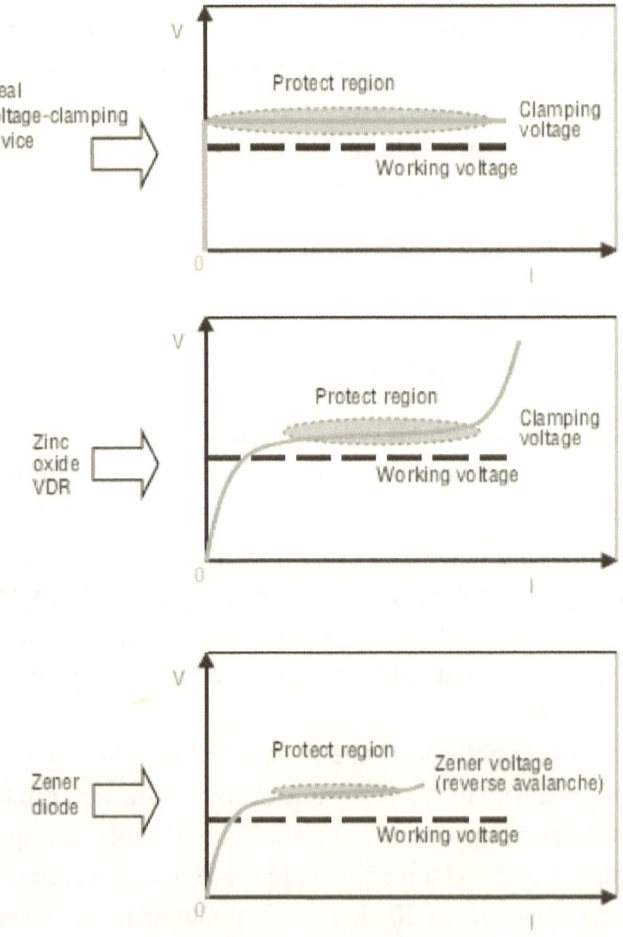

Fig 4.14: The V-I behavior of a varistor

4.10.4 Selection Criteria

For most applications, you can determine the selection by assessing four aspects of the desired application:

(1) **The normal operating conditions of the apparatus or system, and whether ac or dc voltage is applied** Fig4.15 shows a flowchart that may be used to determine steady-state voltage rating or working voltage. You can find VDRs. in various sizes and voltages ranging from 8 V up to 1000 Vrms or more. The higher the nominal voltage of the selected varistor compared with the normal circuit operating voltage, the better its reliability is over time, as the device is able to withstand more surge currents without degrading performance. The disadvantage is a reduction in the level of protection offered by an over-specified varistor. Hence, you should maintain the following relation:

Thumb Rule: Maximum withstand voltage of a protected device > max varistor clamping voltage > max continuous operating voltage.

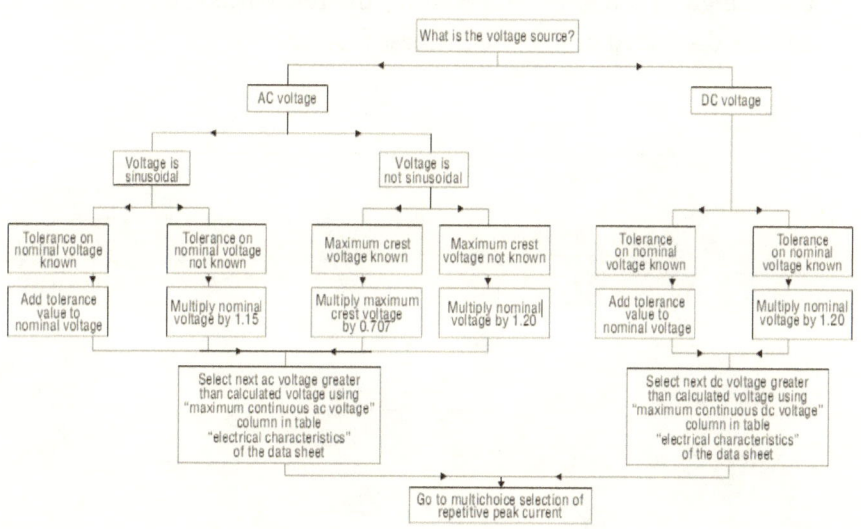

Fig 4.15: Flowchart used to determine the necessary steady-state voltage rating or working voltage

(2) **Determine the repetitive peak current:** Fig 4.16 shows a flowchart that may be used to determine the repetitive peak current. Maximum surge currents are related to the size of the component and start from a few hundred amperes up to several tens of kiloamperes (at standard

waveforms of 8 / 20 μs). Once the repetitive peak current is known, then you can calculate the necessary energy absorption, in Joules (Watt. second or Ws), for the varistor.

(3) **Calculate the energy absorption:** There are two cases-: one for dc and one for ac energy. Energy ratings for available varistors start at a few Joules up to several hundred Joules.

Case 1 Calculating dc Dissipation: The power dissipated in a varistor is equal to the product of the volt-age and current, and may be written:

$$W = I \times V = C \times I^{\beta+1}$$

When the coefficient $\alpha = 30$ ($\beta = 0.033$), the power dissipated by the varistor is proportional to the 31st power of the voltage. A voltage increase of only 2.26% will, in this case, double the dissipated power. Consequently, it's important that the applied voltage doesn't rise above a certain maximum value, or the permissible rating will be exceeded. Moreover since varistors have a negative temperature coefficient at a higher dissipation (and accordingly at a higher temperature) the resistance value will decrease and the dissipated power will increase further.

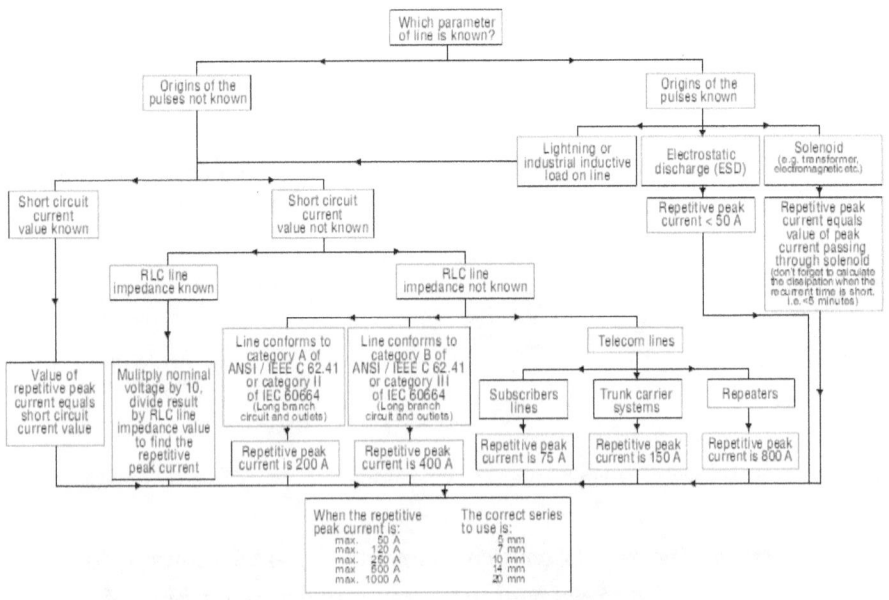

Fig 4.16: Flowchart used to determine the repetitive peak current

Case 2 Calculating ac Dissipation: When a sinusoidal alternating voltage is applied to a varistor, the dissipation is calculated by integrating the VI product. A suitable expression is as follows:

$$P = 1/\pi \times 2^{(a+1)/2} \times \int_0^\pi (\sin\omega t)^{a+1} \times dt$$

Transient energy ratings are quoted in Joules. It's important to ensure the varistor is able to absorb this energy throughout the planned product lifetime or replacement interval without failing. When the device is being used to protect against transients resulting from an inductive or capacitive discharge, such as switching a motor, the transient energy is easily calculated. However, if the varistor is expected to protect against transients originating from external sources, the magnitude of the transient is typically unknown and an approximation technique must be applied.

This involves calculating the energy absorbed after finding the transient current and voltage applied to the varistor. The following equation may be applied:

$$E = \text{integral of (everything upto the Vc(t) I(t)} \Delta t$$
$$\text{from 0 to } \tau = KVc\, I\tau$$

Where I is the peak current, Vc is the resulting clamp voltage, τ (tau) is the impulse duration, and K is an energy form factor constant dependent on the current waveform.

(4) **Package size and style:** Electrical and mechanical considerations must be taken into account when selecting the package size and style. This includes determining the required energy rating and surge current amplitudes, and whether the device is intended to protect against exceptional surges or those caused by repetitive events will feed into the selection process.

The amount of energy expected to be dissipated will also influence this, and designers must ensure the package dimensions are appropriate to the physical and mechanical design of the product. Conventional form factors typically range from disc types of a few millimeters in diameter up to 50 mm, or block and rectangular types for high-energy handling parts.

Other important selection considerations are the effects of lead inductance and device capacitance, which also impact the performance of the varistor in circuit, and must be considered when choosing to use a varistor. In conventional leaded devices, the inductance of the lead can slow the fast action of the varistor to the extent that protection is negated. Modeling the varistor presents a shunt capacitance that may range from a few tens of pF up to several nF, depending on size and voltage range of the device. Depending on the application, the presence of this capacitance can be of little consequence, a desirable property, or, at worst, problematical.

For example, in dc applications a large capacitance is desirable and can provide a degree of filtering and transient suppression. On the other hand, it may preclude the use of a varistor to protect high-frequency circuits.

4.10.5 Sample Applications

Fig 4.17 shows how varistors may be used to suppress internally generated spikes in a TV application. This is a typical application and there can be many more like this.

In Fig 4.18, you can see how a varistor may be used to protect a generic load against power surges originating from the supply. The power supply's own output impedance combines with that of the varistor to create a potential divider whose ratio varies with the varistor impedance to protect the load.

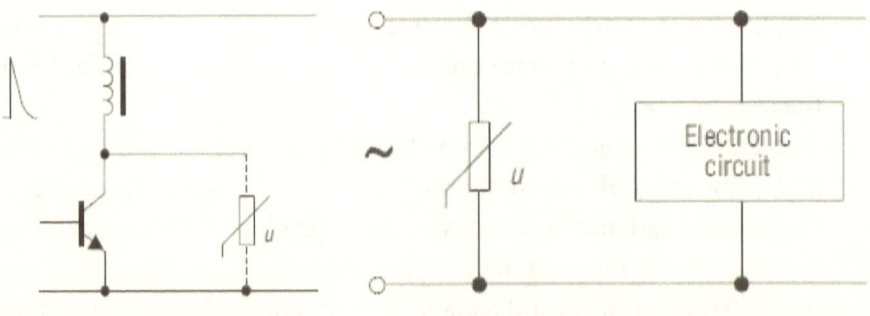

Fig 4.17: Varistor in a TV application

Fig 4.18: suppression directly across mains

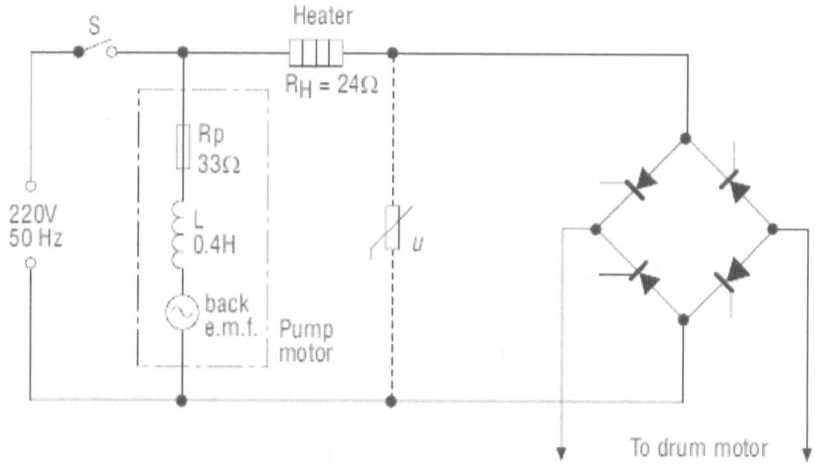

Fig 4.19: Protection of a Thyristor bridge in a washing machine

You can see an alternative application in Fig 4.19. Without varistor protection, the measured peak current through the pump motor when S is closed is 1A.

The energy expended in establishing the electromagnetic field in the inductance of the motor is therefore:

$$I^2 \times L/2 = 0.4/2 = 200 \text{ mJ}$$

Without varistor protection, an initial current of 1A will flow through the thyristor bridge when S is opened, and a voltage sufficient to damage or destroy the thyristors will be developed. Arcing will occur across the opening contacts of the switch. But with a varistor inserted in the circuit, the peak voltage developed across the varistor on opening switch S is:

$$V = C_{MAX} \times I\beta = 600V$$

The thyristors in the bridge can withstand this voltage without damage. The total energy returned to the circuit is 200 mJ. Of this 200 mJ, 15.1mJ is dissipated in the heater, and 184.3 mJ is dissipated in the varistor. The varistor can withstand more than 105 transients containing this amount of energy.

4.11 Implementing some Practical Solutions

The saying, "an ounce of prevention is worth a pound of cure", is highly applicable to lightning protection. It is strongly recommended that comprehensive preventive measures be taken by installing devices and systems for lightning protection on all transmitter sites. Antenna arrestors and an effective antenna ground are mandatory for any site if for no other reason than the safety of associated personnel. The configuration of individual transmitter sites would seldom be identical to ideal layout as shown earlier in fig 4.11. It is however exppected that this model would provide the readers a better understanding of the underlying principles and the ability to design a satisfactory protective scheme for a particular site. It is worth noting that amount of potentially destructive lightning energy and hence the cost of protecting a transmitter site, is not related to the size of the transmitter. The amount of money worth spending at a particular site is, however, related to the cost of the equipment being protected and to its statistical probability of experiencing lightning strikes as discussed earlier.

4.11.1 Rack Cabinet Grounding

The single point grounding technique can also be effective to protect multiple pieces of equipment installed inside an equipment rack cabinet. Treat the rack same as you would a building and mount a panel on the cabinet to act as both an entrance panel and reference ground for all conductors entering and leaving the rack. Install AC supply surge protectors at this point in shunt to ground, and install a series impedance between the panel and the equipment. Don't count on the metal cabinet itself to serve as a ground conductor as paint and oxidation prevents a good connection. A copper strap should be run along the inside of the cabinet, bonded to the cabinet along its length and also bonded to the access panel. The chassis of each equipment is then bonded to this busbar with a single copper braid or strap. Redundant ground connections by means of AC cable and shields of audio cables should be avoided when possible. Finally, connect the rack's access panel to the building reference ground.

4.11.2 An Effective Earth Ground

Once all connections have been made to the master ground point in the building, it must be bonded to an effective earth ground system outside the building. Four-inch or larger copper strap or #2 AWG multi-strand wire is recommended, with short, straight connections. Corrosion will dramatically increase the resistance of a connection, so use silver soldering or cadwelding for all connections exposed to the weather. A network of four or more 10-ft ground rods driven at least 20 feet apart makes the best earth-ground interface.

A commonly used grounding method for transmitter building is the perimeter ground, where a conductor is run around the plinth of building and bonded to ground rods spaced at even intervals around the building. In a large building housing several independent systems in it, each system can have its own reference ground established, which is in turn connected to the perimeter ground at the closest point. This is preferable because it avoids long runs of ground strip inside the building. It's important that any ground connections between these separate systems be avoided, and torroids or surge arrestors be installed on the interconnections that cannot be avoided.

In case of AM transmitters, be aware that there are two separate types of grounds; RF ground and earth ground. While some conductors may serve both functions, the facility should be analyzed separately with regard to its effectiveness for both functions. For instance, it should not be assumed that a tower is adequately grounded for lightning protection because it is connected to earth radial system. The radials are usually ineffective as an earth ground because they lie just below the surface and do not make contact with the water table. If possible, ground rods should be driven into the soil to at least the depth of the water table and bonded to the radial system.

In cases where there may not be an accessible ground water table, such as a mountain top, desert or permafrost site, chemical ground rods have proven effective. These are hollow copper rods filled with earth salts, which slowly leak conductive salt water into the soil, dramatically improving its contact with the soil. However, they are more complicated to install, as a well drilling rig is often required.

At mountain top installations where it may be impossible to install driven or buried ground rods of any kind into solid rock, it has been shown effective to place long lengths of copper strips on the surface, run along the ground in several directions from the transmitter building or tower. This has the effect of capacitively coupling the ground connection to the rock below.

4.11.3 At the Tower

Some important steps to be taken at tower base to maximize protection against a lightning strike are as below:

(1) Tack weld all tower sections together running down at least in one leg, to provide corrosion-free electrical continuity to ground. Also may use a copper strap of size 4 inches (100 x 2mm).
(2) Drive 4 or more ground rods at ten-foot intervals around the base of the tower, and ground these to the tower.
(3) Install a ground rod at each guy wire anchor and connect them to the guy wires with a short jumper cable.

The following additional steps apply to base-insulated AM towers:

(4) Adjust the ball gap located across the base insulator. Make sure the contacts are close enough to be effective – just beyond the point of flashover at modulation peaks. Typically, they should be set for 0.02 inch (1.25 mm) per 1,000 volts peak. Align the terminals horizontally to keep rainwater out of the gap, and to ensure that the gap is self-quenching. Periodically clean any corrosion from the spark balls, which would otherwise increase the resistance of the gap.
(5) There should be an inductive loop in the connection from the antenna tuning unit (ATU) to the tower, usually fabricated from hollow copper tubing. This creates an inductive reactance in the path, causing the voltage to momentarily build up behind the inductor and encouraging the spark gap to ionize and conduct. Run tower lighting conductors, if any, through the center of the tubing to give equal inductive reactance to all paths.
(6) Install an air spark gap between the RF conductor and ground at the tower side of the tuning hut. There should also be a static drain choke in parallel across this path to bleed off atmospheric-generated static from the tower. A 100K ohm non-inductive resistor can also be used, rated at about 200 watts. It should be noted that these devices are

made to carry only small currents caused by static electricity, and will frequently be destroyed in the event of a direct lightning strike.
(7) Bond the radial system to earth ground rods driven at the base of the tower and at each guy anchor point.
(8) If an iso-coupler is used for auxiliary antennas mounted on the tower (FM, STL, cellular, etc.), their coaxial cable shields should be bonded to tower at several locations, plus grounded at tower base below the isocoupler. There are several companies marketing in-line coaxial surge arrestors, and these can be connected in series with the coaxial cable at this point, or at the building entrance.
(9) If a tower lighting system is in use, bond the lighting conduit to the tower at several points as well.
(10) Another area to consider is arcing across the guy insulators, which can develop several hundred volts across them with an electrical storm in vicinity and can damage the insulators. This problem can be tolerated in most cases, but in extreme conditions a viable solution is a static dissipation resistor made by Racal-Decca. It consists of a Voltage Dependent Resistor in series with a fixed resistor, designed to be connected across the insulator. As a voltage builds up, its resistance lowers and drains off the static charge, preventing arcing. This is not an inexpensive solution, and they can be subject to damage during direct lightning strikes.

4.11.4 Ground Rods

It cannot be assumed that the antenna ground mat or radials provides a low impedance to the ideal ground. In poor soil conductivity or in frozen soil, it may function as a good counterpoise, but shall have a high resistance to ground. The ground mat or radials must be supplemented by a ring of at least four driven ground rods which should be as long as possible (typically 3 to 4.5 metres -10 to15 feet). They must penetrate well below the deepest frost level. Separation of the rods should be at least 2 to 3 times their length. They should connect separately via heavy strap to the antenna base's ground terminal together with the mat. The aim should be to obtain the lowest possible inductance as well as the lowest possible resistance.

4.11.4.1 Ground Rod Depth

The total depth each ground rod must be driven into the soil depends on the soil conductivity. Soil resistivities vary greatly depending on the content, quality and the distribution of both the water and natural salts in the soil. It is beneficial to reach the water table, but it is not necessary in all cases. In higher latitudes, rods should be long enough to penetrate below the maximum frost depth. In most cases a total depth of 12 metres (40 feet) or less is necessary, with the average being 4.6 metres (15 feet). Depth also depends on the number of rods and the distance between them. Table 4–1 provides typical resistive measurements for different soils. The resistivity of a given soil will vary with moisture.

Table 4.1: Soil Resistivities

Type of soil	Resitivity (Ω/m)	Type of soil	Resitivity (Ω/m)	Type of soil	Resitivity (Ω/m)
Sea Water	1 to 2	Clay (mixed Sand & Gravel)	10 to 1,350	Sand	90 to 800
Marsh	2 to 3	Chalk	60 to 400	Sand and Gravel	300 to 5,000
Clay	3 to 160	Shale	100 to 500	Rock (normal crystalline)	500 to 10,000

4.11.5 Diversion of Transients on RF Feed Coaxial cables

Even with an excellent low resistance ground at the base of the tower antenna, some current will flow through the shield of the antenna's coaxial RF feed cable, towards the transmitter. It is vital that this current be diverted to ground, away from the transmitter, at the point where the coaxial cable enters the transmitter building. The coaxial shield must be connected directly to the common ground point by a low inductance, low impedance, ground strap. This is the reason the entry point for the RF feed's coaxial cable is recommend to be in close proximity to the station reference ground's common point.

4.11.6 Diversion/Suppression of Transients on AC Power Wiring

Another point of entry for damaging transients is the wiring from the AC power source. A direct lightning hit to the wiring is possible. Common mode transients can also be induced into underground wiring. The defence here is a line transient suppressor system consisting of shunt varistor elements and often, series impedances in the form of air-cored line inductors (around 15 uH). The building entry point for the AC power wiring and the location of the transient suppressor system must be in close proximity to the station reference ground's common point. The AC ground and the ground return for the transient suppressor system should be connected directly to the common ground point by a low inductance, low impedance, ground strap. This will ensure transients are shunted to the station reference ground by the line transient suppressor system, and diverted away from the transmitter where the wiring enters the building.

4.11.7 Shielded Isolation Transformer

An additional protection scheme, which has recently been shown to be effective, is the use of a shielded, isolation transformer between the AC power source and the transmitter. They are normally 1:1 power transformers which employ a shielding technique that enables them to block common mode transients over a broad range of frequencies without impeding the normal AC signal. To be effective the shield must be connected to the station reference ground at the common point by a low inductance, low impedance connection. This is normally achieved by using a wide copper strap that is as short as possible. Failure to provide a low impedance, non-inductive ground will allow the shield potential to rise and couple transients to the secondary. Ordinary transformers have a high primary to secondary capacitance (2000 pF) and offer much less protection than once thought. They can easily couple the fast edge of a common mode transient from their primaries to their secondaries.

4.12 How to set Spark ball gap at tower base

As already explained static discharge ball gaps at tower base, are the first line of defense, and it performs two valuable functions. First, they limit the static charge build up on the tower by providing an easily ionized path for accumulated static discharge, thus preventing potential strike buildup in most cases. Second, they bypass most of the lightning current flow directly to ground at the tower base reducing the risk that a lightning strike will defeat other mitigation measures and protective devices in the transmitting plant.

Almost a century ago Mr. Frank William Peek determined a spherical shape had the most repeatable arcing characteristics. Thus a ball gap could be set reliably as a static discharge lightning arrester, and the tower base static ball gap was born based on Peek's formulas. At that time it was thought a gap between two balls would be relatively insensitive to humidity but modern research has shown a significant humidity effect.

4.12.1 How Right is Ball Gap

If correctly set, the ball gap spacing should be adjusted for arcing in humid air at a voltage just above the maximum voltage that will be developed by transmitter operation – with positive modulation exceeding 125% modulation limit at tower base to accommodate brief overshoots that are not prevented by normal audio processing techniques. Thus, the aim is arcing to take place at the lowest possible voltage that will never be achieved during a station's normal operation i.e. when transmitter is operating under average modualtion with processor in the chain. One of the most common errors with ball gaps is setting them too far apart, making them far less effective, especially on low power installlations.

Another important factor is the correct placement (orientation) of the balls forming the gap. Both balls should be located in the same horizontal plane so that liquids or semisolids do not drip across the gap, thus shorting the tower base to ground.

4.12.2 Procedure to set the Spark Gap

The correct adjustment of spark gaps is the beginning of lightning protection system. As a static charge builds up on the tower accumulating voltage above the maximum operating potential, a static arc ionizes the air in the ball gap briefly acting as a very low resistance, discharging the accumulated static charge of the tower. Minimum spacing of the static ball gap at the tower base and secondary gaps at other locations limits the potential in the rare event of a direct lightning strike on the tower.

Again, the objective is getting the gap as small as possible yet wide enough that any arc that occurs will self-extinguish once the tower's static charge has drained away. There are many ways to actually calculate the voltages and resistances encountered at the tower base.

4.12.3 Calculation of Base Voltage

The voltage across base insulator depends upon base impedance and the current flowing at base as per the power applied to the tower. The voltage across the base insulator of a tower antenna (series fed mast) therefore can be readily calculated by following ohm's law as below, provided the resistance and reactance component of the base impedance of the antenna is known:

The voltage across base insulator is:

$$V_b = I_b Z_b \qquad \rightarrow (1)$$

Here I_b = Current in amp at the base
Z_b = Impedance at base in Ohms and is = $\sqrt{R^2_b + X^2_b}$
R_b = Antenna resistance at base in (Ohms), and X_b = Antenna reactance at base in (Ohms)
For a given power in watts:

$$W = I_b^2 R_b \qquad \rightarrow (2)$$
$$I_b = \sqrt{W/R_b} \qquad \rightarrow (3)$$

Therefore

$$V_b = \sqrt{W/R_b} \times \sqrt{R^2_b + X^2_b} \qquad \rightarrow (4)$$

This value of voltage is root mean square (rms) value for an un-modulated carrier. To obtain peak voltage at 100% modulation, this rms value of equation (4) is to be multiplied by $2\sqrt{2}$. The voltage, so calculated is basis for, selecting the electrical rating of the base insulator and to assess the factor of safety of the selected base insulator and also for setting the spark ball gap.

A caution: Positive modulation limit norms in broadcasting industry world wide is 125%, instant peaks may often exceed this value for very brief intervals in normal operation as a result of processsor overshoot. Thus is it wise to assume a positive peak of 140% or 150% and allow a safety factor for preventing arcs during normal operation. This is especially true with talk program operations where sharper, more energetic peaks may be more common.

The resistance and reactance of towers of various shape and size or diameter can be calculated from emperical formula and also can be found from standard graph avialable in all hand books on broadcasting like NAB etc.

Note: For more details on base impedance and antenna potentials, readers may refer to chapter-4, section 4.5 to 4.7 page 83 to 90 of my book, "AM Radio Tower Antenna".

4.12.4 Break down voltage of Spark ball Gap

The breakdown strength of air at 1.0 MHz is about 80% of its low frequency value, which is 30 kV/cm. The breakdown voltage will vary with the (a) altitude of the site (or air pressure), (b) humidity in the air and also (c) the pollution level of the area. Rapidly rising potential increases the breakdown voltage greatly. At Sea level for a large sphere (ball), the breakdown voltage at 1MHz is approximately 9.4 KVpeak per 1/8 inch air gap. A rough rule of thumb for setting the gap is to allow 0.020 inch per peak KV voltage at the antenna base. This should be increased at high altitude.

The peak antenna voltage may be calculated from;

$$V peak = 2.83 \times Za \times Ia,$$

Here Za is antenna base impedance in ohms and Ia antenna base current in ampere. The air gap should be set as below from the charts.

(a) Uniform Field Gap Spacing graphs

Graphs of Fig 4.20 and 4.21 can be used for determination of the air gap spacing for a uniform field at a frequency of 1.0 MHz, when the peak voltage is known.

Fig 4.20 is intended for use at higher peak voltages (up to 50 kVpeak), while Fig 4.21 will provide a more accurate reading at lower peak voltages (up to 13 kVpeak). The air gap spacing is determined by plotting the maximum peak voltage against the line representing the air breakdown voltage at 1.0 MHz.

Lightning Hazards 255

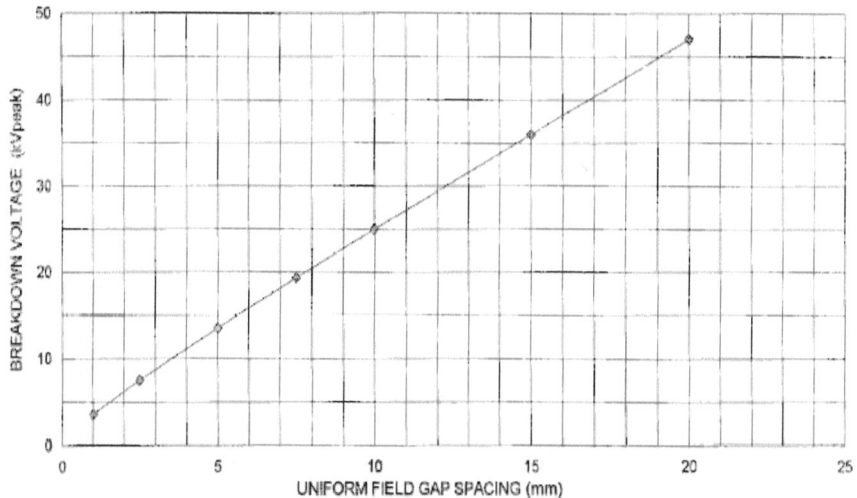

Fig 4.20: Uniform field Gap spacing (High voltage-up to 50KVpeak)

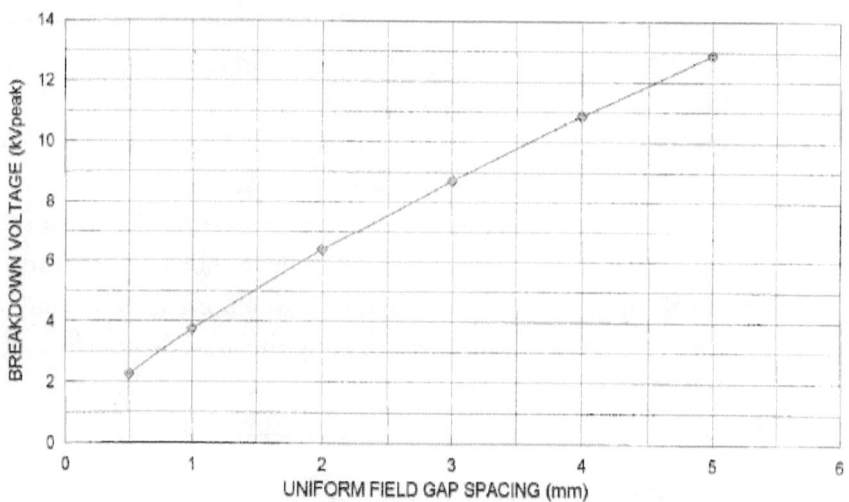

Fig 4.21: Uniform field Gap spacing (Low voltage – less than 14KVpeak)

(b) Air Gap Correction Factors
(1) **For Spark Ball Diameter;** Graph of Fig 4.22 provides a factor that is determined by the diameter of the spark balls when the uniform field air gap spacing is known.

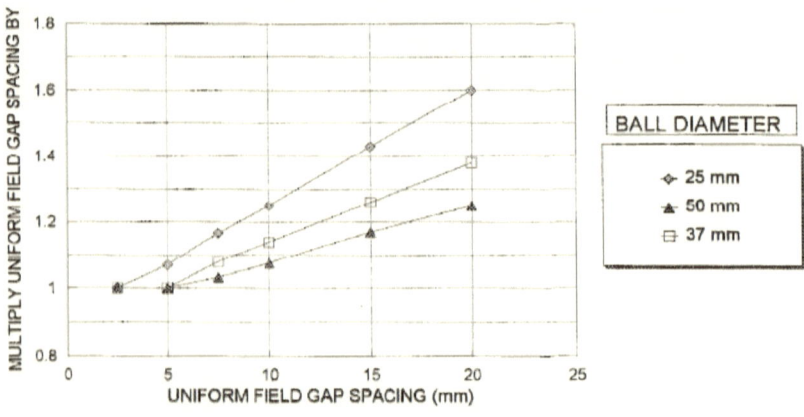

Fig 4.22: Spark Ball Diameter Correction Factor

Spark balls with a diameter of 37.5 to 50 mm provide a uniform field for air gap settings of 5.0 mm or less. If a larger gap is required, the gap spacing multiplier will provide a field enhancement for the curvature of the balls. The field enhancement factors for other geometries can be determined, but are not recommended for use.

(2) **For Altitude;** Fig 4.23 provides a correction factor determined by the site altitude. To determine the Altitude Correction Factor: Enter the bottom graph in Fig 4.23 with the site altitude and draw a vertical line on the 'X' axis at the corresponding point on the altitude in thousands of ft scale. Determine the altitude correction factor. Draw a horizontal line on the 'Y' axis where the vertical line representing the altitude intersects the line representing the altitude correction factor and read the correction factor on the multiply field enhanced gap spacing by scale.

(c) Final air gap setting

Determine the final air gap setting for antenna spark balls by multiplying the field enhanced air gap setting due to spark ball diameter and, by the altitude correction factor as in (1) and (2) above.

Note: The air gap calculations do not contain provision to set the gap at a larger setting than is absolutely essential. The only gap increasing factors that were considered, were the simultaneous worst case condition of 140% modulation peaks and a VSWR of 1.5:1, the geometry of the spark gap balls and the site altitude. It is a natural tendency on the part of operator to increase the spark gap setting to avoid nuisance tripping of the transmitter caused by weather conditions, insects, etc. But minimizing risks to the transmitter and associated electronics should be the highest priority. For maximum protection of the transmitter, it is advisable to be aggressive and set the gap to the minimum setting that will permit normal operation without arcing.

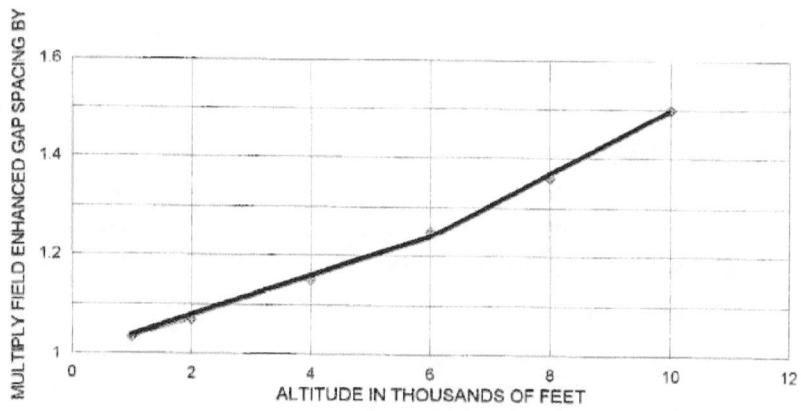

Fig 4.23: Altitude correction Factor

References

1. Lightning Phenomenon – Introduction and Basic Information to Understand the Power of Nature Łukasz Staszewski University of Technology Wybrzeze Wyspianskiego 27 Wroclaw, Poland
2. Recommendations for Transmitters Site Preparation – Doc – issue upto: 22Sept 2004; M/S Nautel Limited, 10089 Peggy's Cove Road, Hackett's Cove, NS, Canada B3Z 3J4, *info@nautel.com* e-mail: *support@nautel.com www.nautel.com*
3. Information Booklet – Lightning Protection for Radio Transmitter Stations,-November 1985, M/S M/S Nautel Limited, 10089 Peggy's Cove Road, Hackett's Cove, NS, Canada B3Z 3J4
4. Varistors: Ideal Solution to Surge Protection – By Bruno van Beneden, Vishay BC components, Malvern, Pa. Application Note: Power Electronics Technology – May 2003 *www.powerelectronics.com*

5. Technical Note – VISHAY BCCOMPONENTS – Varistors Introduction, Revision: 04-Sep-13 Document Number: 29079, For technical query contact: *nlr@vishay.com*; website; www. vishay.com
6. The life and scientific work of Professor Dr. Jovan Surutka: The Elimination of Static Discharges on the Stays of High-Power Antennas: Facta Universitatis ser.: Elec. & Energ. vol. 14, No. 2, August 2001
7. Installation Methods for Protecting Solid State Broadcast Transmitters Against Damage from Lightning and AC Power Surges By John F. Schneider Broadcast Electronics, Inc. Quincy, Illinois USA
8. Radio World September 1991 – Static Protection for Radio Towers- Part-III, PP-23 & 31
9. Radio News – Thomson Broadcast & Multimedia issue-24, autumn 2006 –pp 2&3, Lightning and Broadcast Solid State Systems
10. Handbook for Radio Engineering Mangers – Chapter 37, 38 & 39 (pp 563 – 622) – J.F Ross

CHAPTER 5
ELECTRICAL SHOCK HAZARDS

5.1 General

This chapter describes the electrical hazards in general including in the working of broadcasting set up from electrical safety point of view of staff and equipments. It specifically addresses the working in restricted areas, near exposed energized parts; operating electric equipments near radio transmission towers; working on electrical systems; personal protective grounding; temporary wiring; and hazardous locations etc.

A Hazard means any potential or actual threat to the wellbeing of the operating staff, machinery. Electricity is invisible–this itself makes it dangerous. It has great potential to seriously injure or kill. Everyone is exposed to electrical hazards, not just electricians, so report all electrical shocks and near misses.

As you know electricity seeks an uninterrupted path, or circuit, to follow. If your body becomes part of that circuit, electricity will pass through you and in this process it will harm you.

5.2 Electrical Hazards

An Electrical hazard means hazard due to electricity while working with it or near to it. Electrical safety means: taking precautions to identify and control electrical hazards for safety of specifically operators. Electrical hazards exist in almost every workplace. Common causes of electrocution are: making contact with overhead wires; undertaking maintenance on live equipment; working with damaged electrical equipment, such as extension leads, plugs and sockets, and using equipment affected by rain or water ingress.

Basically, electrical hazards can be categorized into three types as below;

(a) The first and most commonly recognized hazard is an electrical shock. (b) The second type of hazard is electrical burns; arc flash burns, and (c) The third is the effects of arc-blast impacts which include pressure impact, falls, flying particles from vaporized conductors and other potential electrical safety hazards.

5.2.1 Electric shock

Electric shock occurs when a person comes in contact with a live electric circuit and current passes through his body which depends upon the voltage and resistance of body & including the path.

5.2.2 Current and Electrical Shock

Electric shock occurs when a human body becomes part of an electrical circuit and current flow through it. Electric shocks can happen in three ways;

(1) A person may come in contact with both the conductors in a circuit,
(2) A person may provide a path between an ungrounded conductor and the ground, and
(3) A person may become a path between ground and a conducting material that is in contact with an ungrounded conductor.

(a) Voltage and Electric shock

In power supply terminology, "low voltage" is much higher than the 600 volts. At home, you would not think of 600 volts as being low voltage. Even when applied to 120-volt circuits, the term low voltage is deceiving. Actually, low voltage does not necessarily mean low hazard, because potential difference is only one factor making up the dangerous effects of electricity. A general Voltage classification is as;

(1) Extra Low Voltage: a nominal voltage not exceeding 50 Volts AC or 120 Volts DC.
(2) Low Voltage: a nominal voltage exceeding 50 V ac or 120V dc, but not exceeding 1000 V ac or 1500Vdc.
(3) High Voltage: a nominal voltage exceeding 1000 Volts AC or 1500 Volts DC.

For the purposes of electric shock, a "low voltage" is a potential difference between 24 to 600 volts.

The extent of the injury accompanying electric shock due to the voltage depends on three factors;

(1) The amount of the current conducted through the body.
(2) The path of the current through the body.
(3) The duration of the time a person is subjected to the current.

Amount of the current depends on the potential difference and the resistance of the body. The effects of the current on the human body ranges from a temporary mild tingling sensation to death as listed in Table 5.1.

Table 5.1: Effects of the Electrical Current on the Human Body

S No.	Current in milli-amperes	Effects on human body / tissues
1	1 or less	No sensation; probably not noticed, Slight tingling sensation
2	1 to 3	Mild sensation; not painful
3	3 to 10	Painful shock
4	10 to 30	Muscular control could be lost or clamping causing you to freeze
5	30 to 75	Respiratory paralysis; pain; exit burns often visible
6	75 mA to 4 amps	Usually fatal; ventricular fibrillation; entry & exit wounds visible
7	Over 4 amps	Tissue begins to burns. Heart muscles clamp and heart stops beating

An electric shock can injure in either or both of the followings cases;

(1) A severe shock can stop the heart or the breathing muscles, or both.
(2) The heating effects of the current causes severe burns, at entry and exit points of currents in the body.
(3) Severe bleeding, breathing difficulty, and ventricular fibrillation.
(4) In addition, the person may hit something and have some accident as a response to the shock.

Current is the killing factor in electrical shock. But Voltage determines how much current will flow through a given body resistance. The current needed to operate a 10 watt light bulb is 8 to 10 times more than the amount that would kill a person. A voltage of 60 volts is enough to cause a current to flow which is many times greater than that necessary to kill a person.

(b) Resistance of a human body

Human body offers different resistance to electrical currents as it depends upon the molecular construction of the skin cells, level of moisture in the skin at that time, and area of the human body. Table 5.2 gives the values for human resistance to an electrical current. With 120 volts and a skin resistance plus internal resistance totaling 1200 Ohms, there would be 1/10 ampere or 100 mA electric current. If skin contact in the circuit is maintained while the current flows through the skin, the skin resistance gradually decreases. During this time, proper first-aid can mean the difference between life and death. Sufficient circulation can sometimes be maintained by heart compression, which should always be supported with mouth-to-mouth resuscitation, commonly known as CPR.

Table 5.2: Resistance – type of skin or paths of a human body

S No	Type of path or skin	Resistance value
1	Dry skin	100,k to 600,kOhms
2	Wet skin	1,000 Ohms
3	Hand to Foot	400 to 600 Ohms
4	Ear to Ear	100 Ohms

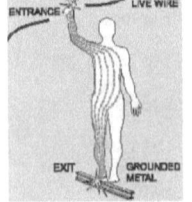

(c) Voltage source and current capacity

The ability of a voltage source to deliver electrical current (and deliver power) depends upon internal 'resistance' of the source plus the resistance of the human body to electron flow. Physically small batteries with high internal resistance suffers voltage 'drop' under load more quickly than larger batteries with less internal resistance. So power supplies with high voltage and high current capability – e.g. mains voltages and large car batteries – are more dangerous. Current capability of few sources is tabulated in Table 5.3 below;

Table 5.3: Current capability of some voltage sources

Voltage source	typical voltage	ability to deliver current	effects
Lightning	100×10^6	very high	death
Static electricity	50,000	very low	brief sharp needle like sensation
Mains electricity	240V AC	high, e.g. 10A	macro shock, cooking, burns
Car battery	12.5–14V DC	very high, e.g. 120A	melts wires, no effect on people

5.2.3 Step and Touch Potential

Voltage is a measure of the potential difference between two points. Voltage means an accumulation of electrons at one point compared to another, and is a measure of the 'force' available to generate current flow.

Voltage sources include static electricity, mains electricity, batteries, right down to bio potentials across cell membranes and small voltages on signal sensors like microphones, pressure sensors etc.

(a) Step Potential

In a ground fault, current flows through grounding system to a ground rod or some type of system ground (steel structure, guy wire) seeking a return to its source. This current flow could possibly exit or in, along the surface of the ground for quite some distance around the point where the earth becomes energized. The current will follow, as nearly as possible, to the conductors supplying the fault current.

Step potential is caused by the flow of fault current through the earth. The closer a person is to the ground rod or grounded device, the greater the concentration of current and the higher the voltage.

This current flow creates a voltage drop as it flows through the earth's surface and a person standing with his feet apart, bridges a portion of this drop thus creating a parallel path for current flow as seen in Figure 5.1.

The wider apart a person's legs are, the larger the voltage difference across his body. Protection from the step potential hazard should be to stay in the zone of equipotential while working. Simply being alert to this hazard is the best defence.

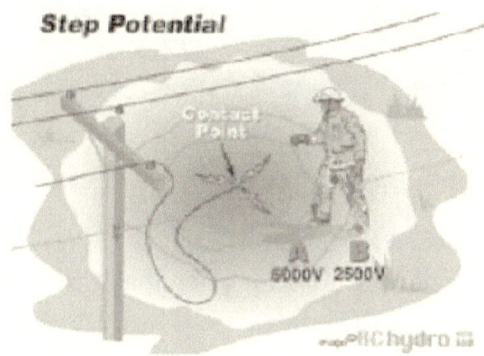

Fig 5.1: Step Potential hazard

For this reason, unqualified personnel standing on the ground are cautioned to stay clear of structures. This means that a person standing near the point where fault current enters the earth may have a large potential difference from foot-to-foot. The potential difference over the same span will be less and less as the span is moved away from either the fault current entry point or the fault current return point at the source.

(b) Touch Potential

Touch potential is a problem similar to step potential, as seen Fig 5.2. It involves a fault current flow in the earth establishing a potential difference between the earth contact point and some nearby conductive structure or hardware.

Protection from step and touch potential is to use switch operating platforms and ground grids. The worker must remain on a local conductive mat as the highest voltage gradient has been moved to the mat's edges.

Sub stations on site have a ground grid located under the rocks, but if an individual is located outside this area and while standing on the earth, touches a ground or a grounded object, a difference in potential may exist during a ground fault.

Fig 5.2: Touch Potential hazard   **Fig 5.3:** Downed Power Line Hazard

(c) Downed Power Lines

It is important to remember that wires installed on utility poles carry electricity. When wires are down, they are dangerous as electricity can still flow through them. Never assume that a downed power supply line is not energized as it still could be "live". See Fig 5.3 Telephone and cable TV wires may be entangled with electric wires and must also be treated as live. Another danger from downed power lines can come in the form of a wire touching or laying on a metal building or metal fence. Again, this could energize the building or fence and kill or injure someone if they were to come in contact with it. Stay at least 50 feet away from all downed lines and keep others from going near them as well.

Be especially careful when driving or parking a vehicle near downed wires. If downed wires are in the street, near the curb, or on the sidewalk, use extreme caution. Never drive over downed power lines. Even if not energized, they can become entangled in your vehicle.

In the event a wire comes down on a vehicle with passengers, the best advice is to stay in the vehicle until professional help arrives to safely remove you from the vehicle. If you must get out of the vehicle because of fire or other life-threatening hazards, jump clear of the vehicle so that you do not touch any part of the car and the ground at the same time. Jump as far as possible away from the vehicle with both feet landing on the ground at the same time. Once you clear the vehicle, shuffle away, with both feet on the ground, or hop away, with both feet landing on the ground at the same time. Do not run away from the vehicle as the electricity forms rings of different voltages. Running may cause your legs to "bridge" current from a higher to a lower voltage ring as shown in Fig 5.4, which would result in a shock. So remain a safe distance away.

The voltage drops as you move away from the point of contact. If one part of your body touches a high voltage zone while another part of your body touches a low-voltage zone, you will become a conductor for electricity. This is why you should shuffle away from the line, keeping your feet close together and hopping or shuffling and not walking as illustrated in Fig 5.5.

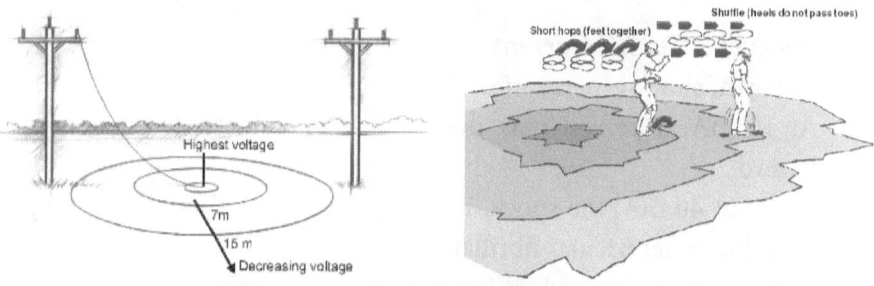

Fig 5.4: Step Potential "Voltage Rings"   **Fig 5.5:** Proper Movement over Energised Grounds

Note; if you must move on energised ground, "Shuffle or Hop", while keeping your feet together and touching each other. Do not take steps.

5.3 Identifying Electrical Hazards

Before taking up a work, including testing and troubleshooting, identify electrical hazard involved in it and also evaluate its effects and risks on the operators / workers.

5.3.1 Hazard Identification and Risk Assessment

Electrical hazard and risk assessment must be done before taking up any electrical work. The hazard assessment includes identifying the followings:
(a) Shock, arc flash, and arc blast hazards;
(b) Non-electrical hazards (e.g. falls, chemical, biological, radiation, and environmental hazards);
(c) Mitigating hazards through engineering and administrative controls, and Personal Protective equipments;
(d) If an Energized Electrical Work Permit is required?

In addition to the electrical hazard assessment, a risk assessment of common electrical tasks must be done by rating the relative hazards. The risk assessment can be used to prioritize the mitigation of hazards.

5.3.2 Hazardous effects of Electricity on Human body

Different types of sources of electricity have different hazardous effects on human body as described below;

(a) Effects of 50Hz AC Current
(1) At 5 mA, shock is perceptible;
(2) At 10 mA, a person may not be able to voluntarily let go of the hazard;
(3) At about 40 mA, the shock, if lasting for 1 second or longer, may be fatal due to ventricular fibrillation.
(4) Increasing current leads to burns and cardiac arrest

(b) Effects of Direct Current (DC)
(1) A DC current of 2 mA is perceptible;
(2) A DC current of 10 mA is the threshold of the let-go current;
(3) A DC current is considered less hazardous than AC wherein muscle contracts and expands alternately.

(c) Effects of 50Hz Voltage amplitude
(1) A voltage of 30 V rms, or 60 V dc, is considered safe except when the skin is broken;
(2) But the internal body resistance can be as low as 500 ohms, so fatalities can occur.

(d) Effects of Short Contact time
(1) For contact less than 0.1 second and with currents just greater than 0.5 mA, ventricular fibrillation may occur only if the shock is in a vulnerable part of the cardiac cycle.
(2) For contact of less than 0.1 second and with currents of several amperes, ventricular fibrillation may occur if the shock is in a vulnerable part of the cardiac cycle.
(3) For contact of greater than 0.8 second and with currents just greater than 0.5 A, cardiac arrest (reversible) may occur.
(4) For contact greater than 0.8 second and with currents of several amperes, burns and death are probable.

(e) Effects of AC at 100 Hz and above
(1) Human body Nerve stimulator tetany is at 50Hz, but the pulses are 200us wide, whereas mains pulses are effectively 10ms wide; so the same amount of mains current feels 50 times stronger at 100 Hz than at 50Hz.
(2) As sensation depends on current density, small currents over large areas are not felt as well as the same current over a smaller area. The myocardium is most sensitive to 30–100Hz electricity, so mains at 50Hz are ideal for inducing fibrillation.
(3) When the threshold of perception increases from 10 kHz to 100 kHz, the threshold of let-go current increases from 10 mA to 100 mA.
(4) Higher frequencies (i.e. diathermy), DC electrical current, and AC which does not pass through the heart do not cause fibrillation but rather heat up and burn the muscle they flow through, sparing skin and fat.

(f) Effects of Wave shape

Contact with voltages from phase controls (square, triangular or pulse train) usually cause effects between those of ac and dc sources. Heating of tissue happens due to average amplitude of the wave shape.

(g) Effects of Capacitive Discharge

(1) A circuit of capacitance of 1 microfarad having a 10 kV capacitor charge may cause ventricular fibrillation.
(2) A circuit of capacitance of 20 microfarad having a 10 kV capacitor charge may be dangerous and probably cause ventricular fibrillation.

5.3.3 Hazards of Power Electronic Equipment

Employer and employees should be aware of the hazardous effects of electricity on the human body and the hazards associated with power electronic equipment used in broadcasting and in control systems of ancillary and industrial devices. Power electronic equipment includes the following types of devices:

(1) Electric arc welding equipment;
(2) High-power radio, radar, and television transmitting towers and antenna;
(3) Industrial dielectric and radio frequency (RF) induction heaters;
(4) Shortwave or RF diathermy devices;
(5) Process equipment that includes rectifiers and inverters such as: Motor drives, Uninterrupted power supply systems, and Lighting controllers.

(a) Type of Hazards

The hazards associated with the following:
(a) High voltages within power supplies;
(b) RF energy–induced high voltages;
(c) RF fields near antennas and antenna transmission lines can cause electrical shock and burns
(d) Radiation; types
   (1) Ionizing; X-radiation (X-ray) hazards from magnetrons, klystrons, thyratrons, cathode-ray tubes etc.
   (2) Non-ionizing RF radiation from: Industrial microwave heaters and diathermy radiators, Industrial scientific and medical

equipment, Radar equipment, Radio communication equipment, including broadcast transmitters, RF induction heaters and dielectric heaters, and Satellite–earth-transmitters.

Effects of RF radiations on the human body, controlled (intentional) and uncontrolled (unintentional) have already been described in details in earlier charter on RF Hazards.

5.3.4 Minimizing Electrical Hazards

To minimize electrical hazards the first goal is to identify high hazard equipment and reduce the hazards of maintaining it through engineering controls. Arc flash analysis will identify the equipments that have a greater potential for arc flash as well as help ensure the safety of qualified electrical workers who frequently use hazard / risk category Personal Protective Equipments (PPE).

(a) Arc Flash Hazard Analysis Implementation

Arc flash hazard analysis consists of collecting data on the power distribution system. The arrangement of components is documented on a one-line diagram with nameplate specifications of every device and lengths and cross-section area of all cables.

Electric utility is consulted to get information about minimum and maximum fault currents that can be expected at the input to their power supply equipments and switchgears supplied by them. In case of departmental power supply equipments and switchgears this data is available from supplied equipment manuals. Subsequently, a short circuit analysis followed by a coordination study is completed. SKM Arc Flash Analysis software then uses the resultant data (using 2012 edition of NFPA 70E or IEEE Standard 1584 2002 equations) to calculate the arc flash protection boundary distances and incident energies. These calculated boundary distances and energies are then used to determine minimum PPE requirements.

For systems of 600 volts and less where an arc flash analysis has not been performed, NFPA 70E Hazard/Risk Category Classification tables 130.7 (C)(15)(a) and 130.7 (C)(15)(b) will provide arc flash boundary distance, provided it meets the maximum short circuit current and fault clearing time criteria. For other fault currents and clearing times greater than those listed in the NFPA 70E tables, an arc flash analysis must be performed as well.

The Arc Flash Analysis Team staff shall complete an arc flash hazard analysis as required by NFPA 70E. Arc flash hazard analysis is done under the supervision of a licensed electrical engineer; for all major electrical system upgrades or renovations; and for all new electrical system installations.

Until an arc flash hazard analysis is done, a qualified electrical worker will use NFPA 70E Hazard/Risk Category Classification tables 130.7 (C) (15) (a) and 130.7 (C) (15) (b) to determine the hazard/risk category; voltage rated (V-rated) gloves use; and V-rated tools use. Warning labels will be eventually placed on all equipments at all locations in power supply system.

(b) Electrical Equipment Labeling

Power supply sub-station transformers, HT/LT Switchgears, industrial control panels, disconnect and any other equipment posing an arc flash hazard will be field marked to warn workers of potential electric arc flash hazards. When arc flash and shock data are available for industrial control panels, labels shall include information on arc flash hazard boundary, the hazard category, required PPE, minimum arc rating, limited approach, restricted approach and prohibited approach distances.

Labeling is intended to reduce the occurrence of serious injury or death due to arcing faults to staffs working on or near energized electrical equipment. Labels shall be located so as to be visible to the staff before examination, adjustment, servicing, or maintenance of the equipment. Arc Flash and Shock Hazard labels are color coded to reflect the hazard/risk category range of calculated incident energy as shown in Table 5.4.

Table 5.4: Arc Flash and Shock Hazard Label Legend

Hazard/Risk Category	Incident Energy Range (cal/cm^2)	Hazard/Risk (Top Color)	Type of Label (Bottom Color)
0	Less than 1.2	White	Warning
1	1.2 – 4	Green	Warning
2	4 – 8	Yellow	Warning
3	8 – 25	Orange	Warning
4	25 – 40	Red	Warning
Dangerous	Greater than 40	Red	Danger

5.4 General Electrical Safety Requirements

All electrical work practices must comply with applicable sections of the Indian Electricity Rules-1956 (IER), National Electrical Code (NEC), National Electrical Safety Code (NESC), or the State adopted electrical codes, as applicable. Chapter–Rule; 29 to 46 of IER-1956 describe in details the General Safety Requirements in India: these are described later under relevant head in this chapter.

These Electrical Safety Rules provide a uniform set of safe work requirements which must be complied with. These rules apply to all persons (Employees, Contractors and Accredited Service Providers) working on or near high voltage and low voltage electrical apparatus associated with power supply system.

5.4.1 Use only Approved Items

Use electrical wire, conduit, apparatus, and equipment only for the specific application that is approved by the Electrical Standards or listed by Underwriters Laboratories (UL). Install and use listed, labeled, or certified equipment according to the instructions included in the listing, labeling, or certification.

5.4.2 Work by Qualified Persons Only

Only qualified personnel familiar with code requirements, safety standards, and experienced in the type of job should work on the electrical circuits and the equipment. Details about qualified persons and training requirements shall be described in the later part of the chapter.

5.4.3 Identify Item Specific Safety Requisites

The employer will determine, by inquiry, direct observation, or instruments, the location of any part of an energized electric power circuit, exposed or concealed before Work is taken up. If the work is likely to cause any person, tool, or machine to violate the boundaries as set forth in table 5.5, de-energize the circuit(s) and ground them, as appropriate. Additionally, all of the following must be required:

(a) Underground Lines

Protect, all underground lines with surface signs and a longitudinal warning tape buried 12 to 18 inches above the lines. Do not perform drilling, or material excavating operation within 6 feet of underground lines unless the lines have been de-energized.

(b) Job Briefing

The supervisor must conduct a job briefing with affected workers. They must hold additional job briefings if significant changes occur during the course of work.

The briefing must cover the following:

(1) **Job Hazard Analysis (JHA);** Identify all the hazards associated with the job in written and discuss them.
(2) **Nonelectrical Hazards;** Identify, in a written JHA hazards not associated with the electrical work but expected to be encountered as by-product of the work, and discuss them.
(3) **Personal Protective Equipment (PPE);** Provide, and use the appropriate PPE needed to do the job safely. Use flash protection Clothing in accordance with NFPA 70E if the job requires operating, racking, circuit breakers with the doors open, or, working within reaching distances of exposed energized parts. Employees working on energized lines and equipment rated at 440 volts or greater must use rubber gloves, hard hats, safety boots, and other approved protective equipment or hot-line tools that meet American Society of Testing and Materials (ASTM) standards.

5.4.4 Restricted Areas

Provide effective barriers or other means to ensure that people do not use areas with electrical circuits or equipments as passage when energized lines or equipments are exposed. Effectively guard live parts of wiring or equipment to protect persons or objects from harmful contact.

Use special tools insulated for the voltage while installing or removing fuses with one or both terminals energized.

5.4.5 Isolate High-Voltage Equipment (over 600 volts)

Isolate exposed high-voltage equipment, such as open terminal transformer banks, HT power supply Filter condenser banks of a high power radio transmitter, air break switches, and similar equipment with exposed energized parts to prevent unauthorized access. Isolation must consist of locked rooms, fences or screened enclosures with door interlocks, partitions, or elevated locations. Keep entrances to isolated areas locked when not under constant observation. Post "DANGER-HIGH VOLTAGE" warning signs at entrances to these areas and properly ground all the conductive components, fences, guardrails, screens, partitions, and equipment frames and enclosures.

5.4.6 Temporary Fences

When extending a fence or removing it for work on high voltage equipment, erect a temporary fence of comparable construction and protection. Electrically bond the temporary fence to the existing fence. If the fence is more than 40 feet long, bond posts to the ground mat at not more than 40-foot intervals. Bond posts at each side of gates or openings to the ground mat/grid and install a bonding jumper across all gate hinges. Bond all corner posts to the ground mat.

5.4.7 Perimeter Markings

Use approved perimeter markings to isolate restricted areas from designated work areas and entries. Erect them before work begins and maintain them for the duration of work. Approved perimeter marking are;

(a) Barrier Tapes
Erect red barrier tape printed with the words, "DANGER—HIGH VOLTAGE", around the perimeter of the work area and access ways approximately 42 inches above the floor or work surface.

(b) Synthetic Rope Barrier
Erect synthetic yellow or orange rope barrier at 36 to 45 inches above the floor with signs of, "DANGER–HIGH VOLTAGE" on a non-conductive material attached at 10-foot intervals.

5.5 Working near Exposed Energized Overhead Lines or Parts

For troubleshooting and testing purposes only, qualified persons using proper test equipment and personal protective equipment (PPE) must adhere to the boundaries (distance from live part to worker) shown in table 5.5. For adjusting, tightening, calibrating or any other work, the circuits must be de-energized, or employees must use voltage-rated gloves and voltage-rated insulated tools.

Table 5.5: Approach boundaries to exposed energized conductors/parts for qualified person's

Nominal voltage phase to phase, or single phase (1)	Limited approach boundaries		Restricted approach boundary includes Inadvertent movement (4)	Prohibited approach boundary (5)
	Moveable conductor (2)	fixed conductor (3)		
0 to 50 V	not specified	not specified	not specified	not specified
51 to 300 V	10 ft	3ft-6in	avoid contact	avoid contact
301 to 750 V	10 ft	3 ft-6in	1 ft	1 inch
751 to 15 kV	10 ft	5 ft	2ft – 2in	7in
15.1 kV to 36 kV	10ft	6ft	2 ft – 7in	10 in
36.1 kV to 46 kV	10ft	8ft	2ft – 9in	1ft – 5in
46.1 kV to 72.5 kV	10ft	8ft	3ft – 3in	2ft – 1in
72.6 kV to 121 kV	10ft – 8in	8ft	3ft – 2in	2ft – 8in
138 kV to 145 kV	11ft	10ft	3ft – 7in	3ft – 1in
161 kV to 169 kV	11-ft 8-in	11-ft 8-in	4-ft	3-ft 6-in
230 kV to 242 kV	13-ft	13-ft	5-ft 3-in	4-ft 9-in
345 kV to 362 kV	15-ft 4-in	15-ft 4-in	8-ft 6-in	8-ft
500 kV to 550 kV	19-ft	19-ft	11-ft 3-in	10-ft 9-in
765 kV to 800 kV	23-ft 9-in	23-ft 9-in	14-ft 11-in	14-ft 5-in

Notes: This table is taken from NFPA 70E table 2–1.3.4 and OSHA 29 CFR, 1910.269 tables R6.

Limited Approach Boundaries; shock protection boundary to be crossed only by qualified persons. Unqualified persons must not cross this boundary unless accompanied by a qualified person.

Restricted approach Boundary; A shock protection boundary to be crossed only by qualified persons. The boundary's proximity to a shock hazard requires the use of shock protection techniques and equipment when crossed.

Prohibited Approach Boundary; A shock protection boundary to be crossed only by qualified persons. When crossed by a body part or object, this boundary requires the same protection as if direct contact is made with a live part (i.e., requires voltage rated tools and voltage rated gloves and, in some cases, other voltage rated clothing).

5.5.1 Low Voltage Testing

For low voltage troubleshooting and testing purposes (under 480 volts) a qualified person may penetrate the prohibited approach boundary shown in table 5.5, column 5, with test instrument probes, leads, and CT's, etc. The qualified person must wear Class 00 (500 volt-rated) gloves.

5.5.2 Unqualified Person Restrictions

When a person without electrical training works on the ground or in an elevated position near overhead lines or any other exposed live parts, supervisors must ensure that the unqualified person and the longest conductive object he might contact or handle, should never come closer to any live line or part than those distances listed in table 5.5, column 2, for energized lines or column 3 for other exposed live parts.

5.5.3 Equipment Transit Clearances

A flag person must guide cranes for loading unloading of equipment / stores, JCBs, Earth Excavators for construction work and other equipment in transit near exposed energized lines or parts at all times. Do not move any equipment or machinery under energized overhead high-voltage lines or near exposed energized parts of RF feeder lines, unless clearances listed in table 5.6 are maintained.

Table 5.6: Equipment in transit clearances

Up to 50 kV	4 feet
50 kV up to and including 345 kV	10 feet
Over 345 kV up to 750 kV	16 feet

Lower any boom or mast to transport position and Ground the equipment while it is being transported. Two grounds must be leap-frogged as the vehicle is moved or the vehicles must be treated as energized.

5.5.4 Sign Posting

Place sign posts at all crossings where equipment will be moved under energized high-voltage line(s) with appropriate signs 50 feet from and on both sides of the line(s). They must be large enough to be easily read from moving equipment. The sign must include the following information:

a) Warning of high-voltage line; b) Line voltage; c) Maximum height of equipment to pass under the line.

Determine the maximum height of the equipment by subtracting the clearance distance shown in table 5.6 from the actual line to ground distance during maximum sag conditions.

5.6 Operating Equipment near Radio and Microwave Transmission Towers

Because of high frequency, low power output, and point-to point microwave transmissions do not present an induced RF hazard. However, on many microwave towers, VHF radio transmission antennas are also mounted, so it will pose RF hazards. Near a high power self radiating MW towers also there will be heavy RF pickup and static charge will be developed in all the equipments.

Vehicles however, will rarely need to be grounded near transmission towers. Tires contain carbon compounds and so are conductive or semi conductive and static charges will bleed off through tires and/or outriggers. However, voltage could build up if all tires were insulated from the earth by dry rip-rap or other insulation. Therefore, the following safety precautions apply to all transmission towers;

(a) Shut down Transmitter

Shut down the transmitter or ground and test the equipment (in case of low power transmitter) to determine if a hazard exists before working near any radio tower where an electrical charge may be induced in the equipment or materials being handled. To conduct a test, connect an insulated wire to the vehicle and touch it to the tower base. If you see or hear the spark, you must ground the vehicle.

(b) Ground Mobile Equipment near Transmission Towers

If needed, ground the equipment to dissipate static electrical charge. On equipment with a rotating boom, attach a ground wire to the structure supporting the boom. Place and remove ground wires using hot-sticks or voltage-rated gloves. Attach the ground connection first (if possible, to the tower ground), then attach the other end to the equipment.

These ground wires do not have to be sized to carry fault current. They need only to carry low level current to bleed off static voltage charges induced on the vehicle or lifted materials. Any convenient wire size that will mechanically withstand the service will be sufficient. A smaller conductor would carry the current, but an insulated #2 copper conductor is recommended for mechanical strength.

(c) Material Ground Wire
Also, attach a ground wire to conductive materials handled by hoisting equipment. Attach the ground connection first, and then attach the other end on the materials.

Alternatively, provide a ground jumper from the load to the required grounding conductor installed on the structure.

5.7 Compliance of Electrical Installation

Electrical installations must comply with the applicable provisions of the current National Electrical Safety standards, National Electrical Code, Safety and Health Standards.

The Underwriters Laboratories or other nationally recognized testing laboratory must approve or list electrical wire, conduit, apparatus, power tools and equipment, for the specific application. This approval / listing must appear on each piece of equipment or tool as part of the "marking or labeling" required below.

5.7.1 Marking of Equipments

Do not use electrical equipment unless the manufacturer's name, trademark, and other descriptive marking by which the manufacturer may be identified, is located on the equipment.

Markings must also provide voltage, current, wattage, approvals / listings, and ratings as required by the National Electric Code in effect at the time of purchase. Markings must be sufficiently durability to withstand the environment.

5.7.2 Working Space between Electrical Equipments

(1) Provide access and working space around electrical equipments and enclosures, e.g., panel boards, motor controls, disconnects, etc., to permit ready and safe operation and maintenance as shown in Fig 5.6 and table 5.7. Keep working space clear at all the times without exception.

(2) Provide a working space of at least 30 inches (750 mm) horizontally where the rear or the side access is required to work on de-energized parts of the enclosed equipment (see fig 5.7).

Table 5.7: working spaces around equipments

Nominal Voltage	Minimum clear distance (ft)		
	Condition1	Condition2	Condition3
0–150	3	3	3
151–600	3	3.5	4
601–2500	3	4	5
2501–9000	4	5	6
9001–25000	5	6	6

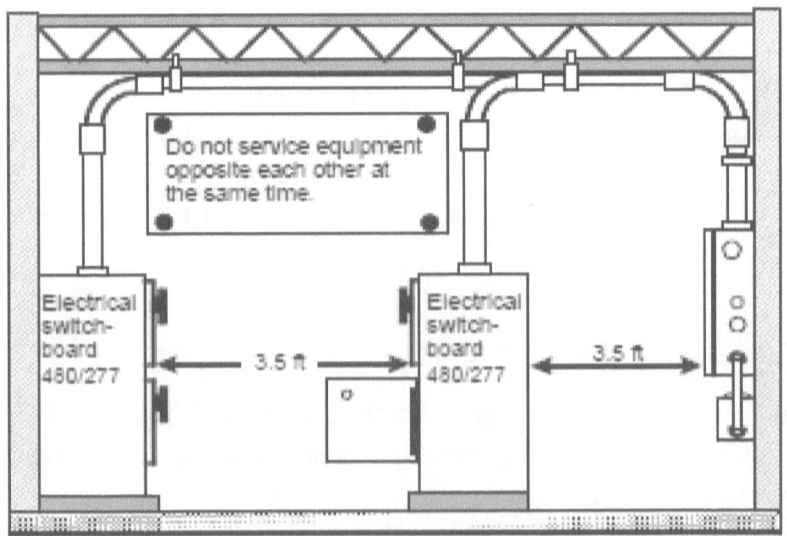

Fig 5.6: Access and working space around equipments

Condition 1 – Exposed live parts on one side and no live parts or grounded parts on the other side of the working space, or exposed live parts on both sides effectively guarded by suitable wood or other insulating materials. Insulated wire or insulated busbars operating at not over 300 volts to ground shall not be considered live parts.

Condition 2 – Exposed live parts on one side and grounded parts on the other side. Consider concrete, brick, or tile walls grounded.

Condition 3 – Exposed live parts on both sides of work space (not guarded or enclosed, as in Condition1) with the worker between.

(3) Doors and hinged panels: Doors and hinged panels must have at least at least a 90⁰ opening. Keep working space clear at all times. Do not store parts, tools, and equipment in the clear space (see figure 5.8).

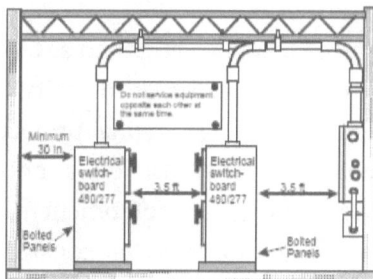

Fig 5.7: working space for rear or side access

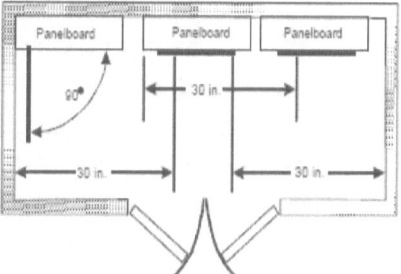

Fig 5.8: working space for doors and hinged panels

5.7.3 Passage Barriers

Provide effective barriers or other means (barrier tape) to ensure that areas containing electrical circuits or equipment are not used as passageways when energized parts or equipment are exposed for testing or maintenance. This includes open doors on motor control centers, and switchgear.

5.8 Personal Protective Grounding

Qualified persons must comply with applicable provisions of "Personal Protective Grounding." Include written grounding procedures in all clearances, special work permits etc. The JHA must include the procedures, and employees must discuss them before beginning work.

5.8.1 Over 600 Volts

Put ground closest to work and within the sight of workers for all circuits and equipment in excess of 600V. The clearance holder is personally responsible for proper placement and removal of protective grounds.

5.8.2 Personal Protective Ground Cables

Personal protective grounds and clamps must be capable of conducting the calculated maximum fault current available for the time necessary to clear the fault. They must be sized in accordance with FIST 5–1.

5.8.3 Prior to Applying Grounds

After implementing hazardous energy control, use a hot-stick "noise tester" or similar approved device of sufficient insulating capacity to verify that the circuit or equipment is de-energized before placing personal protective grounds. Test the voltage tester immediately before use on a known energized source of similar voltage before testing the equipment to be worked on. The circuit/equipment to be worked on must be considered energized while conducting the test.

5.8.4 Placement and Removal of Personal Protective Grounds

After de-energization, install personal protective grounds so that all phases of lines and equipment are visibly and effectively bonded together in a multi-phase short and connected to ground at one point. Do not use single-phase personal protective grounds or grounding chains. Install personal protective grounds using a hot-stick or voltage-rated gloves on both sides of the work area, if possible. This precaution prevents a possible back feed, especially when working on transformers and related equipment. When attaching grounds, attach the ground end first, and then attach the other end to the de-energized circuit. When removing personal protective grounds, first remove the grounding clamp from the de-energized circuit using a hot stick or voltage rated gloves, and then remove the other end from the ground connection.

5.9 Training Requirements for Electrical Operations

Before discussing the training requirements for operation of electrical equipments, machinery, and switchgear or on power supply lines we will define the type of manpower who may work on electrical system.

An Ordinary Person; A person, without sufficient knowledge about electricity, training or experience to enable him to avoid the dangers / hazards which electrical apparatus may impose if he operates them such workers are; Fitter, Carpenters, Cellotex cutters cum decorators, Painters, Housekeeping and Janitors staff.

An Qualified Person; A person, who is trained professionally and competent to use appropriate skills to complete a given task in electric operation without hazard, and

An Electrically qualified: A person, who is trained and competent in an electrical trade or profession such as a Power Line Worker, Electrical technician or Fitter or Mechanic, Electrical Engineer or a Cable Jointer. All countries have their own standards broad based on some already established norms and tested standard for training. NFPA70E, 2000 Edition, and OSHA29CFR 1910.269 contain references for training requirements for electrically qualified persons. Training Requirements for Electrical persons as below;

(a) Normal Training
A person must have all the training listed below to be a qualified person for electrical job operations.

(1) Required training must be of classroom type and also on-the-job training.
(2) Qualified persons must be trained in and made familiar with "Safety-Related Work Practices," safety procedures, and other safety requirements pertaining to their work. Qualified persons must be trained in first-aid and CPR and be familiar with applicable emergency procedures. They must be trained in any other safety practices, including those not specifically addressed in this section such as confined space entry, and pole-top rescue, fall protection, personal protective equipment, etc.

(3) Qualified persons must be trained and knowledgeable in Job Hazard Analysis (JHA). This training and knowledge includes recognizing work hazards, doing the work safely, writing a JHA, and communicating hazards and safety work practices to fellow employees.
(4) Qualified persons must be trained and knowledgeable in the construction, operation, and maintenance of equipment and specific work methods. They must be trained to recognize and avoid hazards with respect to equipment or work methods and must be familiar with applicable codes and standards. They shall be familiar with the proper use of special precautionary techniques, personal protective equipment, insulating and shielding materials, and insulated tools and test equipment.

(b) Additional Training

Qualified persons permitted to work within limited approach boundaries (table 5.5) of exposed conductors and parts must, at a minimum, be additionally trained in all of the following:

(1) Skills and techniques necessary to distinguish exposed energized parts from other parts.
(2) Skills and techniques necessary to determine the nominal voltage of exposed energized parts. These skills and techniques include those necessary to safely use high and low-voltage meters, test instruments, and personal protective equipment while performing measurements and testing.
(3) The approach distances specified in table 5.5 and corresponding voltages to which the qualified person will be exposed. (Paste this table in the Electric Shop and hand it out to each team member before beginning work on a project that involves work near exposed energized lines or other equipment).
(4) The decision-making process to determine the degree and extent of the hazard and the personal protective equipment necessary to perform the task safely. For example, clothing that would increase injury by fire is not permitted. *Note*; Clothing made of acetate, nylon, polyester, and rayon is prohibited. (Refer; OSHA 29 CFR 1910.269 on apparel).

(c) In-Training

A person who is undergoing on-the-job training and who, in the course of this training, has demonstrated the ability to perform specific duties safely at his or her level of training, and is under the direct supervision of a qualified person, is considered a qualified person for the performance of those specific duties only. For qualified persons, the employer must determine by regular supervision and inspections of the employee's work and his/her on-the-job work practices, at least annually, that each qualified person is complying with the safety-related-work practices required.

(d) Additional training (or re-training)

An employee must receive additional training (or re-training) under any of the following conditions;

(1) If supervision and/or annual inspections of the employees work and on the-job, safety-related work practices indicate the employee is not knowledgeable or complying with the requirements of this section.
(2) If new technology, new type equipment, or changes in procedures dictate the use of safety-related work practices that are different from those which the employee would normally use.
(3) If the worker must use safety-related practices not normally used during normal job duties.
(4) If the worker has not performed this specific task within 1 year or feels a need for additional training to perform the job safely.
(5) If the worker's other qualifications have expired, such as First Aid and CPR.

Note: Employee who performs a task less than once a year must receive hazard retraining before the employee may perform the task again. Retraining may be done during the JHA, but must also include a jobsite visit to discuss hazards.

Performing a task less than once a year is not considered a part of normal job duties.

The Employer must keep a record of training of its each employee including the retraining, refresher training, special training, and update etc.

5.10 Indian Electricity Rules 1956 – an Overview

Ministry of Irrigation and Power, Govt of India in exercise of the powers conferred by Section 37 of the Indian Electricity Act, 1910 (IX of 1910), the Central Electricity Board hereby makes the following rules, the same having been previously published as required by subsection (1) of Section 38 of the said Act, namely: – THE INDIAN ELECTRICITY RULES 1956, as notified and as published in Gazette of India, Pt. II (3), dt 26.6.1956.

(a) Brief of rules-an abstract

These rules enacted in 1956 based on 1910 Act and amended thereafter on many occasions (amendment upto 29.4.1995 included), describes all about Electricity usage, rules, regulations, implementations etc in various sectors as applicable in India.

Chapter I; Preliminary: It contains Rule: 1, 2 and 3 and describes Short title and commencement, Definitions and Authorisation.

Chapter II; Inspectors: It contains Rule: 4; Qualification of Inspectors, 4A; Appointment of officers to assist the Inspectors, 4B; Qualification of officers appointed to assist the Inspectors,5; Entry and inspection,6; Appeals, 7; Amount of fees, 8; Incidence of fees recoverable in the cases of dispute, 9; Submission of records, and Rule 10; List of consumers.

Chapter III; Licence: It contains Rule: 11; Application for licence, 12; Copies of maps and draft licence for public inspection, 13; Contents of draft licence, 14; Form of draft licence,15; Advertisement of application and contents thereof,16; Amendment of draft licence,17; Local enquiries,18; Approval of draft licence,19; Notification of grant of licence,20; Date of commencement of licence,21; Deposit of maps,22; Deposit of printed copies,23; Application for written consent of State Government in certain cases,24; Amendment of licence,25; Sale of Plans,26; Preparations and submission of accounts,27; Model conditions of supply, and 28; Forms of requisitions.

Chapter IV; General Safety Requirements: Rule; 29 to 46 described in detail later under this head being relevant to chapter.

Chapter V; General Conditions Relating to Supply and Use of Energy: Rule; 47 to 59 described in detail later under this head being relevant to chapter.

Chapter VI; Electricity Supply Lines, Systems and Apparatus for Low and Medium Voltages: Rule; 60 to 62 described in details later under this head being relevant to chapter.

Chapter VII; Electricity Supply Lines, Systems and Apparatus for High and Extra-High Voltages: Rule; 63 to 73 described in details later under this head being relevant to chapter.

Chapter VIII; Overhead Lines, Under Ground Cables and Generating Stations: Rule 74 to 93 described in details later under this head being relevant to chapter.

Chapter IX; Electric Traction: It contains Rule: 94; Additional rules for electric traction, 95; Voltage of supply to vehicle,96; Insulation of lines,97; Insulation of returns,98; Proximity to metallic pipes etc.,99; Difference of potential on return,100; Leakage on conduit system,101; Leakage on system other than conduit system,102; Passengers not to have access to electric circuit,103; Current density in rails,104; Isolation of sections,105; Minimum size and strength of trolley-wire,106; Height of trolley-wire and length of span,107; Earthing of guard wires,107A; Proximity to magnetic observatories and laboratories, and Rule:108; Records.

Chapter X; Additional Precautions to be adopted in Mines and Oil-Fields: It contains Rule:109; Application of Chapter, 110; Responsibility for observance, 111; Notices, 112; Plans, 113; Lighting, communications and fire precautions,114; Isolation and fixing of transformer, switchgear, etc,115; Method of earthing,116; Protective equipment,117; Earthing metal, etc,118; Voltage limits,119; Transformers,120; Switchgear and terminals,121; Disconnection of supply,122; Cables,123; Flexible cables,124. Portable and transportable machines,125; Sundry precautions,126; Precautions where gas exists,127; Shot-firing,128; Signalling,129; Haulage,130; Earthing of neutral points, 131; Supervision, and Rule: 132; Exemptions

Chapter XI; Miscellaneous: It contains Rule:133; Relaxation by Government,134; Relaxation by Inspector,135; Supply and use of energy by non-licensees and others,136; Responsibility of Agents and Managers,137; Mode of entry,138; Penalty for breaking seal,138A; Penalty for breach of rule 44A,139; Penalty for breach of rule 45,140; Penalty for breach of rule 82,140A; Penalty for breach of rule 77, 79 or 80,141; Penalty for breach of rules,142; Application of rules, and 143; Repeal.

Annexure-I to XIV: Some typical specs relating to electrolysis etc, Forms, log sheets, formats for various returns, requisitions, compliances & guidelines for training under rule 3(2A).

Appendix-I: Syllabus for Operating & Supervisory (O&S) staff of thermal power plants.

Appendix-II: Specialised mechanical & electrical course for O&S staff of power plants.

Appendix-III: Syllabus for skilled helpers for O&M of Thermal power plants.

Appendix-IV: Specialised electrical & mechanical course for skilled operators of respective power plants.

Appendix-V: Syllabus for O&M staff of Hydro power generating plants.

Appendix-VI: Syllabus for O&M staff of EHT sub station.

Appendix-VII: Assessment forms for O&M, O&S, and skilled helpers to various staff of power plants etc for use by training dept.

(b) Description of relevant technical chapters
Chapter IV to VIII, about various electrical works related to general buildings and broadcasting set ups including its office establishments having Rule: 29 to 93 (of Indian Electricity Rules 1956 (IE-1956), amended to date (upto 29.4.1995) shall be described, as below;

Chapter IV; General Safety Requirements:
Rule 29; For Construction, installation, protection, operation and maintenance of electric supply lines and apparatus –

(1) All electric supply lines and apparatus shall be of sufficient ratings for power, insulation and estimated fault current and of sufficient mechanical strength, for the duty which they may be required to perform under the environmental conditions of installation, and shall be constructed, installed, protected, worked and maintained in such a manner as to ensure safety of human beings, animals and property.

(2) Save as otherwise provided in these rules, the relevant code of practice of the Bureau of Indian Standards BIS, including International code, regd broadcasting equipments and buildings housing such equipments, also National Electrical Code, if any may be followed to carry out the purposes of this rule and in the event of any inconsistency, the provision of these rules shall prevail.

(3) The material and apparatus used shall conform to the relevant specifications of the Bureau of Indian Standards, where such specifications have already been laid down.

Rule 30; For erecting Service lines and apparatus on consumer's premises –

(1) The supplier shall ensure that all electric supply lines, wires, fittings and apparatus belonging to him or under his control, which are on a consumer's premises, are in a safe condition and in all respects fit for supplying energy and the supplier shall take due precautions to avoid danger arising on such premises from such supply lines, wires, fittings and apparatus.

(2) Service-lines placed by the supplier on the premises of a consumer which are underground or which are accessible shall be so insulated and protected by the supplier as to be secured under all ordinary conditions against electrical, mechanical, chemical or other injury to the insulation.

(3) The consumer shall, as far as circumstances permit, take precautions for the safe custody of the equipment on his premises belonging to the supplier.

(4) The consumer shall also ensure that the installation under his control is maintained in a safe condition.

Rule 31; For Cut-out on consumer's premises –
(1) The supplier shall provide a suitable cut-out in each conductor of every service-line other than an earthed or earthed neutral conductor or the earthed external conductor of a concentric cable within a consumer's premises, in an accessible position. Such cut-out shall be contained within an adequately enclosed fireproof receptacle. If more than one consumer is supplied through a common service-line, each such consumer shall be provided with an independent cut-out at the point of junction to the common service.
(2) Every electric supply line other than the earth or earthed neutral conductor of any system or the earthed external conductor of a concentric cable shall be protected by a suitable cut-out by its owner.
(3) -Deleted

Rule 32; Identification of earthed and earthed neutral conductors and position of switches and cut-outs therein –

Where the conductors include an earthed conductor of a two-wire system or an earthed neutral conductor of a multi-wire system or a conductor which is to be connected thereto, the following conditions shall be complied with;

(1) An indication of a permanent nature shall be provided by the owner of the earthed or earthed neutral conductor, or the conductor which is to be connected thereto, to enable such conductor to be distinguished from any live conductor. Such indication shall be provided – (a) Where the earthed or earthed neutral conductor is the property of the supplier, at or near the point of commencement of supply; (b) Where a conductor forming part of a consumer's system is to be connected to the supplier's earthed or earthed neutral conductor, at the point where such connection is to be made; (c) In all other cases, at a point corresponding to the point of commencement of supply or at such other points as may be approved by an Inspector or any officer appointed to assist the Inspector and authorised under sub-rule (2) of rule 4A.
(2) No cut-out, link or switch other than a linked switch arranged to operate simultaneously on the earthed or earthed neutral conductor and live conductors shall be inserted or remain inserted in any earthed

or earthed neutral conductor of a two wire-system or in any earthed or earthed neutral conductor of a multi-wire system or in any conductor connected thereto with the following exceptions: –

(a) A link for testing purposes, or (b) A switch for use in controlling a generator or transformer

Rule-33; Earthed terminal on consumer's premises –

(1) The supplier shall provide and maintain on the consumer's premises for the consumer's use a suitable earthed terminal in an accessible position at or near the point of commencement of supply as defined under rule 58. Provided that in the case of medium, high or extra-high voltage installation the consumer shall, in addition to the afore mentioned earthing arrangement, provide his own earthing system with an independent electrode. Provided further that the supplier may not provide any earthed terminal in the case of installations already connected to his system on or before the date to be specified by the State Government in this behalf if he is satisfied that the consumer's earthing arrangement is efficient.

(2) The consumer shall take all reasonable precautions to prevent mechanical damage to the earthed terminal and its lead belonging to the supplier. The supplier may recover from the consumer the cost of installation on the basis of schedule of charges notified in advance and where such schedule of charges is not notified, the procedure prescribed; in sub-rule (5) of rule 82 will apply.

Rule-34; Accessibility of bare conductors –

Where bare conductors are used in a building, the owner of such conductors shall: – (a) Ensure that they are inaccessible; (b) Provide in readily accessible position switches for rendering them dead whenever necessary; and (c) Take such other safety measures as are considered necessary by the Inspector.

Rule-35; Danger Notices-

The owner of every medium, high and extra-high voltage installation shall affix permanently in a conspicuous position a danger notice in Hindi or English and the local language of the district, with a sign of skull and bones of a design as per the relevant ISS No. 2551– [inserted vide GSR 512, dt. 29.6.1983, w.e.f. 16.7.1983]

(a) Every motor, generator, transformer and other electrical plant and equipment together with apparatus used for controlling or regulating the same;

(b) All supports of high and extra-high voltage overhead lines which can be easily climb-upon without the aid of ladder or special appliances;
Explanation— Rails, tubular poles, wooden supports, reinforced cement concrete poles without steps, I-sections and channels, and shall be deemed as supports which cannot be easily climbed upon for the purposes of this clause.

(c) Luminous tube sign requiring high voltage supply-ray and similar high-frequency installations; Provided that where it is not possible to affix such notices on any generator, motor transformer of other apparatus, they shall be affixed as near as possible thereto; or the word 'danger' and the voltage of the apparatus concerned shall be permanently painted on it. Provided further that where the generator, motor, transformer of other apparatus is within an enclosure one notice affixed to the said enclosure shall be sufficient for the purposes of this rule.

Rule-36; Handling of electric supply lines and apparatus –

(1) Before any conductor or apparatus is handled adequate precautions shall be taken, by earthing or other suitable means, to discharge electrically such conductor or apparatus, and any adjacent conductor or apparatus if there is danger there from, and to prevent any conductor or apparatus from being accidentally or inadvertently electrically charged when persons are working thereon. Every person who is working on an electric supply line or apparatus or both shall be provided with tools and devices such as gloves, rubber shoes, and safety belts, ladders, earthing devices, helmets, line testers, hand lines and the like for protecting him from mechanical and electrical injury. Such tools and devices shall always be maintained in sound and efficient working conditions:

(2) No person shall work on any live electric supply line or apparatus and no person shall assist such person on such work, unless he is authorised in that behalf, and takes the safety measures approved by the Inspector.

(3) Every telecommunication line on supports carrying a high or extra-high voltage line shall, for the purpose of working thereon, be deemed to be a high voltage line.

Rule 37; Supply to vehicles, cranes, etc. –
Every person owning a vehicle, travelling crane or the like to which energy is supplied from an external source shall ensure that it is efficiently controlled by a suitable switch enabling all voltage to be cut off in one operation and, where such vehicle, travelling crane or the like runs on metal rails, the owner shall ensure that the rails are electrically continuous and earthed.

Rule-38; Cables for portable or transportable apparatus-
(1) Flexible cables shall not be used for portable or transportable motors, generators, transformer rectifiers, electric drills, electric sprayers, welding sets or any other portable or transportable apparatus unless they are heavily insulated and adequately protected from mechanical injury.
(2) Where the protection is by means of metallic covering, the covering shall be in metallic connection with the frame of any such apparatus and earth.
(3) The cables shall be three core type and four-core type for portable and transportable apparatus working on single phase and three phases supply respectively and the wire meant to be used for ground connection shall be easily identifiable.

Rule-39; Cables protected by bituminous materials –
(1) Where the supplier or the owner has brought into use an electric supply line (other than an overhead line) which is not completely enclosed in a continuous metallic covering connected with earth and is insulated or protected in situ by composition or material of a bituminous character –
   (a) Any pipe, conduit or the like into which such electric supply line may have been drawn or placed shall, unless other arrangements are approved by the Inspector in any particular case, be effectively sealed at its point of entry into any street box so as to prevent any flow of gas to or from the street box; and
   (b) Such electric supply line shall be periodically inspected and tested where accessible, and the result of each such inspection and test shall be duly recorded by the supplier or the owner.

(2) It shall not be permissible for the supplier or the owner after the coming into force of these rules, to bring into use any further electric supply line as aforesaid which is insulated or protected in situ by any composition or material known to be liable to produce noxious or explosive gases on excessive heating.

Rule; 40; Street boxes –

(1) Street boxes shall not contain gas pipes, and precautions shall be taken to prevent, as far as reasonably possible, any influx of water or gas.
(2) Where electric supply lines forming part of different systems pass through the same street box, they shall be readily distinguishable from one another and all electric supply lines at high or extra-high voltage in street boxes shall be adequately supported and protected to as to prevent risk of damage to or danger from adjacent electric supply lines.
(3) All street boxes shall be regularly inspected for the purpose of detecting the presence of gas and if any influx or accumulation is discovered, the owner shall give immediate notice to any authority or company who have gas mains in the neighborhood of the street box and in cases where a street box is large enough to admit the entrance of a person after the electric supply lines or apparatus therein have been placed in position, ample provision shall be made – (a) To ensure that any gas which may by accident have obtained access to the box shall escape before a person is allowed to enter; and (b) For the prevention of danger from sparking.
(4) The owners of all street boxes or pillars containing circuits or apparatus shall ensure that their covers and doors are so provided that they can be opened only by means of a key or a special appliance.

Rule-41; Distinction of different circuits –

The owner of every generating station, sub-station, junction-box or pillar in which there are any circuits or apparatus, whether intended for operation at different voltages or at the same voltage, shall ensure by means of indication of a permanent nature that the respective circuits are readily distinguishable from one another.

Rule-41A; Distinction of the installations having more than one feed –
The owner of the every installation including sub-station, double pole structure, four pole structure or any other structure having more than one feed, shall ensure by means of indication of a permanent nature, that the installation is readily distinguishable from other installations. Ref GSR 529, dt 11.7.1986, w.e.f. 19.7.1986.

Rule-42; Accidental charging –
The owners of all circuits and apparatus shall so arrange them that there shall be no danger of any part thereof becoming accidentally charged to any voltage beyond the limits of voltage for which they are intended. Where A.C. and D.C. circuits are installed on the same support they shall be so arranged and protected that they shall not come into contact with each other when live.

Rule-43; Provisions applicable to protective equipment –
(1) Fire buckets filled with clean dry sand and ready for immediate use for extinguishing fires, in addition to fire extinguishers suitable for dealing with electric fires, shall be conspicuously marked and kept in all generating stations, enclosed sub-stations and switch stations in convenient situation. The fire extinguishers shall be tested for satisfactory operation at least once a year and record of such tests shall be maintained. (2) First-aid boxes or cupboards conspicuously marked and equipped with such contents as the State Government may specify shall be provided and maintained in every generating station, enclosed sub-station and enclosed switch station so as to be readily accessible during all working hours. All such boxes and cupboards shall, except in the case of unattended sub-stations and switch stations, be kept in charge of responsible persons who are trained in first-aid treatment and one of such person shall be available during working hours.
(3) Two or more gas masks shall be provided conspicuously and installed and maintained at accessible places in every generating station with capacity of 5 MW and above and enclosed sub-station with transformation capacity of 5 MVA and above for use in the event of fire or smoke. Provided that where more than one generator with capacity of 5 MW and above is installed in a power station, each generator would be provided with at least two separate gas masks in accessible

and conspicuous position. Provided further that adequate number of gas masks would be provided by the owner of every generating station and enclosed sub-station with capacity less than 5MW and 5MVA respectively (if so desired by the Inspector).

Rule-44; Instructions for restoration of persons suffering from electric shock –
(1) Instructions, in English or Hindi and the local language of the district and where Hindi is the local language, in English and Hindi for the restoration of persons suffering from electric shock, shall be affixed by the owner in a conspicuous place in every generating station, enclosed sub-station, enclosed switch-station and in every factory as defined in clause (m) of section 2 of the Factories Act, 1948 (63 of 1948) in which electricity is used and in such other premises where electricity is used as the Inspector or any officer appointed to assist the Inspector may, by notice in writing served on the owner, direct.
(2) Copies of the instructions shall be supplied on demand by an officer or officers appointed by the Central or the State Government in this behalf at a price to be fixed by the Central or the State Government.
(3) The owner of every generating station, enclosed sub-station, enclosed switch-station and every factory or other premises to which this rule applies, shall ensure that all authorised persons employed by him are acquainted with and are competent to apply the instructions referred to in sub-rule (1).
(4) In every manned high voltage or extra-high voltage generating station, sub-station or switch station, an artificial respirator shall be provided and kept in good working condition.

Rule-44A; Intimation of Accident-
If any accident occurs in connection with the generation, transmission, supply or use of energy in or in connection with, any part of the electric supply lines or other works of any person and the accident results in or is likely to have resulted in loss of human or animal life or in any injury to a human being or an animal, such person or any authorised person of the State Electricity Board/Supplier, not below the rank of a Junior Engineer or equivalent shall send to the Inspector a telegraphic report within 24 hours of the knowledge of the occurrence of the fatal

accident and a written report in the form set out in Annexure XIII within 48 hours of the knowledge of occurrence of fatal and all other accidents. Where practicable a telephonic message should also be given to the Inspector immediately the accident comes to the knowledge of the authorised officer of the State Electricity Board/ Supplier or other person concerned.

Rule-45; Precautions to be adopted by consumers owners occupiers, electrical contractors, electrical workmen and suppliers –

(1) No electrical installation work, including additions, alterations, repairs and adjustments to existing installations, except such replacement of lamps, fans, fuses, switches, low voltage domestic appliances and fittings as in no way alters its capacity or character, shall be carried out upon the premises of or on behalf of any consumer, supplier, owner or occupier] for the purpose of supply to such consumer, supplier, owner or occupier except by an electrical contractor licensed in this behalf by the State Government and under the direct supervision of a person holding a certificate of competency and by a person holding a permit issued or recognised by the State Government.

Provided that in the case of works executed for or on behalf of the Central Government and in the case of installations in mines, oil fields and railways, the Central Government and in other cases the State Government may, by notification in the Official Gazette, exempt, on such conditions as it may impose, any such work described therein either generally or in the case of any specified class of consumers, suppliers, owners or occupiers] from so much of this sub-rule as requires such work to be carried out by an electrical contractor licensed by the State Government in this behalf.

(2) No electrical installation work which has been carried out in contravention of sub-rule (1) shall either be energised or connected to the works of any supplier.

Rule-46; Periodical inspection and testing of consumer's installation –

(1) (a) Where an installation is already connected to the supply system of the supplier, every such installation shall be periodically inspected and tested at intervals not exceeding five years either by the Inspector or any officer appointed to assist the Inspector

or by the supplier as may be directed by the State Government in this behalf or in the case of installations belonging to, or under the control of the Central Government, and in the case of installation in mines, oilfields and railways by the Central Government.

(b) Where the supplier is directed by the Central or the State Government as the case may be to inspect and test the installation he shall report on the condition of the installation to the consumer concerned in a form approved by the Inspector and shall submit a copy of such report to the Inspector or to any officer appointed to assist the Inspector and authorised under sub-rule (2) of rule 4A.

(c) Subject to the approval of the Inspector, the forms of inspection report contained in Annexure IX-A may, with such variations as the circumstances of each case require, be used for the purposes of this sub-rule. (2) (a) The fees for such inspection and test shall be determined by the Central or the State Government, as the case may be, in the case of each class of consumers and shall be payable by the consumer in advance. (b) In the event it of the failure of any consumer to pay the fees on or before the date specified in the fee-notice, supply to the installation of such consumer shall be liable to be disconnected under the direction of the Inspector. Such disconnection, however, shall not be made by the supplier without giving to the consumer seven clear days' notice in writing of his intention so to do. (c) In the event of the failure of the owner of any installation to rectify the defects in his installation pointed out by the Inspector or by any officer appointed to assist him and authorised under sub-rule (2) of rule 4A in the form set out in Annexure-IX and within the time indicated therein, such installation shall be liable to be disconnected under the directions of the Inspector after serving the owner of such installation with a notice. Provided that the installation shall not be disconnected in case an appeal in made under rule 6 and the appellate authority has stayed the orders of disconnection. Provided further the time period indicated in the notice shall not be less than 48 hours in any case. Provided also nothing contained in this clause shall have any effect on the application of rule 49.

(3) Notwithstanding the provisions of this rule, the consumer shall at all times be solely responsible for the maintenance of his installation in such condition as to be free from danger.

Chapter V; General Conditions Relating to Supply and Use of Energy:
Rule-47; Testing of consumer's installation –

(1) Upon receipt of an application for a new or additional supply of energy and before connecting the supply or reconnecting the same after a period of six months, the supplier shall inspect and test the applicants' installation. The supplier shall maintain a record of test results obtained at each supply point to a consumer, in a form to be approved by the Inspector.

(2) If as a result of such inspection and test, the supplier is satisfied that the installation is likely to constitute danger, he shall serve on the applicant a notice in writing requiring him to make such modifications as are necessary to render the installation safe. The supplier may refuse to connect or reconnect the supply until the required modifications have been completed and he has been notified by the applicant.

Rule-47A; Installation and Testing of Generating Units –

Where any consumer or occupier installs a generating plant, he shall give a thirty days' notice of his intention to commission the plant to the supplier as well as the Inspector. Provided, that no consumer or occupier shall commission his generating plant of a capacity exceeding 10KW without the approval in writing of the Inspector.

Rule-48; Precautions against leakage before connection –

(1) The supplier shall not connect with his works the installation or apparatus on the premises of any applicant for supply unless he is reasonably satisfied that the connection will not at the time of making the connection cause a leakage from that installation or apparatus of a magnitude detrimental to safety. Compliance with this rule shall be checked by measuring the insulation resistance (IR) as provided below:
   (i) High Voltage Equipments installations – (a); High Voltage equipments shall have the IR value as stipulated in the relevant Indian Standard. (b) At a pressure of 1000 V applied between

each live conductor and earth for a period of one minute the insulation resistance of HV installations shall be at least 1 Mega ohm or as specified by the Bureau of Indian standard (BIS) from time to time.

(ii) Medium and Low Voltage Installations – At a pressure of 500 V applied between each live conductor and earth for a period of one minute, the insulation resistance of medium and low voltage installations shall be at least 1 Mega ohm or as specified by the BIS from time to time.

(2) If the supplier declines to make a connection under the provisions of sub-rule (1), he shall serve upon the applicant a notice in writing stating his reason for so declining.

Rule-49; Leakage on consumer's premises –

(1) If the Inspector or any officer appointed to assist the Inspector and authorised under sub-rule (2) of rule 4A or the supplier had reason to believe that there is in the system of a consumer leakage which is likely to affect injuriously the use of energy by the supplier or by other persons, or which is likely to cause danger, he may give the consumer reasonable notice in writing that he desires to inspect and test the consumer's installation.

(2) If on such notice being given – (a) The consumer does not give all reasonable facilities for inspection and testing of his installation, or (b) When an insulation resistance at the consumer's installation is so low as to prevent safe use of energy. The supplier may, and if directed so to do by the Inspector shall discontinue the supply of energy to the installation but only after giving to the consumer 48 hours notice in writing of disconnection of supply and shall not recommence the supply until he or the Inspector is satisfied that the cause of the leakage has been removed.

Rule-50; Supply and use of energy –

(1) The energy shall not be supplied, transformed, converted or used or continued to be supplied, transformed, converted or used unless provisions as set out below are observed: –

(a) The following controls of requisite capacity to carry and break the current are placed after the point of commencement of supply

as defined in rule 58 so as to be readily accessible and capable of being easily operated to completely isolate the supply to the installation such equipment being in addition to any equipment installed for controlling individual circuits or apparatus: –
   (i) A linked switch with fuse(s) or a circuit breaker by low and medium voltage consumers.
   (ii) A linked switch with fuse(s) or a circuit breaker by HV consumers having aggregate installed transformer/apparatus capacity up to 1000 KVA to be supplied at voltage upto 11 KV and 2500 KVA at higher – voltages (above 11 KV and not exceeding 33 KV).
   (iii) A circuit breaker by HV consumers having an aggregate installed transformer/apparatus capacity above 1000 KVA and supplied at 11 KV and above 2500 KVA supplied at higher voltages (above 11 KV and not exceeding 33 KV).
   (iv) A circuit breaker by EHV consumer; Provided that where the point of commencement of supply and the consumer apparatus are near each other one linked switch with fuse(s) or circuit breaker near the point of commencement of supply as required by this clause shall be considered sufficient for the purpose of this rule;
(b) In case of every transformer the following shall be provided: –
   (i) On primary side for transformers a linked switch with fuse(s) or circuit breaker of adequate capacity: Provided that the linked switch on the primary side of the transformer may be of such capacity as to carry the full load current and to break only the magnetising current of the transformer: Provided further that for transformers of capacity 5000 KVA and above a circuit breaker shall be provided: Provided further that the provision of linked switch on the primary side of the transformer shall not apply to the unit auxiliary transformer of the generator. (ii) On the secondary side of transformers of capacity 100 KVA and above transforming HV to MV or LV, a linked switch with fuse(s) or circuit breaker of adequate capacity capable of carrying and breaking full load current and for transformers transforming HV to EHV as the case may be, a circuit breaker: Provided that where the transformer

capacity exceeds 630 KVA a circuit breaker of adequate capacity shall be installed on the secondary side;

(c) Except in the case of composite control gear designed as a unit distinct circuit is protected against excess energy by means of suitable cut-out or a circuit breaker of adequate breaking capacity suitably located and, so constructed as to prevent danger from overheating, arcing or scattering of hot metal when it comes into operation and to permit for ready renewal of the fusible metal of the cut-out without danger;

(d) The supply of energy of each motor or a group of motors or other apparatus meant for operating one particular machine is controlled by a suitable linked switch or a circuit breaker or an emergency tripping device with manual reset of requisite capacity placed in such a position as to be adjacent to the motor or a group of motors or other apparatus readily accessible to and easily operated by the person incharge and so connected in the circuit that by its means all supply of energy can be cut off from the motor or group of motors or apparatus from any regulating switch, resistance of other device associated therewith;

(e) All insulating materials are chosen with special regard to the circumstances of its proposed use and their mechanical strength is sufficient for its purpose and so far as is practicable of such a character or so protected as to maintain adequately its insulating property under all working conditions in respect of Temperature and moisture; and

(f) Adequate precautions shall be taken to ensure that no live parts are so exposed as to cause danger.

(2) Where energy is being supplied, transformed, converted or used the consumer, supplier or the owner of the concerned installation shall be responsible for the continuous observance of the provisions of sub-rule (1) in respect of his installations.

(3) Every consumer shall use all reasonable mean to ensure that where energy is supplied by a supplier no person other than the supplier shall interfere with the service lines and apparatus placed by the supplier on the premises of the consumer.

Rule-50A; Additional provisions for supply and use of energy in multi-storey building (more than 15 meters in height) –

(1) Before making an application for commencement of supply or recommencement of supply after an installation has been disconnected for a period of six months or more the owner/occupier of a multi-storey building shall give not less than 30 days' notice in writing to the Inspector together with particulars. The supply of energy shall not be commenced or recommenced within this period, without the approval or otherwise in writing of the Inspector.

(2) The supplier/owner of the installation shall provide at the point of commencement of supply a suitable isolating device with cut out or breaker to operate on all phases except neutral in the 3 phase 4 wire circuit and fixed in a conspicuous position at not more than 2.75 meters above the ground so as to completely isolate the supply to the building in case of emergency.

(3) The owner/occupier of a multi-storey building shall ensure that electrical installations/works inside the building are carried out and maintained in such a manner as to prevent danger due to shock and fire hazards' and the installation is carried out in accordance with the relevant codes of practices.

(4) No other service pipes shall be taken along the ducts provided for laying power cables. All ducts provided for power cables and other services shall be provided with fire-barrier at each floor crossing.

Rule-51; Provisions applicable to medium, high or extra-high voltage installations –

The following provisions shall be observed where energy at medium, high or extra-high voltage is supplied, converted, transformed or used;

(1) (a) All conductors (other than those of overhead lines) shall be completely enclosed in mechanically strong metal casting or metallic covering which is electrically and mechanically continuous and adequately protected against mechanical damage unless the said conductors are accessible only to an authorised person or are installed and protected to the satisfaction of the Inspector so as to prevent danger; Provided that non-metallic conduits conforming

to the relevant Indian Standard Specifications may be used for medium voltage installations, subject to such conditions as the Inspector or Officer appointed to assist an Inspector may think fit to impose.

(b) All metal works, enclosing, supporting or associated with the installation, other than that designed to serve as a conductor shall be connected with an earthing system as per standards laid down in the Indian Standards in this regard and in also accordance with rule 61(4).

(c) Every switchboard shall comply with the following provisions, namely: – (i) A clear space of not less than 1 meter in width shall be provided in front of the switchboard; (ii) If there are any attachments or bare connections at the back of the switchboard, the space (if any) behind the switchboard shall be either less than 20 centimeters or more than 75 centimeters in width, measured from the farthest outstanding part of any attachment or conductor; (iii) If the space behind the switchboard exceeds 75 centimeters in width, there shall be a passage-way from either end of the switchboard clear to a height of 1.8 meters.

(d) In case of installations provided in premises where inflammable materials including gases and/or chemicals are produced, handled or stored, the electrical installations, equipment and apparatus shall comply with the requirements of flame proof, dust tight, totally enclosed or any other suitable type of electrical fittings depending upon the hazardous zones as per relevant Indian Standard Specification.

(2) Where an application has been made to a supplier for supply of energy to any installation, he shall not commence the supply or where the supply has been discontinued for a period of one year and above, recommence the supply unless he is satisfied that the consumer has complied with, in all respects the conditions of supply set out in sub-rule (1), of this rule, rules 50, 63 and 64.

(3) Where a supplier proposes to supply or use energy at a medium voltage or to recommence supply after it has been discontinued for a period of six months, he, shall, before connecting or reconnecting the supply, give notice in writing of such intention to the Inspector

or any officer of specified rank and class appointed to assist the Inspector.

(4) If at any time after connecting the supply, the supplier is satisfied that any provision of the sub-rule (1) of this rule or of rules 50 and 64, is not being observed he shall give notice of the same in writing to the consumer and the inspector, specifying how the provisions has not been observed and to rectify such defects in a reasonable time and if the consumer fails to rectify such defects pointed out, he may discontinue the supply after giving the consumer a reasonable opportunity of being heard and recording reasons in writing, unless the inspector directs otherwise. The supply shall be discontinued only on written orders of an officer duly notified by the supplier in this behalf. The supply shall be restored with all possible speed after such defects is rectified by the consumer to the satisfaction of the supplier.

Rule-52; Appeal to Inspector in regard to defects –

(1) If any applicant for a supply or a consumer is dissatisfied with the action of the supplier in declining to commence, to continue or to recommence the supply of energy to his premises on the grounds that the installation is defective or is likely to constitute danger, he may appeal to the Inspector to test the installation and the supplies shall not, if the Inspector or under his orders, any other officer appointed to assist the Inspector, is satisfied that the installation is free from the defect or danger complained of, be entitled to refuse supply to the consumer on the grounds aforesaid, and shall, within twenty-four hours after the receipt of such intimation from the Inspector, commence, continue or recommence the supply of energy.

(2) Any test for which application has been made under the provision of sub-rule (1) shall be carried out within seven days after the receipt of such application.

(3) This rule shall be endorsed on every notice given under the provisions of rules 47, 48, and 49.

Rule-53; Cost of inspection and test of consumer's installation –

(1) The cost of the first inspection and test of consumer's installation carried out in pursuance of the provisions of rule 47 shall be borne by the supplier and the cost of every subsequent inspection and test

shall be borne by the consumer, unless in the appeal under rule 52, the Inspector directs otherwise.
(2) The cost of any inspection and test made by the Inspector or any officer appointed to assist the Inspector, at the request of the consumer or other interested party, shall be borne by the consumer or other interested party, unless the Inspector directs otherwise.
(3) The cost of each and every such inspection and test by whomsoever borne shall be calculated in accordance with the scale specified by the Central or the State Government as the case may be in this behalf.

Rule-54; Declared voltage of supply to consumer –

Except with the written consent of the consumer or with the previous sanction of the State Government a supplier shall not permit the voltage at the point of commencement of supply as defined under rule 58 to vary from the declared voltage –
(i) In the case of low or medium voltage, by more than 6 per cent, or;
(ii) In the case of high voltage, by more than 6 per cent on the higher side or by more than 9 per cent on the lower side, or;
(iii) In the case of extra-high voltage, by more than 10 per cent on the higher side or by more than 12.5 per cent on the lower side.

Rule-55; Declared frequency of supply to consumer –

Except with the written consent of the consumer or with the previous sanction of the State Government a supplier shall not permit the frequency of an alternating current supply to vary from the declared frequency by more than 3 per cent.

Rule-56; Sealing of meters and cut-outs –

(1) A supplier may affix one or more seals to any cut-out and to any meter, maximum demand indicator, or other apparatus placed upon a consumer's premises in accordance with section 26, and no person other than the supplier shall break any such seal.
(2) The consumer shall use all reasonable means in his power to ensure that no such seal is broken otherwise than by the supplier.
(3) The word 'supplier' shall for the purpose of this rule include a State Government when any meter, maximum demand indicator or other apparatus is placed upon a consumer's premises by such Government.

Rule-57; Meters, maximum demand indicators and other apparatus on consumer's premises –

(1) Any meter or maximum demand indicator or other apparatus placed upon a consumer's premises in accordance with section 26 shall be of appropriate capacity and shall be deemed to be correct if its limits of error are within the limits specified in the relevant Indian Standard Specification and where no such specification exists, the limits of error do not exceed 3 per cent above or below absolute accuracy at all loads in excess of one tenth of full load and up to full load.

Provided, that for extra high voltage consumers the limit of error shall be 1 per cent.

(2) No meter shall register consumption at no load.

(3) Every supplier shall provide and maintain in proper condition such suitable apparatus as may be prescribed or approved by the Inspector for the examination, testing and regulation of meters used or intended to be used in connection with the supply of energy. Provided that the supplier may with the approval of the Inspector and shall, if required by the Inspector, enter into a joint arrangement with any other supplier for the purpose aforesaid.

(4) Every supplier shall examine, test and regulate all meters, maximum demand indicators and other apparatus for ascertaining the amount of energy supplied before their first installation at the consumer's premises and at such other intervals as may be directed by the State Government in this behalf.

(5) Every supplier shall maintain a register of meters showing the date of the last test, the error recorded at the time of the test, the limit of accuracy after adjustment and final test, the date of installation, withdrawal, reinstallation, etc., for the examination of the Inspector or his authorised representative.

(6) Where the supplier has failed to examine, test and regulate the meters and keep records thereof as aforesaid, the Inspector may cause such meters to be tested and sealed at the cost of the owner of the meters in case these are found defective.

Rule-58; Point of commencement of supply –

The point of commencement of supply of energy to a consumer shall be deemed to be the point at the incoming terminal of the cut-outs installed by the consumer under rule 50.

Rule-59; Precautions against failure of supply: Notice of failures –

(1) The layout of the electric supply lines of the supplier for the supply of energy throughout his area of supply shall under normal working conditions be sectionalised and so arranged, and provided with cut-outs or circuit-breakers so located, as to restrict within reasonable limits the extent of the portion of the system affected by any failure of supply.

(2) The, supplier shall take all reasonable precautions to avoid any accidental interruptions of supply, and also to avoid danger to the public or to any employee or authorised person when engaged on any operation during and in connection with the installation, extension, replacement, repair and maintenance of any works. (3) The supplier shall send to the Inspector or any officer of a specified rank and class appointed to assist the Inspector notice of failure of supply of such kind as the Inspector or any officer of specified rank and class to assist the Inspector] may from time to time require to be notified to him, and such notice shall be sent by the earliest practicable post after the failure occurs or after the failure becomes known to the supplier and shall be in such form and contain such particulars as Inspector may from time to time specify.

(4) For the purpose of testing or for any other purpose connected with the efficient working of the undertaking, the supply of energy may be discontinued by the supplier for such period as may be necessary, subject (except in cases of emergency) to not less than 24 hours notice being given by the supplier to all consumers likely to be affected by such discontinuance: Provided that the supply of energy shall be discontinued during such hours as are likely to interfere the least with the use of energy by consumers and the energy shall not be discontinued if the Inspector so directs.

Chapter VI; Electric Supply Lines, Systems and Apparatus for Low and Medium Voltages:
Rule-60; Test for resistance of insulation –

(1) Where any electric supply line for use at low or medium voltage has been disconnected from a system for the purpose of addition, alteration or repair, such electric supply line shall not be reconnected to the system until the supplier or the owner has applied the test prescribed under rule 48.

(2) The provision of sub-rule (1) shall not apply to overhead lines except, overhead insulated cables unless the Inspector otherwise directs in any particular case.

Rule-61; Connection with earth –
(1) The following provisions shall apply to the connection with earth of systems at low voltage in cases where the voltage normally exceeds 125 volts and of systems at medium voltage: –
   (a) Neutral conductor of a phase, 4 wire system and the middle conductor of a 2 phase, 3-wire system shall be earthed by not less than two separate and distinct connections with a minimum of two different earth electrodes of such large number as may be necessary to bring the earth resistance to a satisfactory value both at the generating station and at the sub-station. The earth electrodes so provided, may be interconnected to reduce earth resistance. It may also be earthed at one or more points along the distribution system or service line in addition to any connection with earth which may be at the consumer's premises.
   (b) In the case of a system comprising electric supply lines having concentric cables, the external conductor of such cables shall be earthed by two separate and distinct connections with earth.
   (c) The connection with earth may include a link by means of which the connection may be temporarily interrupted for the purpose of testing or for locating a fault.
   (d) (i) In a direct current three wire system the middle conductor shall be earthed at the generating station only, and the current from the middle conductor to earth shall be continuously recorded by means of a recording ammeter, and if any time the current exceeds one-thousandth part of the maximum supply-current immediate steps shall be taken to improve the insulation of the system. (ii) Where the middle conductor is earthed by means of a circuit breaker with a resistance connected in parallel, the resistance shall not exceed 10 ohms and on the opening of the circuit breaker, immediate steps shall be taken to improve the insulation of the system, and the circuit-breaker shall be reclosed as soon as possible. (iii) The resistance shall be used only as a

protection for the ammeter in case of earths on the system and until such earths are removed. Immediate steps shall be taken to locate and remove the earth.

(e) In the case of an alternating current system, there shall not be inserted in the connected with earth any impedance (other than that required solely for the operation of switch-gear of instruments), cut-out or circuit-breaker, and the result of any test made to ascertain whether the current (if any) passing through the connection with earth is normal shall be duly recorded by the supplier.

(f) No person shall make connection with earth by the aid of, nor shall be keep it in contact with, any water main not belonging to him except with the consent of the owner thereof and of the Inspector.

(g) Alternating current systems which are connected with earth as aforesaid may be electrically interconnected: Provided that each connection with earth is bonded to the metal sheathing and metallic armouring (if any) of the electric supply lines concerned.

(2) The frame of every generator, stationary motor, portable motor, and the metallic parts (not intended as conductors) of all transformers and any other apparatus used for regulating or controlling energy and all medium voltage energy consuming apparatus shall be earthed by the owner by two separate and distinct connections with earth.

(3) All metal castings or metallic coverings containing or protecting any electric supply-line or apparatus shall be connected with earth and shall be so joined and connected across all junction boxes and other openings as to make good mechanical and electrical connection throughout their whole length.

Provided that where the supply is at low voltage, this sub-rule shall not apply to isolated wall tubes or to brackets, electroliers, switches, ceiling fans or other fittings (other than portable hand lamps and portable and transportable apparatus) unless provided with earth terminal and to class-II apparatus/ appliances. Provided further that where the supply is at low voltage and where the installations are either new or renovated all plug sockets shall be of the three-pin type, and the third pin shall be permanently and efficiently earthed. Explanation-The words "Class-II apparatus/appliance" will have the same meaning as assigned to these words in the relevant ISS.

(4) All earthing systems shall-(a) Consist of equipotential bonding conductors capable of carrying the prospective earth fault current and a group of pipe/rod/plate electrodes for dissipating the current to the general mass of earth without exceeding the allowable temperature limits as per relevant Indian Standards in order to maintain all non-current carrying metal works reasonably at earth potential and to avoid dangerous contact potentials being developed on such metal works; (b) Limit earth resistance sufficiently low to permit adequate fault current for the operation of protective devices in time and to reduce neutral shifting; (c) Be mechanically strong, withstand corrosion and retain electrical continuity during the life of the installation. All earthing systems shall be tested to ensure efficient earthing, before the electric supply lines or apparatus are energised.

(5) All earthing systems belonging to the supplier shall in addition, be tested for resistance on dry day during the dry season not less than once every two years.

(6) A record of every earth test made and the result thereof shall be kept by the supplier for a period of not less than two years after the day of testing and shall be available to the Inspector or any officer appointed to assist the Inspector and authorised under sub-rule (2) of rule 4A when required.

The supply of Energy to every electrical installation other than low voltage installation below 5 KW and those low voltage installations which do not attract provisions of section 30 of the Indian Electricity Act, 1910, shall be controlled by an earth leakage protective device so as to disconnect the supply instantly on the occurrence of earth fault or leakage of current.

Provided that the above shall not apply to overhead supply lines having protective devices which are effectively bonded to the neutral of supply transformers and conforming to rule 91 of IE – Rules1956.

Rule-62; Systems at medium voltage –

Where a medium voltage supply system is employed, the voltage between earth and any conductor forming part of the same system shall not, under normal conditions, exceed low voltage.

Chapter VII; Electric Supply Lines, Systems and Apparatus for High and Extra-High Voltages:

Rule-63; Approval by Inspector –

(1) Before making an application to the Inspector for permission to commence or recommence supply after an installation has been disconnected for one year and above at high or extra-high voltage to any person, the supplier shall ensure that the high or extra-high voltage electric supply lines or apparatus belonging to him are placed in position, properly joined and duly completed and examined. The supply of energy shall not be commenced by the supplier unless and until the Inspector is satisfied that the provisions of rules 65 to 69 both inclusive have been complied with and the approval in writing of the Inspector have been obtained by him. Provided, that the supplier may exercise the aforesaid electric supply lines or apparatus for the purpose of tests specified in rule 65.

(2) The owner of any high or extra-high voltage installation shall, before making application to the Inspector for approval of his installation or additions thereto, test every high or extra-high voltage circuit or additions thereto, other than an overhead line, and satisfy himself that they withstand the application of the testing voltage set out in sub-rule (1) of rule 65 and shall duly record the results of such tests and forward them to the Inspector. Provided that an Inspector may direct such owner to carry out such tests as he deems necessary or, if he thinks fit, accept the manufacturer's certified tests in respect of any particular apparatus in place of the tests required by this sub-rule.

(3) The owner of any high or extra-high voltage installation who makes any additions or alterations to his installation shall not connect to the supply his apparatus or electric supply lines, comprising the said alterations or additions unless and until such alterations or additions have been approved in writing by the Inspector.

Rule-64; Use of energy at high and extra-high voltage –

(1) The Inspector shall not authorise the supplier to commence supply or where the supply has been discontinued for a period of one year and above, to commence the supply at high or extra-high voltage to any consumer unless; (a) All conductors and apparatus situated on the

premises of the consumer are so placed as to be inaccessible except to an authorised person and all operations in connection with the said conductors and apparatus are carried out by an authorised person; (b) The consumer has provided and agrees to maintain a separate building or a locked weather-proof and fire-proof enclosure of agreed design and location, to which the supplier at all times have access for the purpose of housing his apparatus and metering equipment, or where the provision for a separate building or enclosure is impracticable, the consumer has segregated the aforesaid apparatus of the supplier from any other part of his own apparatus; Provided that such segregation shall be by the provision of fire proof walls, if the Inspector considers it to be necessary; Provided further that in the case of an out-door installation consumer shall suitably segregate the aforesaid apparatus belonging to the supplier from his own to the satisfaction of the Inspector; (c) All pole type sub-stations are constructed and maintained in accordance with rule 69.

(2) The following provisions shall be observed where energy at high or extra-high voltage is supplied, converted, transformed or used:

   (a) (i) Clearances as per Indian Standard Code shall be provided for electrical apparatus so that sufficient space is available for easy operation and maintenance without any hazard to the operating and maintenance personnel working near the equipment and for ensuring adequate ventilation.(ii) The following minimum clearances shall be maintained for bare conductors or live parts of any apparatus in out-door substations, excluding overhead lines, of HV and EHV installations: –

Voltage Class		Ground clearance (Meters)	Sectional clearance (Meters)
Not exceeding	11KV	2.75	2.6
-do-	33KV	3.7	2.8
-do-	66KV	4.0	3.0
-do-	132KV	4.6	3.5
-do-	220KV	5.5	4.3
-do-	400KV	8.0	6.5

(b) The windings of motors or other apparatus within reach from any position in which a person may require to be shall be suitably protected so as to prevent danger.

(c) Where transformer or transformers are used, suitable provision shall be made, either by connecting with earth a point of the circuit at the lower voltage or otherwise, to guard against danger by reason of the said circuit becoming accidentally charged above its normal voltage by leakage from or contact with the circuit at the higher voltage.

(d) A sub-station or a switch station with apparatus having more than 2000 litres of oil shall not be located in the basement where proper oil draining arrangement cannot be provided.

(e) Where a sub-station or a switch station with apparatus having more than 2000 litres of oil is installed, whether indoors or out-doors, the following measures shall be taken, namely: – (i) The baffle walls of 4 hour fire rating shall be provided between the apparatus in the following cases: – (A) Single phase banks in the switch-yards of generating stations and sub-stations; (B) On the consumer premises; (C) Where adequate clearance between the units is not available. (ii) Provisions shall be made for suitable oil soak pit and where use of more than 9000 litres of oil in any one oil tank, receptacle or chamber is involved, provision shall be made for the draining away or removal of any oil which may leak or escape from the tanks receptacles or chambers containing the same, special precautions shall be taken to prevent the spread of any fire resulting from the ignition of the oil from any cause and adequate provision shall be made for extinguishing any fire which may occur. Spare oil shall not be stored in a sub-station or switch station.

(f) (i) Without prejudice to the above measures, adequate fire protection arrangement shall be provided for quenching the fire in the apparatus; (ii) Where it is necessary to locate the sub-station/switch station in the basement following measures shall be taken: – (a) The room shall necessarily be in the first basement at the periphery of the basement; (b) The entrances to the room shall be provided with fire resisting doors of 2 hours fire rating.

A curb (sill) of a suitable height shall be provided at the entrance in order to prevent the flow of oil from a ruptured transformer into other parts of the basement. Direct access to the transformer room shall be provided from outside. (c) The transformer shall be protected by an automatic high velocity water spray system or by carbon dioxide or BCF (Bromo Chlorodifeuro methane) or BTM (Bromo trifluoro methane) fixed installation system; and (iii) Oil filled transformers installed indoors shall not be on any floor above the ground or below the first basement.

(g) Cable trenches inside the sub-stations and switch stations containing cables shall be filled with sand, pebbles or similar non-inflammable materials or completely cover with non-inflammable slabs.

(h) Unless the conditions are such that all the conductors and apparatus may be made dead at the same time for the purpose of cleaning or for other work, the said conductors and apparatus shall be so arranged that these may be made dead in sections, and that work on any such section may be carried on by an authorised person without danger.

(i) Only persons authorised under sub-rule (1) of rule 3, shall carry out the work on live lines and apparatus. (3) All EHV apparatus shall be protected against lightning as well as against switching over voltages. The equipment used for protection and switching shall be adequately coordinated with the protected apparatus to ensure safe operation as well as to maintain the stability of the inter-connected units of the power system.

Rule-64 A; Additional provisions for use of energy at high and extra-high voltage –

The following additional provisions shall be observed where energy at high or extra-high voltage is supplied, converted, transferred or used, namely: -

(1) Inter-locks-Suitable inter-locks shall be provided in the following cases: –

    (a) Isolators and the controlling circuit breakers shall be interlocked so that the isolators cannot be operated unless the corresponding breaker is in open position;

(b) Isolators and the corresponding earthing switches shall be interlocked so that no earthing switch can be closed unless and until the corresponding isolator is in open position;

(c) Where two or more supplies are not intended to be operated in parallel, the respective circuit breakers or linked switches controlling the supplies shall be inter-locked to prevent possibility of any inadvertent paralleling or feedback;

(d) When two or more transformers are operated in parallel, the system shall be so arranged as to trip the secondary breaker of a transformer in case the primary breaker of that transformer trips;

(e) All gates or doors which give access to live parts of an installation shall be inter-locked in such a way that these cannot be opened unless the live parts are made dead. Proper discharging and earthing of these parts should be ensured before any person comes in close proximity of such parts;

(f) Where two or more generators operate in parallel and neutral switching is adopted, inter-lock shall be provided to ensure that generator breaker cannot be closed unless one of the neutrals is connected to the earthing system.

(2) Protection – All systems and circuits shall be so protected as to automatically disconnect the supply under abnormal conditions. The following protection shall be provided, namely: –

(a) Over current protection to disconnect the supply automatically if the rated current of the equipment, cable or supply line is exceeded for a time which the equipment, cable or supply line is not designed to withstand;

(b) Earth-fault/earth leakage protection to disconnect the supply automatically if the earth fault current exceeds the limit of current for keeping the contact potential within the reasonable values;

(c) Gas pressure type protection to given alarm and tripping shall be provided on all transformers of ratings 1000 KVA and above;

(d) Transformers of capacity 10 MVA and above shall be protected against incipient faults by differential protection; and

(e) All generators with rating of 100 KVA and above shall be protected against earth fault/leakage. All generators of rating 1000KVA and above shall be protected against faults within the generator winding using restricted earth fault protection or differential protection or by both.

Rule-65; Testing, Operation and Maintenance –
(1) Before approval is accorded by the Inspector under rule 63, the manufacturer's test certificates shall, if required, be produced for all the routine tests as required under the relevant Indian Standard.
(2) No new HV or EHV apparatus, cable or supply line shall be commissioned unless such apparatus, cable or supply line are subjected to site tests as per relevant code of practice of the Bureau of Indian Standards. (3) No HV of EHV apparatus, cable or supply line which has been kept disconnected, for a period of 6 months or more, from the system for alterations or repair shall be corrected to the system until such apparatus, cable or supply line are subjected to the relevant tests as per code of practice of Bureau of Indian Standards.
(4) Notwithstanding the provisions of sub-rules (1) to (3) (both inclusive) the Inspector may require certain additional tests to be carried out before charging the installations or subsequently.
(5) All apparatus, cables and supply lines shall be maintained in healthy conditions and tests shall be carried out periodically as per the relevant codes of practice of the Bureau of Indian Standards.
(6) Records of all tests, trappings, maintenance works and repairs of all equipments, cables and supply lines shall be duly kept in such a way that these records can be compared with earlier ones.
(7) It shall be the responsibility of the owner of all HV and EHV installations to maintain and operate the installations in a condition free from danger and as recommended by the manufacturer and/or by the relevant codes of practice of the Bureau of Indian Standards] and/ or by the Inspector.

Rule-66; Metal sheathed electric supply lines –
Precautions against excess leakage –
(1) The following provisions shall apply to electric supply lines (other than overhead lines for use at high or extra-high voltage: –
    (a) The conductors shall be enclosed in metal sheathing which shall be electrically continuous and connected with earth, and the conductivity of the metal sheathing shall be maintained and

reasonable precautions taken where necessary to avoid corrosion of the sheathing;

Provided that in the case of thermoplastic insulated and sheathed cables with metallic armour the metallic wire or tape armour shall be considered as metal sheathing for the purpose of this rule; Provided, further that this rule shall not apply to cable with thermoplastic insulation without any metallic screen or armour.

(b) The resistance of the earth connection with metallic sheath shall be kept low enough to permit the controlling circuit breaker or cut-out to operate in the event of any failure of insulation between the metallic sheath and the conductor.

(c) Where an electric supply-line as aforesaid has concentric cables and the external conductor is insulated from an outer metal sheathing and connected with earth, the external conductor may be regarded as the metal sheathing for the purposes of this rule provided that the foregoing provisions as to conductivity are complied with.

(2) Nothing in the provisions of sub-rule (1) shall preclude the employment in generating stations, sub-stations and switch-stations (including outdoor substations and outdoor switch-stations) of conductors for use at high or extra-high voltages which are not enclosed in metal sheathing or preclude the use of electric supply lines laid before the prescribed date to which the provisions of these rules apply

Rule-67; Connection with earth –

(1) All non-current carrying metal parts associated with HV/EHV installation shall be effectively earthed to a grounding system or mat which will: – (a) Limit the touch and step potential to tolerable values; (b) Limit the ground potential rise to tolerable values so as to prevent danger due to transfer of potential through ground, earth wires, cables sheath fences, pipe lines, etc.; (c) Maintain the resistance of the earth connection to such a value as to make operation of the protective device effective.

(1-A) In the case of star-connected system with earthed neutrals or delta connected system with earthed artificial neutral point: –

(a) The neutral point of every generator and transformer shall be earthed by connecting it to the earthing system as defined in rule 61(4) and hereinabove by not less than two separate and distinct connections; Provided that the neutral point of a generator may be connected to the earthing system through an impedance to limit the fault current to the earth; Provided further that in the case of multi-machine system neutral switching may be resorted to, for limiting the injurious effect of harmonic current circulation in the system; (b) In the event of an appreciable harmonic current flowing in the neutral connection so as to cause interference, with communication circuits, the generator or transformer neutral, shall be earthed through a suitable impedance; (c) In case of the delta connected system the neutral point shall be obtained by the insertion of a grounding transformer and current limiting resistance or impedance wherever considered necessary at the commencement of such a system.

(2) Single-phase high or extra-high voltage systems shall be earthed in a manner approved by the Inspector. (3) In the case of a system comprising electric supply lines having concentric cables, the external conductor shall be the one to be connected with earth.

(4) Where a supplier proposes to connect with earth an existing system for use at high or extra-high voltage which has not hitherto been so connected with earth he shall give not less than fourteen days' notice in writing together with particulars to the telegraph-authority of the proposed connection with earth.

(5) (a) Where the earthing lead and earth connection are used only in connection with earthing guards erected under high or extra-high voltage overhead lines where they cross a telecommunication line or a railway line, and where such lines are equipped with earth leakage relays of a type and setting approved by the Inspector, the resistance shall not exceed 25 ohms.

(b) Every earthing system belonging to either the supplier or the consumer shall be tested for its resistance to earth on a dry day during dry season not less than once a year. Records of such tests shall be maintained and shall be produced, if required before the Inspector or any officer appointed to assist him and authorised under sub-rule (2) of rule 4A.

(6) In so far as the provisions of rule 61 are consistent with the provisions of this rule, all connections with earth shall also comply with the provisions of that rule.

Rule-68; General conditions as to transformation and control of energy –

(1) Where energy at high or extra-high voltage is transformed, converted, regulated or otherwise controlled in sub-stations or switch-stations (including outdoor substations and out-door switch-stations) or in street boxes constructed underground, the following provisions shall have effect: – (a) Sub-stations and switch-stations shall preferably be erected above ground, but where necessarily constructed underground due provisions for ventilation and drainage shall be made and any space housing switchgear shall not be used for storage of any materials especially inflammable and combustible materials or refuse. (b) Outdoor sub-station is except pole type sub-stations and. outdoor switch – stations shall (unless the apparatus is completely enclosed in a metal covering connected with earth, the said apparatus also being connected with the system by armoured cables) be efficiently protected by fencing not less than 1.8 meters in height or other means so as to prevent access to the electric-supply lines and apparatus therein by an unauthorised person. (c) Underground street boxes (other than sub-stations) which contain transformers shall not contain switches or other apparatus, and switches, cut-outs or other apparatus required for controlling or other purposes shall be fixed in separate receptacles above ground wherever practicable.

(2) Where energy is transformed, suitable provisions shall be made either by connecting with earth a point of the system at the lower voltage or otherwise to guard against danger by reason of the said system becoming accidentally charged above its normal voltage by leakage from a contact with the system at the higher voltage.

Rule-69; Pole type sub-stations –

Where platform type construction is used for a pole type sub-station and sufficient space for a person to stand on the platform is provided a substantial hand rail shall be built around the said platform and if the hand rail is of metal, it shall be connected with earth; Provided that in the case of pole type sub-station on wooden supports and wooden platform the metal hand-rail shall not be connected with earth.

Rule-70; Condensers

Suitable provision shall be made for immediate and automatic discharge of every static condenser on disconnection of supply.

Rule-71; Additional provisions for supply to high voltage luminous tube sign installation –

(1) Any person who proposes to use or who is using energy for the purpose of operating a luminous tube sign installation, or who proposes to transform or who is transforming energy to a high voltage for any such purpose shall comply with the following conditions:

    (a) All live parts of the installation (including all apparatus and live conductors in the secondary circuit, but excluding tubes except in the neighborhoods of their terminals) shall be inaccessible to unauthorised persons and such parts shall be effectively screened.

    (b) Irrespective of the method of obtaining the voltage of the circuit which feeds the luminous discharge tube sign, no part of any conductor of such circuit shall be in metallic connection (except in respect of its connection with earth) with any conductor of the supply system or with the primary winding of the transformer.

    (c) All live parts of an exterior installation shall be so disposed as to protect them against the effects of the weather and such installation shall be so arranged and separated from the surroundings as to limit, as far as possible, the spreading of fire.

    (d) The secondary circuit shall be permanently earthed at the transformer and the core of every transformer shall be earthed.

    (e) Where the conductors of the primary circuit are not in metallic connection with the supply conductors, (e.g., where a motor-generator or a double-wound convertor is used), one phase of such primary circuit shall be permanently earthed at the motor generator or convertor, or at the transformer. An earth leakage circuit breaker of sufficient rating shall be provided on the low voltage side to detect the leakage in such luminous tube sign installations.

    (f) A final sub-circuit which forms the primary circuit of a fixed luminous discharge tube sign installation shall be reserved solely for such purpose.

- (g) A separate primary final sub-circuit shall be provided for each transformer or each group of transformers having an aggregate input not exceeding 1000 VAs, of a fixed luminous-discharge tube sign installation.
- (h) An interior installation shall be provided with suitable adjacent means for disconnecting all phases of the supply except the "neutral" in a three-phase four wire circuit.
- (i) For installations on the exterior of a building a suitable emergency fire-proof linked switch to operate on all phases except the neutral in a three phase four wire circuits shall be provided and fixed in a conspicuous position at not more than 2.75 meters above the ground.
- (j) A special "caution" notice shall be affixed in a conspicuous place on the door of every high voltage enclosure to the effect that the low voltage supply must be cut off before the enclosure is opened.
- (k) Where static condensers are used, they shall be installed on the load side of the fuses and the primary (low voltage) side of the transformers.

(2) Where static condensers are used on primary side, means shall be provided for automatically discharging the condensers when the supply is cut off; Provided that static condensers or any circuit interrupting devices on the high or extra-high voltage side shall not be used without the approval in writing of the Inspector.

(3) The owner or user of any luminous tube sign or similar high voltage installation shall not bring the same into use without giving notice to the Inspector not less than 14 days in writing of his intention so to do.

Rule-72; Additional provisions for supply to high voltage electrode boilers –

(1) Where a system having a point connected with earth is used for supply of energy at high or extra-high voltage to an electrode boiler which is also connected with earth, the following conditions shall apply: –
- (a) The metal work of the electrode boiler shall be efficiently connected to the metal sheathing and metallic armouring (if any) of the high voltage electric supply line whereby energy is supplied to the electrode boiler.

(b) The supply of energy at high or extra-high voltage to the electrode boiler shall be controlled by a suitable circuit-breaker so set as to operate in the event of the phase currents becoming unbalanced to the extent of 10 per cent of the rated current consumption of the electrode boiler under normal conditions of operation; Provided that if in any case a higher setting is essential to ensure stability of operation of the electrode boiler, the setting may be increased so as not to exceed 15 per cent of the rated current consumption of the electrode boiler under normal conditions of operation.

(c) An inverse time element device may be used in conjunction with the aforesaid circuit breaker to prevent the operation thereof unnecessarily on the occurrence of unbalanced phase currents of momentary or short duration.

(d) The supplier shall serve a notice in writing on the telegraph-authority at least seven days prior to the date on which such supply of energy is to be afforded specifying the location of every point (including the earth connection of the electrode boiler) at which the system is connected with earth.

(2) The owner or user of any high or extra-high voltage electrode boiler shall not bring the same into use without giving notice of not less than 14 days' to the Inspector in writing of his intention so to do.

Rule-73; Supply to X-ray and high frequency installation –

(1) Any person who proposes to employ or who is employing energy for the purpose of operating an X-ray or similar high-frequency installation, shall comply with the following conditions: –

(a) Mechanical barriers shall be provided to prevent too close an approach to any high-voltage parts of the X-ray apparatus, except the X-ray tube and its leads, unless such high-voltage parts have been rendered shock-proof by being shielded by earthed metal or adequate insulating material.

(b) Where extra-high voltage generators operating at 300 Peak KV or more are used, such generators shall be installed in rooms separate from those containing the other equipment and any step-up transformer employed shall be so installed and protected as to prevent danger.

(c) A suitable switch shall be provided to control the circuit supplying a generator, and shall be so arranged as to be open except while the door of the room housing the generator is locked from the outside.

(d) X-ray tubes used in therapy shall be mounted in an earthed metal enclosure.

(e) Every X-ray machine shall be provided with a millimeter or other suitable measuring instrument, readily visible from the control position and connected, if practicable, in the earthed lead, but guarded if connected in the high-voltage lead. Notwithstanding the provisions of clause (e), earth leakage circuit breaker of sufficient rating shall be provided on the low voltage side to detect the leakage in such X-ray installations.

(f) This sub-rule shall not apply to shock-proof portable units or shockproof self contained and stationary units. Note: – The expression "shock-proof", as applied to X-ray and high-frequency equipment, shall mean that such equipment is guarded with earthed metal so that no person may come into contact with any live part.

(2) (a) In the case of non-shock-proof equipment, overhead high-voltage conductors, unless suitably guarded against personal contact, shall be adequately spaced and high-voltage leads on tilting tables and fluoroscopes shall be adequately insulated or so surrounded by barriers as to prevent inadvertent contact.

(b) The low voltage circuit of the step up transformer shall contain a manually operated control device having overload protection, in addition to the over-current device for circuit protection, and these devices shall have no exposed live parts and for diagnostic work there shall be an additional switch in the said circuit, which shall be of one of the following types: (i) A switch with a spring or other mechanism that will open automatically except while held close by the operator, or (ii) A time switch which will open automatically after a definite period of time for which it has been set.

(c) If more than one piece of apparatus is operated from the same high or extra-high voltage source each shall be provided with a high or extra-high voltage switch to give independent control.

(d) Low frequency current-carrying parts of a machine of the quenched gap or open gap type shall be so insulated or guarded that they cannot be touched during operation, the high frequency circuit proper which delivers high-frequency current normally for the therapeutic purposes, being exempted.

(e) All X-ray generators having capacitors shall have suitable means for discharging the capacitors manually. (f) Except in the case of self-contained units, all 200 Peak KV or higher, X-ray generators shall have a sphere gap installed in the high-voltage system adjusted so that it will break down on over-voltage surges.

(3) (a) All non-current carrying metal parts of tube stands, fluoroscopes and other apparatus shall be properly earthed and insulating floors, mats or platforms shall be provided for operator in proximity to high or extra-high voltage parts unless such parts have been rendered shock proof.

(b) Where short wave therapy machines are used, the treatment tables and examining chairs shall be wholly non-metallic.

(4) The owner of any X-ray installations or similar high frequency apparatus shall not bring the same into use without giving notice to the Inspector not less than 14 days in writing of his intention to do so. Provided that the aforesaid notice shall not be necessary in the case of shockproof portable X-ray and high-frequency equipment which have been inspected before the commencement of their use and periodically, thereafter.

Chapter VIII; Overhead Lines, Under Ground Cables and Generating Stations:
Rule-74; Material and strength –

(1) All conductors of overhead lines other than those specified in sub-rule (1) of rule 86 shall have a breaking strength of not less than 350 kg.

(2) Where the voltage is low and the span is of less than 15 meters and is on the owner's or consumer's premises, a conductor having an actual breaking strength of not less than 150 kg may be used.

Rule-75; Joints –

Joints between conductors of overhead lines shall be mechanically and electrically secure under the conditions of operation. The ultimate strength of the joint shall not be less than 95 per cent of that of the conductor, and the electrical conductivity not less than that of the conductor. Provided, that no conductor of an overhead line shall have more than two joints in a span.

Rule-76; Maximum stresses: Factors of safety –

(1) (a) The owner of every overhead line shall ensure that it has the following minimum factors of safety: –
   (i) For metal supports: 1.5,
   (ii) For mechanically processed concrete supports: 2.0,
   (iii) For hand-moulded concrete supports: 2.5, and (iv) for wood supports: 3.0.

The minimum factors of safety shall be based on such load as would cause failure of the support to perform its function (assuming that the foundation and other components of the structure are intact). The aforesaid load shall be-(i) Equivalent to the yield point stress or the modulus of rupture, as the case may be, for supports subject to bending and vertical loads, (ii) The crippling load for supports used struts. The said owner shall also ensure that the strength of the supports in the direction of the line is not less than one-fourth of the strength required in the direction transverse to the line. Provided that in the case of latticed steel or other compound structures, factors of safety shall not be less than 1.5 under such broken wire conditions as may be specified by the State Government in this behalf.

(b) The minimum factor of safety for stay-wires, guard-wires or bearer wires shall be 2.5 based on the ultimate tensile strength of the wire.

(c) The minimum factor of safety for conductors shall be 2, based on their ultimate tensile strength. In addition, the conductors' tension at 32°C, without external load, shall not exceed the following percentages of the ultimate tensile strength of the conductor: –
Initial unloaded tension 35 per cent

Final unloaded tension 25 per cent

Provided that in the case of conductors having a cross section of a generally triangular shape, such as conductors composed of 3-wires, the final unloaded tension at 32°C shall not exceed 30 per cent of the ultimate tensile strength of such conductor.

(2) For the purpose of calculating the factors of safety prescribed in sub-rule (1) –
- (a) The maximum wind pressure shall be such as the State Government may specify in each case;
- (b) For cylindrical bodies the effective area shall be taken as two-thirds of the projected area exposed to wind pressure;
- (c) For latticed steel or other compound structures the wind pressure on the lee side members shall be taken as one-half of the wind pressure on the windward side members and the factors of safety shall be calculated on the crippling load of struts and upon the elastic limit of tension members;
- (d) The maximum and minimum temperatures shall be such as the State Government may specify in each case.

(3) Notwithstanding anything contained in sub-rules (1) and (2), in localities where overhead lines are liable to accumulations of ice or snow the State Government may, by order in writing, specify the loading conditions for the purpose of calculating the factor of safety.

Rule-77; Clearance above ground of the lowest conductor –

(1) No conductor of an overhead line, including service lines, erected across a street shall at any part thereof be at a height of less than –
- (a) For low and medium voltage lines 5.8 meters
- (b) For high voltage lines 6.1 meters

(2) No conductor of an overhead line, including service lines, erected along any street shall at any part thereof be at a height less than –
- (a) For low and medium voltage lines 5.5 meters
- (b) For high voltage lines 5.8 meters

(3) No conductor of in overhead line including service lines, erected elsewhere than along or across any street shall be at a height less than –

(a) For low, medium and high voltages lines upto and including 11,000 volts, if bare ———— 4.6 meters
(b) For low, medium and high voltage lines upto and including 11,000 volts, if insulated ———— 4.0 meters
(c) For high voltage lines above 11,000 volts ———— 5.2 meters

(4) For extra-high voltage lines the clearance above ground shall not be less than 5.2 meters plus 0.3 meter for every 33,000 volts or part thereof by which the voltage of the line exceeds 33,000 volts.

Provided that the minimum clearance along or across any street shall not be less than 6.1 meters.

Rule-78; Clearance between conductors and trolley wires –

(1) No conductor of an overhead line crossing a tramway or trolley bus route using trolley wires shall have less than the following clearances above and trolley wire –
   (a) Low and medium voltage lines ————-; 1.2 meters
       Provided that where an insulated conductor suspended from a bearer wire crosses over a trolley wire the minimum clearance for such insulated conductor shall be 0.6 meter.
   (b) High voltage lines up to and including 11,000 volts; 1.8 meters
   (c) High voltage lines above 11,000 volts; 2.5 meters
   (d) Extra-high voltage lines; 3.0 meters

(2) In any case of a crossing referred to in sub-rule (1), whoever lays his line later in time, shall provide the clearance between his own line and the line which will be crossed in accordance with the provisions of said sub-rule.

Provided that if the later entrant is the owner of the lower line and is not able to provide adequate clearance, he should bear the cost of modification of the upper line so as to comply with this rule.

Rule-79; Clearances from buildings of low and medium voltage lines and service lines –

(1) Where a low or medium voltage, overhead line passes above or adjacent to or terminates on any building, the following minimum clearances from any accessible point, on the basis of maximum sag, shall be observed: –
   (a) For any flat roof, open balcony, verandah roof and lean-to-roof –

(i) When the line passes above the building a vertical clearance of 2.5 meters from the highest point, and

(ii) When the line passes adjacent to the building a horizontal clearance of 1.2 meters from the nearest point, and

(b) For pitched roof-(i) When the line passes above the building a vertical clearance of 2.5 meters immediately under the lines, and

(ii) When the line passes adjacent to the building a horizontal clearance of 1.2 meters.

(2) Any conductor so situated as to have a clearance less than that specified in sub-rule (1) shall be adequately insulated and shall be attached at suitable intervals to a bare earthed bearer wire having a breaking strength of not less than 350 kg.

(3) The horizontal clearance shall be measured when the line is at a maximum deflection from the vertical due to wind pressure.

Explanation – For the purpose of this rule, expression "building" shall be deemed to include any structure, whether permanent or temporary

Rule-80; Clearances from buildings of high and extra-high voltage lines –

(1) Where a high or extra-high voltage overhead line passes above or adjacent to any building or part of a building it shall have on the basis of maximum sag a vertical clearance above the highest part of the building immediately under such line, of not less than –

(a) For high voltage lines upto and including 33,000 volts ––– - 3.7 meters

(b) For extra-high voltage lines —-3.7 meters plus 0.30 meter for every additional 33,000 volts or part thereof.

(2) The horizontal clearance between the nearest conductor and any part of such building shall, on the basis of maximum deflection due to wind pressure, be not less than –

(a) For high voltage lines upto and including 11,000 volts——— 1.2 meters

(b) For high voltage lines above 11,000 volts and up to and including 33,000 volts ——-2.0 meters

(c) For extra-high voltage lines—— 2.0 meters plus 0.3 meter for every additional 33,000 volts for part thereof.

Rule-81; Conductors at different voltages on same supports –
Where conductors forming parts of systems at different voltages are erected on the same supports, the owner shall make adequate provision to guard against danger to linesman and others from the lower voltage system being charged above its normal working voltage by leakage from or contact with the higher voltage system and the methods of construction and the clearances between the conductors of the two systems shall be subject to the prior approval of the Inspector.

Rule-82; Erection of or alternation to buildings, structures, flood banks and elevation of roads –
(1) If at any time subsequent to the erection of an overhead line (whether covered with insulating material or bare), any person proposes to erect a new building or structure or flood bank or to raise any road level or to carry out any other type of work whether permanent or temporary or to make in or upon any building or structure of flood bank or road, any permanent or temporary addition or alternation, he and the contractor whom he employs to carry out the erection, addition or alteration, shall if such work, building, structure, flood bank, road or additions and alterations thereto, would, during or after the construction result in contravention of any of the provisions of rule 77, 79 or 80, give notice in writing of his intention to the supplier and to the Inspector and shall furnish therewith a scale drawing showing the proposed building, structure, flood bank road, any addition or alteration and scaffolding required during the construction.
(2) (a) On receipt of the notice referred to in sub-rule (1) or otherwise, the supplier shall examine whether the line under reference was lawfully laid and whether the person was liable to pay the cost of alteration and if so, send a notice without undue delay, to such person together with an estimate of the cost of the expenditure likely to be incurred to so after the overhead line and require him to deposit, within 30 days of the receipt of the notice with the supplier, the amount of the estimated cost.
(b) If the person referred to in sub-rule (1) disputes the suppliers estimated cost of alteration of the overhead line or even the responsibility to pay such cost the dispute may be referred to the

Inspector by either of the parties whereupon the same shall be decided by the Inspector.

(3) No work upon such building, structure, flood bank, road and addition or alternation thereto shall be commenced or continued until the Inspector has certified that the provisions of rule 77, 79 or 80 are not likely to be contravened either during or after the aforesaid construction. Provided that the Inspector may, if he is satisfied that the overhead line has been so guarded as to secure the protection of persons or property from injury, or risk of injury, permit the work to be executed prior to the alteration of the overhead line or in the case of temporary addition or alteration, without alteration of the overhead line.

(4) On receipt of the deposit, the supplier shall after the overhead line within one month of the date of deposit or within such longer period as the Inspector may allow and ensure that it shall not contravene the provisions of rule 77, 79 or 80 either during or after such construction.

(5) In the absence of an agreement to the contrary between the parties concerned, the cost of such alteration of the overhead line laid down shall be estimated on the following basis, namely: –
   (a) The cost of additional material used on the alteration giving due credit for the depreciated cost of the material which would be available from the existing line;
   (b) The wages of Labor employed in affecting the alteration;
   (c) Supervision charges to the extent of 15 per cent of the wages mentioned in clause (b); and
   (d) Any charges incurred by the supplier in complying with the provisions of section 16 of the Act in respect of such alterations.

(6) Where the estimated cost of the alteration of the overhead line is not deposited the supplier shall be considered as an aggrieved party for the purpose of this rule.

Rule-82A; Transporting and Storing of material near overhead lines –

(1) No rods, pipes or similar materials shall be taken below or in the vicinity of any bare overhead conductors or lines if they are likely to infringe the provisions for clearances under rules 79 and 80, unless such materials are transported under the direct supervision of a competent person authorised in this behalf by the owner of such overhead conductors or lines.

(2) Under no circumstances rods, pipes or other similar materials shall be brought within the flash over distance of bare live conductors or lines; and
(3) No material or earth work or agricultural produce shall be dumped or stored or trees grown below or in the vicinity of bare overhead conductor's lines so as to reduce the requisite safety clearances specified under rules 79 and 80.

Rule-83; Clearances: General-
For the purpose of computing the vertical clearance of an overhead line, the maximum sag of any conductor shall be calculated on the basis of the maximum sag in still air and the maximum temperature as specified by the State Government under rule 76(2) (d). Similarly, for the purpose of computing any horizontal clearance of an overhead line the maximum deflection of any conductor shall be calculated on the basis of the wind pressure specified by the State Government under rule 76(2)(a) or may be taken as 350, whichever is greater.

Rule-84; Routes: Proximity to aerodromes –
Overhead lines shall not be erected in the vicinity of aerodromes until the aerodrome authorities have approved in writing the route of the proposed lines.

Rule-85; Maximum interval between supports –
All conductors shall be attached to supports at intervals not exceeding the safe limits based on the ultimate tensile strength of the conductor and the factor of safety prescribed in rule 76.

Provided that in the case of overhead lines carrying low or medium voltage conductors, when erected in, over, along or across any street, the interval shall not, without the consent in writing of the Inspector, exceed 65 meters.

Rule-86; Conditions to apply where telecommunication lines and power lines are carried on same supports –
(1) Every overhead telecommunication line erected on supports carrying a power line shall consist of conductors each having a breaking strength of not less than 270 kg.
(2) Every telephone used on a telecommunication line erected on supports carrying a power line shall be suitably guarded against lightning and shall be protected by cutouts.

(3) Where a telecommunication line is erected on supports carrying a high or extra-high voltage power line arrangement shall be made to safeguard any person using the telephone against injury resulting from contact, leakage or induction between such power and telecommunication lines.

Rule-87; Lines crossing or approaching each other –
(1) Where an overhead line crosses or is in proximity to any telecommunication line, either the owner of the overhead line or the telecommunication line, whoever lays his line later, shall arrange to provide for protective devices or guarding arrangements, in a manner laid down in the Code of Practice or the guidelines prepared by the Power and Telecommunication Coordination Committee and subject to the provisions of the following sub-rules: –
(2) When it is intended to erect a telecommunication line or an overhead line which will cross or be in proximity to an overhead line or a telecommunication line, as the case may be, the person proposing to erect such line shall give one month's notice of his intention so to do along with the relevant details of protection and drawings to the owner of the existing line.
(3) Where an overhead line crosses or is in proximity to another overhead line, guarding arrangements shall be provided so as to guard against the possibility of their coming into contact with each other. Where an overhead line crosses another overhead line, clearances shall be as under: –

S. No	Nominal System Voltage	11-66 KV	110-132 KV	220 KV	400KV	800 KV
1	Low & Medium	2.44	3.05	4.58	5.49	7.94
2	11-66 KV	2.44	3.05	4.58	5.49	7.94
3	110-132 KV	3.05	3.05	4.58	5.49	7.94
4	220 KV	4.58	4.58	4.58	5.49	7.94
5	400 KV	5.49	5.49	5.49	5.49	7.94
6	800 KV	7.94	7.94	7.94	7.94	7.94

Table: Minimum clearances in meters between lines crossing each other

Provided that no guarding are required when an extra high voltage line crosses over another extra-high voltage, high voltage, medium or low voltage line or a road or a tram subject to the condition that adequate clearances are provided between the lowest conductor of the extra-high voltage line and the top most conductor of the overhead line crossing underneath the extra-high voltage line and the clearances as stipulated in rule 77 from the topmost surface of the road is maintained.

(4) A person erecting or proposing to erect a line which may cross or be in proximity with an existing line, may normally provide guarding arrangements on his own line or require the owner of the other overhead line to provide guarding arrangements as referred to in sub-rule, (3).

(5) In all cases referred to in the preceding sub-rules the expenses of providing the guarding arrangements or protective devices shall be borne by the person whose line was last erected.

(6) Where two lines cross, the crossing shall be made as nearly at right angles the nature of the case admits and as near the support of the line as practicable, and the support of the lower line shall not be erected below the upper line.

(7) The guarding arrangements shall ordinarily be carried out by the owner of the supports on which it is made and he shall be responsible for its efficient Maintenance.

(8) All work required to be done by or under this rule shall be carried out to the satisfaction of the Inspector.

Rule-88; Guarding –

(1) Where guarding is required under these rules the provisions of sub-rules (2) to (4) shall apply.

(2) Every guard-wire shall be connected with earth at each point at which its electrical continuity is broken. (3) Every guard-wire shall have an actual breaking strength of not less than 635 kg and if made of iron or steel, shall be galvanised.

(4) Lines crossing trolley-wires – In the case of a crossing over a trolley-wire the guarding shall fulfill the following conditions, namely: –
   (a) Where there is only one trolley-wire, two guard-wires shall be erected as in diagram A;

(b) Where there are two trolley-wires and the distance between them does not exceed 40 cms, two guard-wires shall be erected as in diagram B;

(c) Where there are two trolley wires and the distance between them exceeds 40 cms but does not exceed 1.2 meters, three guard-wires shall be erected as in diagram C;

(d) Where there are two trolleys—wires and the distance between them exceeds 1.2 meters, each trolley-wire shall be separately guarded as in diagram D;

(e) The rise of the trolley boom shall be so limited that the trolley leaves the trolley-wire, it shall not foul the guard-wires; and

(f) Where a telegraph-line is liable to fall or be blown down upon an arm, stay-wire or span-wire and so slide down upon a trolley-wire, guard hooks shall be provided to prevent such sliding.

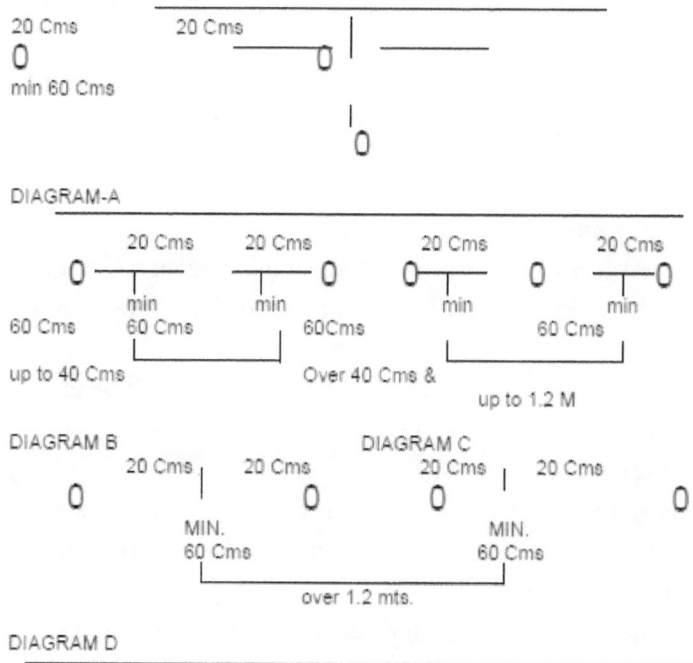

Rule-89; Service-lines from Overhead lines –

No Service-line or tapping shall be taken off an overhead line except at a point of support.

Provided, that the number of tapings per conductor shall not be more than four in case of low and medium voltage connections.

Rule-90; Earthing –

(1) All metal supports and all reinforced and prestressed cement concrete supports of overhead lines and metallic fittings attached thereto shall be permanently and efficiently earthed. For this purpose a continuous earth wire shall be provided and securely fastened to each pole and connected with earth ordinarily at three points in every km, the spacing between the points being as nearly equidistance as possible. Alternatively, each support and the metallic fitting attached thereto shall be efficiently earthed.

(1-A) Metallic bearer wire used for supporting insulated wire of low and medium voltage overhead service lines shall be efficiently earthed or insulated.

(2) Each stay-wire shall be similarly earthed unless insulator has been placed in it at a height not less than 3.0 meters from the ground.

Rule-91; Safety and protective devices –

(1) Every overhead line, (not being suspended from a dead bearer wire and not being covered with insulating material and not being a trolley-wire) erected over any part of street or other public place or in any factory or mine or on any consumers' premises shall be protected with a device approved by the Inspector for rendering the line electrically harmless in case it breaks.

(2) An Inspector may by notice in writing require the owner of any such overhead line wherever it may be erected to protect it in the manner specified in sub-rule (l).

(3) The owner of every high and extra-high voltage overhead line shall make adequate arrangements to the satisfaction of the Inspector to prevent unauthorised persons from ascending any of the supports of such overhead lines which can be easily climbed upon without the help of a ladder or special appliances. Rails, reinforced cement concrete

poles and pre-stressed cement concrete poles without steps, tubular poles, wooden supports without steps, I – sections and channels shall be deemed as supports which cannot be easily climbed upon for the purpose of this rule.

Rule-92; Protection against lightning –

(1) The owner of every overhead line, sub-station or generating station which is so exposed as to be liable to injury from lightning shall adopt efficient means for diverting to earth any electrical surges due to lightning.

(2) The earthing lead for any lightning arrestor shall not pass through any iron or steel pipe, but shall be taken as directly as possible from the lightning – arrestor to a separate earth electrode and/or junction of the earth mat already provided for the high and extra-high voltage sub-station subject to the avoidance of bends wherever practicable.

Note – A vertical ground electrode shall be connected to this junction of the earth mat.

Rule-93; Unused overhead lines –

(1) Where an overhead line ceases to be used as an electric supply line, the owner shall maintain it in a safe mechanical condition in accordance with rule 76 or shall remove it.

(2) Where any overhead line ceases to be used as an electric supply line, an Inspector may, by a notice in writing served on the owner, require him to maintain it in a safe mechanical condition or to remove it within fifteen days of the receipt of the notice.

References

1. Handbook for Radio Engineering Managers – Chapter – 37-pp 548 to 552 –By J.F. Ross
2. General Electrical Safety Requirements–Section 12, October 2009; Complying with applicable sections of the Occupational Safety and Health Administration (OSHA), National Fire Protection Association (NFPA), National Electrical Code, National Electrical Safety Code, and State adopted electrical codes.

3. Electrical Safety Program – Adopted January 2014 – Purdue University
4. THE INDIAN ELECTRICITY RULES, 1956
5. STUDY GUIDE ELECTRICAL SAFETY HAZARDS AWARENESS – EFCOG ELECTRICAL SAFETY IMPROVEMENT PROJECT
6. Occupational Safety & Health Administration (OSHA) Standards 29 CFR 1910.331 – 1910.335
7. NFPA® 70E, "Electrical Safety Requirements for Employee Workplaces" 2000 Edition
8. OSHA, Electrical Safety Related Work Practices, 29 CFR –Part 1910

CHAPTER 6

EARTHING AND GROUNDING SYSTEMS

6.1 General

Proper earthing techniques are necessary and essential for the safety, operation and correct performance of the equipments as well as the protection of the human beings operating such gadgets. The integrity of grounding and proper operation of the equipment depend on proper bonding of the earth electrode system, proper system grounding of service equipment and separately derived sources and proper equipment grounding for operational frequencies (DC or AC-mains) as well as higher RF frequencies. It is recommended that the earthing design and installation be compliant to all applicable codes and standards. Since grounding or earthing is not designed as an active component of the functioning of any electrical or Electronics devices or systems; so this path must be free of any operational current. However a functional earth always carries the current for the normal operation of a device such as surge suppressor and electromagnetic interference (EMI) filters. Also generally a Protective Earth (PE) is used as a functional earth.

Once the grounding complies with the safety rules, the ground connections have to be improved to obtain a good electrical performance of the system. It is important to keep in mind that for operational reasons one should imagine that at low frequencies the ground system is a kind of low resistive divider in which all noise currents usually flows everywhere. However at high frequency the ground impedances start increasing mainly due to the inductive effect, following the noise only through the lowest impedance path. Therefore, the ground connections should be tackled in two steps. The first step is oriented to low frequency currents where it is important to avoid ground loops because once these currents are established they can flow everywhere, decreasing the performance of the system. The second step is focused on the noise at high frequency, where one could imagine that everything in the system is connected (through real or parasitic impedances) designing the ground path and connections in a specific way that gives a low impedance path to HF noise currents and thus avoiding the flow of these current.

6.2 What is Earthing or Grounding?

Process of connecting equipments, machines, circuits, devices, instruments, metallic enclosures or cabinets or a system for their proper operation or functioning etc to a good ground or earth through a system of earth electrode to extend zero potential is called Earthing or Grounding. Grounding is one of the primary mean to minimise unwanted noise and pick-up and ensure the safe operation and proper electrical performance of the equipment.

6.2.1 Purpose of Earthing or Grounding

Safety of the equipment and operating personnel, system protection and proper Electrical performance are the main reasons to extend earth connection to a system or equipment. However not all electronic equipments needs to be connected to earth for functioning, satellites are such an example. Sometimes wrong grounding configurations, oriented to satisfy the special power and performance requirements of electronic loading equipment, can compromise safety rules generating dangerous situations for operating personnel and equipment. Personnel and equipment safety & performance grounding issues have to be analyzed together.

The use of electronic equipment /ups/computers / CFL lamps / Speed drives etc has increased the possibility of development of voltage difference in the neutral of the power supply. As you know these devices operate at different cut off voltages in the sine wave cycle and hence the sine wave is distorted resulting in the higher potential difference (PD) in the Neutral and this causes the flow of currents in the Neutral Circuit which results in the damage of PCB cards and other sensitive components in the electronic circuit/equipments. Also Industrial loads are in general "Not Balanced Loads" and therefore the line voltages are spilled over to Neutral which results into stray currents thereby damaging the electronic equipments / PCB cards. So to maintain the ZERO PD (potential difference) between the Neutral and earth, and to ground the Spill over voltages, the Neutral is grounded. It is therefore essential to connect Neutral to Earth by Solid permanent contact so that there is no possibility of loose connection at any stage of operation to prevent any electrical mishaps.

The earthing of radio installations plays an important role as regards to the electrical performance of the broadcasting and recording equipments by providing reference (zero) ground. Secondly it plays another very important function of safety & protection of the equipments during fault from heavy currents and burnout as well as damage from lightning and static. Thirdly the earthing of radio installations plays another important role as regards to the safety and protection of the operating & maintenance personnel from electric shocks and burns.

Metal parts of the equipment enclosures, racks, raceways and equipment grounding conductors susceptible of being energized by electrical currents (due to circuit faults, RF induction, electrostatic discharge, and lightning), must be effectively grounded to ensure safety of operating personnel, fire hazard reduction, equipment protection and its electrical performance. Grounding these metallic objects facilitates the operation of over-current protective devices during ground faults and permits return current from EMI filters and surge protective devices, connecting line to ground or line to chassis, to flow in proper sequence. All metallic conduits and raceways in areas containing electronic load equipment have to be carefully bonded to form an electrically continuous conductor.

Summing up the purpose of Earthing;

(a) Safety for Operator and Equipments
1) To save human life from danger of electric shock or burns or death; 2) To protect building, machinery & appliances under fault conditions; 3) To ensure all exposed conductive parts do not reach dangerous potential; 4) To provide safe path to dissipate lightning and short circuit currents, and 5) To provide stable reference for the operation of sensitive electronic equipments

(b) Over Voltage Protection
1) To provide an alternate path to the lightning strikes, line surges or unintentional dangerously high voltage induced in the systems to minimize

the damage to the equipments and the system; 2) To provide lowest impedance path to the static charge induced on a broadcast tower to safe guard its insulators and fixtures, and 4) To provide lowest impedance path to RF pickups on feeder lines of a MF band transmitters.

(c) Voltage Stabilisation

1) To bring all sources of current in equipments, circuits, transformers, Generators etc to a common reference point to facilitate establish their relationship to each other; 2) Earth being the most omnipresent conductive surface is used as universal Earthing Standard since beginning of electrical operations.

(d) Zero Reference or Neutral Grounding

1) All Neutral should have Zero Potential Difference to avoid shock; 2) To establish equal reference for all measurements and control signal; 3) To ensure prevention of harmonics currents and overheating in the conductor, and 4) To avoid spill over or transfer of load unbalance.

(e) To increase conductivity of Ground

1) To increase the conductivity of the soil below a MW self radiating tower, a system of earth radials consisting of radially buried copper wires at an interval of 3 degree and for collecting and bringing back all the RF current induced in the soil to the base of the tower for return path of the RF circuit, 2) To reduce the ground losses for returning the RF current induced below MF band feeder line to RF PA of transmitter,3) To provide low impedance path to lightning strikes and static charges induced on the broadcast tower, and 4) To provide ground potential to charge accumulated on guy wires of a broadcast tower.

6.3 Classifications of Earthing on Generic Basis

Earthing plays very important role in the functioning, performance, safety of operating personnel and also the safety of the equipments of Radio transmitters, big electric plants, and machines. Various types of earthings are:

(a) Signal and Safety Earthing

Signal earthings provides same voltage reference for different parts of the signal system to compare and function. Proper signal grounding will minimize conductive, electromagnetic, magnetic (inductive) or electric (capacitive)

Earthing and Grounding Systems

noise coupling on the signals. Safety earthing eliminates voltage differences between surfaces that may become energized and therefore present a shock hazards to the operating personnel. Also safety grounding protects the equipment by providing a low impedance path to the ground to power surges due to the lightning strikes or in case of faults short circuits or over current.

(b) Safety and Earth Grounding

Safety and Earth Grounding difference is shown in Fig 6.1. In order to remove dangerous voltage due to a ground fault, an over current protection device needs to open quickly. In order to open a fuse or a breaker quickly, the ground connection must have an impedance low enough to permit ground fault current to reach a level several times larger than the rating of over-current protection devices.

Safety ground connection as in Fig 6.1(a) will offer low impedance ground connection. Earth ground provides a connection to the ground of sufficiently low impedance to prevent the destruction of the equipment or electric shock which can occur due to superimposed voltages from lightning strikes. Earth connection however does not have low enough impedance to draw large currents needed to open the circuit breaker. Safety grounding as described above is not required for metal parts of equipment that operates at 50V or less.

(c) Signal Return and Ground

Both signal and safety grounding conductors do not carry any current during the normal operating conditions. This is what distinguishes them from the signal and power return conductors. Fig 6.2 demonstrates the difference as well as the different symbols used.

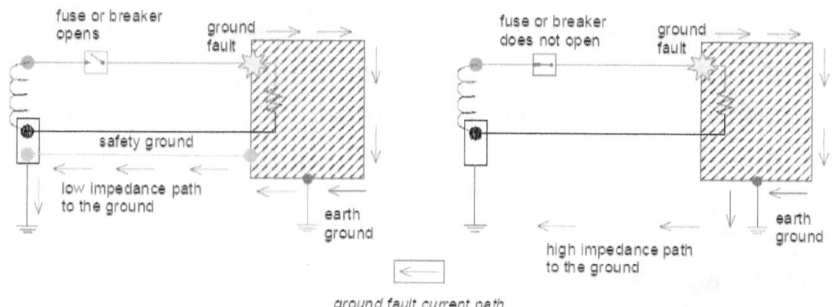

Fig 6.1: Safety and Earth Grounding

The signal return path carries signal current back to its source. The signal ground connection carries no current under normal operation. It provides a common voltage reference point for different circuits inside the enclosure. The neutral line provides a return path for the AC current. Safety ground line carries no current under normal operation. It provides a low impedance path to the ground in case the "hot" line is shorted to the Chassis. This would cause large current to flow through the safety ground line causing the Ground Fault Circuit to trip.

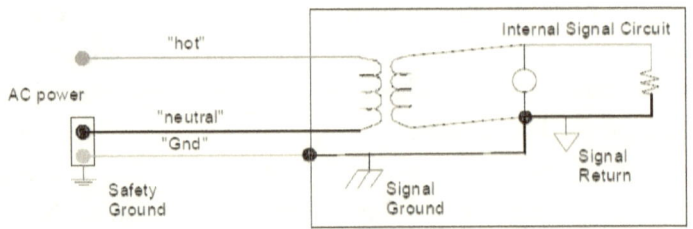

Fig 6.2: Differences between Return and Ground Wires

6.3.1 Single Point versus Multi Point Grounding

It is important to distinguish when single point and when multipoint grounding needs to be used. Fig 6.2 is an example where single point grounding should be used. It is essential to have the Neutral and Ground wires connected at only one point. That point is the Neutral Tie Block. If this connection was made at more than one point then there would be a possibility of current flowing through the Ground wire. As already mentioned there must not be any current flowing through the grounding wires under normal operating conditions.

The same goes true for the Signal Return connection to the Chassis of the instrument. The connection should be made at only one point to prevent any signal currents flowing through the Chassis. In general single point grounding is beneficial if the grounding structure is large in area and there are possibilities of potential differences between different parts of the grounding structure. If this is the case then there may be a current flowing through the grounding structure which may enter as noise through multiple grounding connections of the grounded instruments. Single point grounding would prevent this from happening.

Multipoint grounding is usually used at the remote sites when connecting Nanometrics instruments to the earth ground. Multiple connections have the advantage of reducing the potential differences between the grounded instruments and structures. In addition single point grounding must not be used if there is a distance of over 15m between the two pieces of the equipment. In this case there is going to be resistance in the wires due to the distance and a possibility a current flow in the grounding structure (which is what we are trying to avoid). In this case each instrument would have to be grounded individually using a multipoint grounding scheme.

6.3.2 Ground Loops

One of the greatest misconceptions concerning earthing is ground loops. Ground loops, although a frequent cause of noise, are blamed more for interference problems than they actually cause. Figure 6.3 demonstrates a difference between an actual ground loop and a setup that is often mistaken for a ground loop.

A ground loop is a conducting path that consists partially of the signal conductor and partially of the grounding structure as shown in Fig 6.3(a). The ground loop has to have these two parts in order to bring noise into the system. In this example, the signal loop is grounded at both the source and the load ends. This is usually the case with single ended equipment. The duplicate ground path forms the equivalent of a loop antenna which very easily picks up interference currents. The best way to eliminate this ground loop is to disconnect one of the Chassis connections such as at the point marked as #1 in Fig 6.3 (a).

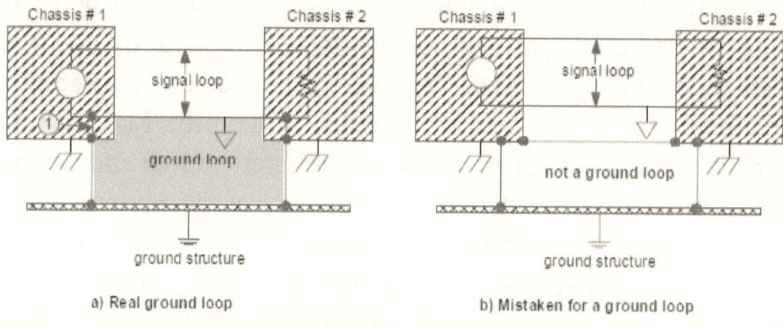

Fig 6.3: Ground Loop Misconceptions

It is important not to remove any of the connections from the Chassis to Ground since that would mean that one of the enclosures would be grounded through the signal return conductor (a bad practice).

Fig 6.3 (b) shows a system setup which is often mistaken for a ground loop. In this case the loop consists of only the grounding wires which under normal operation carry no current. Since no part of this loop is a signal carrying conductor, there is not a ground loop.

6.4 Classification of Earthing on Function basis

All mechanical equipments in the electronic equipment areas should be carefully bonded for electrical safety and noise current control to local building steel using direct grounding and bonding means. When located in the same area as the electronic load equipment, mechanical equipment should be bonded at multiple points to the same ground references as the electronic load equipment. Heating, ventilation, air conditioning, process cooling equipment, related metal piping and electrical conduits are recommended to be bonded to the same ground reference serving the electronic load equipment.

The earthing for radio installations is needed for functioning of the system also. A very elaborate system of earthing is needed for functioning of MW Radio Antennas. Effective and elaborate earthing systems are needed for radio stations especially for transmitters, and it is quite different from other installations.

(a) Equipment or Protective Earthing

A protective earth (PE) ensures all exposed conductive surfaces are kept at earth potential to avoid risk of shock to operator. So Earthing of non-current carrying exposed metal part and conductor is essential for the safety of the operator by eliminating risk of shocks from the voltage induced in their metallic body due to pickup while working in RF field. Main features of equipment earthing system are:(1) To ensure safety of staff and equipment against earth fault; (2) To stabilize the potential of the equipment; (3) To prevent arching between metal parts and earth due to voltage pickup in high RF fields; (4) To ensure operating staff protection from electric shock hazard due to accumulation of static charge on the body of the equipment; (5) To obtain satisfactory electrical performance from the equipment w.r.t common ground; (6) To have common zero-voltage reference point for all the equipment working in a system; (7) To prevent radiation of spurious and harmonics frequencies in metal structures due to poor joints; (8) To provide return path for earth return circuits, and (9) To provide protection against lightening by providing grounding path to lightning currents.

Therefore earthing of a Radio stations is designed to ensure that all apparatus / chassis other than live parts are at zero potential (earth or ground) at all the times.

(b) System Earthing
In circuits where current carrying conductor associated with the safety of equipment or circuits is connected to ground, then it is called System Earthing. This type of earthing involves the ground connection of power services and separately derived systems. The main features of system earthing are; (1) It is used for generator, transformers, uninterruptible power systems (UPS), RF cables, and Microwave towers. The system earthing is the intentional connection of a circuit conductor (typically the neutral on a three phase, four wire system) to earth; (2) the purpose of the system grounding is for electrical safety of personnel and equipment as well as fire safety reasons; (3) Safety is basically governed by the electrical codes and standards as adopted by government agencies and commercial entities; (4) System grounding also impacts the performance of electronic load equipment for reasons related to the control of the common-mode noise and fault currents, however the personnel and the equipment protection is the primary task of the grounding, and (5) The grounding of power systems is from a safety point of view to limit the potential difference between grounded objects, to provide a good operation of over-current protective devices in case of ground fault, to stabilize the phase voltages with reference to ground and to limit transient voltages due to lightning and load switching.

(c) Reference Earthing
Earthing done in the reference circuit for comparing the amplitude of signal generated by over current or overvoltage fault or leakage current of various fault detecting devices to cut off the input power supply source to the device by tripping the incoming Circuit breaker is called Reference Earthing. Its main features are; (1) It is used in the control system of all transmitter plants

individual equipments, auxiliaries as well as ancillary devices, machines for functional controlling, supervising, monitoring system for protecting them from destruction in case of faults; (2) The reference earth is directly connected to main bus or to the earth stud or terminal provided for this purpose in the cabinet; (3) It is implemented by connecting a highly stable very low value resistance (less than an ohm) in the return path of the circuit and ground terminal, and (4) The return common path of the circuit is at lowest potential in the device.

(d) Discharge Earthing
Earth extended for discharge of capacitors, inductors in High Voltage rectifiers circuits, RF tuned tank coils and capacitors of radio transmitters' antenna tuning units, to limit heavy inrush currents during charging and discharging, as well as while switching off for repairs and in case of fault including static build up and lightning is called Discharge Earthing. The main features of Discharge Earthing are; (1) The earthing connection is directly extended to return side of the component from earth buss; (2) There is no sub connection from this to any other component; (3) This connection is extended with a heavy duty copper band or strap or braided wire ribbon capable of taking fault currents without deterioration and fatigue of the metal; (4) The connection up to main buss is run on insulated supports, and (5) To ensure the contact at both ends remains always firm use spring washers.

(e) Earth Radials & Feeder Earth system
This is one type of system earthing where earthing circuit is part of return path. Earthing System in case of MW self radiating tower radiator, is provided for making the earth below it conducting for improving its efficiency by reducing the ground losses as well as to provide return path to radiated RF. It's very elaborate system requiring great attention to details while executing in the field. Main features of such earthing system are; (1) It provides return path to RF currents to antenna through space; (2) Two copper wire conductors are laid below MW feeder line to collect the induced current below it to improve the efficiency of the system, and (3) The above mentioned two wires are then connected to ground terminal of feeder outlet from transmitter for completing RF ground return from antenna to PA amplifier of the transmitter.

(Note: For more details of a typical practical earth radial system, screening and RF earth schematic & sectional view between transmitter and tower, readers may refer to chapter-12, section 12.4.4 fig 12.5 & 12.6 of my book, "AM Radio Tower Antennas" page 350 to 351)

6.5 Characteristics of Grounding Structures and Materials

While choosing grounding materials or structures there are several factors to consider. The grounding materials obviously have to be an excellent conductor. It also has to have a "large" surface area compared to the system that is being grounded. Finally the grounding structure has to be close to the system and not more than 15m away. The actual distance depends on the type and size of the grounding wire. If the distance is larger than 10-15m, then due to the resistance of the grounding wire, there could be a potential difference between the equipment enclosures and the ground. This would defeat the purpose of the grounding connection.

Another important characteristic of the grounding is that it should not be part of any signal current carrying path. Fundamentally any signal current must not flow through a grounding structure under normal system operation.

Acceptable earth grounding electrodes include: a metal underground water pipe in contact with the earth for no less than 10 feet, the metal frame of the building, a minimum 2 AWG (7 mm) bare copper ground ring encircling the building and a ground rod of at least 10 feet in length with no less than 8 feet in contact with the soil. Building sites will often have several of the above mentioned grounding electrodes. It is important that if more electrodes are present they all be bonded together to form a single reference electrode grounding system. A minimum 6 AWG (4.875 mm) wire should be used for interconnection purpose.

Note; In case of Radio transmitter installations, where strong RF field is involved, instead of wires, copper bands or braided strips of size 100x1 or 1.5 mm, 75x2 or 1.5 mm, 50x1.5 or 2 mm are used for required performance. Equivalent area cross sectional GI strips can also be used but there may not be desired results. However for Power supply and Audio systems earthing wires or G I strip is considered more than adequate.

6.5.1 Conductivity of Different Types of Soils

As mentioned in the previous sections, the greatest requirement of the grounding structure is to provide a low impedance path to the ground for any type of stray currents. The impedance of the earth ground depends on the resistance (cross–section) of the grounding wire used, resistance of the electrode conductors (grounding pipe, strip or plate or rod for example), contact resistance between the electrodes and the adjacent soil and the resistance of the body of the soil surrounding the electrodes. The last two factors are a property of the soil of the project site. If the resistance of the soil adjacent to electrode is not good then the soil is treated to make it better for serving the purpose of Earth. Table 6.1 lists some of the common soils with their average approximate resistivity. This table is meant as a rough guide to the resistivity of different types of soil. The soils usually contain more than one of the listed types mixed together.

Table 6.1: Resistivity of Different Types of Soils

S No	Soil Type	Resistivity (Ω-m)	S No	Soil Type	Resistivity (Ω-m)
1	Loam Topsoil[1]	26	7	Clayey gravel, poorly graded gravel	300
2	Clay of high plasticity[2]	33	8	Sands, sandstone	510
3	Fine sands with slight plasticity	55	9	Gravel	800
4	Porous limestone, chalk	65	10	Igneous rocks (volcanic, granite, basalt)	1500
5	Clayey sands, poorly graded sand-clay mixtures[3]	125	11	Crystalline limestone, marble rocks	5500
6	Clay-Sand-Gravel mixtures	145	12		

Notes:
1. Dark top layers of soil (a mixture of clay, silt, sand and organic components).
2. The plasticity of soil increases with the addition of water (high plasticity–high ability to store and hold water).
3. Poorly graded soil means, all particles are of the same approximate size and therefore the naturally low density cannot be increased by compaction.

Soil Resistivity is affected by the mineral content, moisture and temperature. Water is a single most important factor in the conductivity characteristics of a particular soil type. Soils with high plasticity, like clay and topsoil which can hold water, provide excellent grounding. On the other hand porous soils like non-clayey sands, gravel and sandstone, which are common soil types in drier climates, are poor conductors and provide bad grounding. The worst conducting soils are igneous and sedimentary rocks due to their high mineral content and inability to hold water. Soils with high conductivities require different grounding strategies than soils with low conductivities.

6.5.2 Conductivity of Concrete

In some installations satellite antennas are installed with penetrating mounts. These penetrating mounts are buried in concrete to a depth of 0.76m for a 1.2m antenna and up to1.2m for a 3.8m antenna. Concrete is a relatively poor conductor, with electrical characteristics that are quite similar to that of the dry sandy soil (see Sands in Table 6.1). The concrete, which surrounds the metal mount, makes up a non-homogeneous ground with the soil that surrounds it thus increasing the ground resistance. Therefore the satellite antenna should always be grounded to an external grounding structure.

6.5.3 Grounding in High Resistivity Soils

Table 6.1 is a good reference for resistivity of different types of soils. The earthing systems for broadcasting stations should have less than 1 ohm and in particular earthing for Audio Equipments should always. But if the grounding structures have higher resistances in some areas due to higher soil resistivity then there should be some means to lower the resistivity of soil in order to get the desired earth resistance. Some methods used to lower the soil resistivity and improve the resistance of the grounding structure shall be discussed below.

6.5.4 Methods of lowering the soil resistivity

In order to achieve the desired earth resistance at a site for your system from a soil having high resistivity, you may have to provide more earth electrodes, so as to reduce the overall resistance by paralleling number of earth pits. In the following paragraphs we will describe some methods for reducing the resistivity of the soil as well as well resistance by providing more earth electrodes.

(1) **Use Multiple Grounding Rods-Method-1:** It has been proven by experiments that two rods properly spaced and connected to each other shall have a combined resistance of the order of 60% to that of one rod & for 3 rods it is 40%. In general, proper spacing of the rods means placing them at least one rod length apart. They have to be interconnected using copper bands of size100x1mm or a minimum # 6 AWG copper wire. For maximum results if large number of rods is available, they should be arranged in a radial configuration around the site as shown in Fig 6.4.

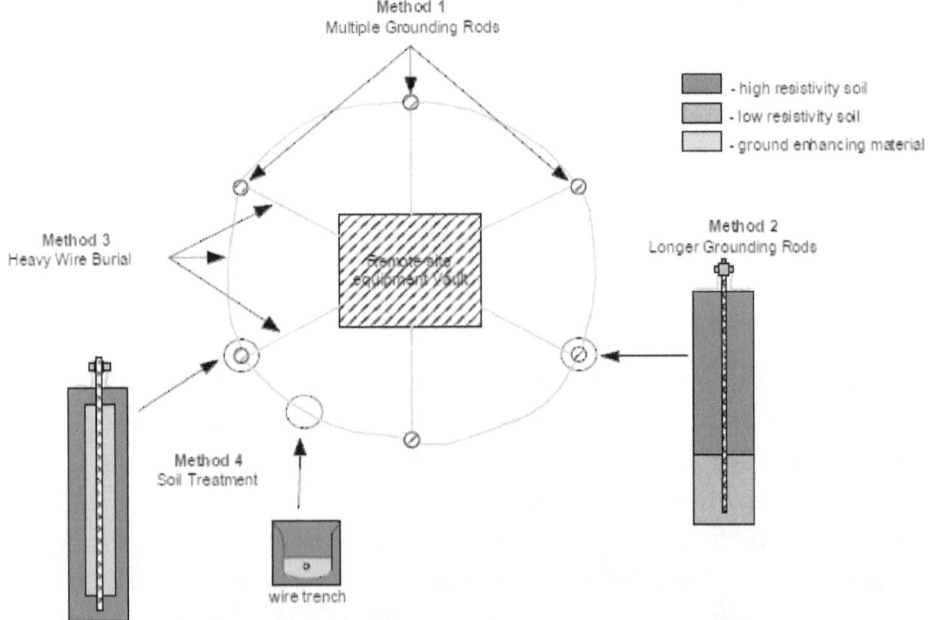

Fig 6.4: Methods of lowering the Ground Resistivity

(2) **Increase Ground Rod Length-Method-2:** Soil resistivity normally decreases with depth as there is an increase in moisture content. If this is the case then rods longer than the standard 10 ft may be required. A rather sharp decrease in the measured ground resistance will occur

when a grounding rod reaches the low resistance soil. This is why the borehole casings are the best possible grounds. This method would yield best results if used with method-1. It is important to note that in very cold climates the grounding rod has to go below the frost depth to have any effect, as the temp below freezing increase soil resistivity (as soon as moisture turns to ice resistivity increases substantially).

(3) **Bury wider size Strip or thicker dia Wire-Method-3:** If site has only a thin layer of topsoil between the surface and the bedrock, then ground rod method would not be practical. The best grounding strategy in this case would be to bury thicker wire like a copper strap of size 50x2 mm or a heavy duty # 4 AWG copper wire in a 12-18 inch deep trench around the site. This ring would then need to be connected in several places to the central grounding element with the wire of the same thickness. This method can be further improved by driving a few short grounding rods as deep as the topsoil around the circle and then bonding them to the copper wire.

(4) **Do Soil Treatment-Method-4:** This method is very effective but rarely practical due to the unavailability of the soil treatment materials. It involves treating the soil to lower its resistivity. One way to lower the resistivity is to increase the moisture content of the soil (increasing the moisture content 5-10% lowers the resistivity significantly). This process would have to be repeated quite often since the problem with the soil in the first place is that it does not retain moisture. Unless the sites had easy accessibility and there is somebody to water the soil around the grounding rods repeatedly this method would not be effective.

Another way to lower the resistivity is to treat the soil with salt (copper sulphate, magnesium sulphate or sodium chloride). Salts leak into the soil to reduce the resistivity. This method is more durable and practical than the previous one since it involves considerably less maintenance work. The downside to the salt treatment is that it may corrode some of the grounding electrodes and it may be against local environmental regulations.

Finally the soil may be treated with ground enhancing materials like bentonite clay or coke powder. These substances absorb the moisture from the surrounding soil and harden keeping the moisture within its structure. To apply such materials to the ground rod dig a hole around the ground rod about 6 inches in diameter. The depth of the hole should be a couple of inches shorter than the rod length. Apply the material to the hole (around the rod) and then bury the top portion of the hole with the soil as shown in Figure 6.6. If a wire method is used for grounding then fill the trench with about 1 inch of the material before placing the wire and then bury it together with the wire as shown in Figure 6.6. The biggest advantage of this method is that most of these materials are permanent and do not require any maintenance.

(5) **Battery Negative Terminal Grounding–Method-5:** If none of the above mentioned methods are feasible then the grounding structure should be connected to the negative terminal of a battery at only one point. Batteries usually have a very low internal resistance which makes them acceptable grounds in such cases. This method would usually be used where no grounding electrodes are available at all. An example of this is the Antarctica system (can't ground in ice with no soil within reachable depth). Another example: system installed on mountains or in rocky areas with very little or no topsoil.

(6) **Use Grid Earthing:** If desired earth resistance is not possible even after adopting all the methods of lowering the earth resistivity explained above, then use grid earthing as stated below to attain the consistent low value of earth resistance:

   (a) To manage very high values of fault currents so that the low ohmic value of earth resistance will reduce the effective watt loss thereby controlling the heat generated in the system.

   (b) To maintain very low levels of ohmic values of earth resistance like 0.1, 0.2 etc. so that the sensitive electronic equipment do not malfunction & there electronics cards & PCB's do not fail & get damage electrically.

As you know from the ohm's law; if there are n resistance connected in parallel, then overall value of the resistance shall be;
$1/R = 1/R_1 + 1/R_2 + 1/R_3 + 1/R_4 + \cdots + 1/R_n$.

Case-I; when three earthings used in a Grid:
Let us Suppose, $R_1=0.6$, $R_2=0.8$, $R_3=0.4$ or $1/R=1/R_1+1/R_2+1/R_3$ or $1/R=1/0.6+1/0.8+1/0.4$ or $1/R=1.66+1.25+2.5$ or $1/R=5.41$ or $R=0.18$ ohms

Case –II; when four earthings are used in a Grid:
Let us suppose, R1=0.4, R2=0.6, R3=0.8, R4=1.2 or
1/R = 1/0.4 + 1/0.6 + 1/0.8 + 1/1.2

$$1/R = 2.5 + 1.67 + 1.25 + 0.833 \text{ or } 1/R = 6.25 \text{ or } R = 0.16 \text{ ohms}$$

So a constant low value of ohmic resistance can be maintained with grid system. Therefore 3 or 4 earth systems (depends upon the budget) to maintain a long term low ohmic values is recommended due to following factors:-

(1) Change of a soil condition at different locations for the earthing system in the same premises.
(2) Change of climatic condition: a) Extremely dry climate – SUMMER; b) Extremely cold climate – WINTER; c) Extremely humid climate- RAINS, and d) Occasionally scanty rain fall which is assumed to be less than 250mm in a year which is unlikely in any geographical location.

6.6 Design features of an Earthing system

Earthing systems are designed to provide required type of earthing for proper functioning, electrical performance of the electronic equipments and also to protect the equipments and personnel operating them in case of faults, lightning and other serious eventualities.

6.6.1 Desirable characteristics of an earth system

The desirable characteristics an earth system should possess are:

(a) The system must carry currents as are likely to flow from the equipment faults and lightening strokes with the minimum potential rise above earth (zero).
(b) The system must carry such current most of the time without overheating or damage to earth pit elements.
(c) The system must have sufficiently low surge impedance to prevent flashover from earthed metalwork to insulated conductors or flammable material during lightening discharges.
(d) Impedance of the system must remain constant throughout the seasonal cycle and during the use.

(e) The material used in the system must have high resistance to soil and atmosphere corrosion.
(f) The cost of the system must be economical within the desired technical parameters.

6.6.2 Features of Earthing Arrangements for Radio transmitters

Following section will list some of the features an earthing system of a radio transmitter installation should have;

(a) There must be two different and distinct earth connections extended up to each equipment or circuit through bonded strips or copper bands of appropriate size as per the power of the transmitter from a two distinct buss.
(b) The two connections from buss to circuit /equipments should be shortest and may be equal or not.
(c) The length of connecting strips should not be more than wavelength α/20 at operating wavelength.
(d) The earth buss – two no's distinct and different should be run in the trenches over insulated supports directly from the link buss in earth entry pit outside transmitter hall building.
(e) The connections to the equipments and buss should be brazed or fastened with star / spring washers if bolt and nut is used to extend the connection.
(f) Normally 2 no's of earth pits make one system. Number of pits in an earth system should be increased to arrive at an earth resistance of less than 1 ohm. So layout of an earth system should facilitate adding of more earth pits, as it may be necessary to add more pits if the required earth resistance is not achieved.
(g) To ensure earth connection is available to each circuit / equipment under all eventuality all earth pit systems should be and are combined together externally only, to have redundancy and availability of ground potential.
(h) Recommended figure is less than1 ohm. But an earth system should never have more than 5 ohm resistance unless otherwise specified.

Short wave transmitters always should be provided with grid type of system. MW transmitter setups are provided with buss system and extension of pig tails from nearest point. FM and TV transmitter can be provided with either depending upon spread of their cabinet &ancillary.

6.6.3 Selection of earthing system

The choice of Earthing System has implication for the safety, functional and electromagnetic compatibility of the electrical and electronics components and the power supply. Earthing System should be selected on the basis of expected fault currents, protections / safety required of the equipments from surges, lightning etc in the electrical and Radio installation and it depends on factors such as;

(1) Power and frequency of the operation of the transmitter and its ancillaries; 2) Type and amount of Connected load; 3) Capacity of DG Set being used; 4) Capacity of the station Transformer; 5) Factor of safety taken into consideration by design engineers & project managers handling the project; 6) Calculate fault current on the type of faults: – a) Fault can be either Phase-to-Phase, and b) Fault can be Phase-to-earth/neutral

Calculations are based on the fact that full load current under normal power factor in a system is 1.4 amps per kVA. Impedance change generally is 5% of the capacity of the transformer. Therefore fault current will be 20 times of transformer capacity on the secondary side e.g. a 100 kVA transformer will have full load current of 140 amp. So current in the worst condition of fault on the secondary side of the transformer will be 2800 amps.

6.6.4 Selection of Metal for earthing electrode

The metal to be used for earthing electrode is selected on number of criteria like; its conductivity, solubility in water /acids, infallibility etc., there are two types of materials, which can be used for earthing and its electrode.

(a) Zinc
Zinc oxide will be formed during the process naturally as the mineral zincite. Practically it is insoluble in water; but soluble in dilute acetic or mineral acids, ammonia, ammonium chloride, ammonium carbonate, and fixed alkali solutions; however it is insoluble in alcohol.

It exhibits piezoelectric characteristics in thin film form. Such thin films can be deposited by process like sputter deposition. Zinc oxide is also luminescent and light sensitive. Zinc oxide is a unique Dielectric Strength material that exhibits semiconducting and piezoelectric dual properties. Zinc oxide is not combustible. It is also used as an extinguishant that is suitable for the materials involved in the surrounding fire. Zinc metal powder (zinc dust) and zinc compounds have long been utilized for their anticorrosive properties in metal-protective coatings, and are basis of important metal primers such as Zinc Chromate primers.

Zinc dust-Zinc Oxide paints are especially useful as primers for new or weathered galvanized Iron. Zinc dust-Zinc Oxide paints however, retain their flexibility and adherence on such surfaces for many years. Zinc dust-Zinc Oxide paints also provides excellent protection to steel structures under normal atmospheric conditions and to steel surfaces in under-water conditions such as dam faces and the interior of fresh water tanks. In view of above facts it is recommended to use Galvanized Iron strips against the conventional copper strips.

(b) Copper

Metal oxides are generally not very good conductors. In fact, most are Dielectrics and hence non-conductors. Certainly, Copper and aluminium are much better conductors than their corresponding Oxides at any temperature. Copper, lead, and aluminium oxides formed by corrosion are decidedly poor Conductors.

Most metals, including copper and aluminium, form thin metal oxide film layers when exposed to air for even a brief time. This make them turn dull after a few days or week's use. Copper is generally used as TINNED COPPER to avoid oxidation effect on the bare copper conductor's when used.

The Termination is generally done by soldering the multi strands of copper wire for effective termination and to avoid oxidation at the point of contact. The termination at bulk loads is normally done by lugs duly filled with soldering compounds or by very effective Crimping.

The copper oxide is poor conductor, is proved by the practice of cleaning the copper contacts in the Switchgears by sand paper periodically to maintain the contact conductivity. The copper get deteriorate due to oxidation and contacts get pitted requiring even replacement. This proves the inefficacy of copper due to oxidation.

Note; The current density of copper with respect to iron or zinc is of course higher and it is taken care by using a better cross section of earthing strips i.e. 40x6 mm (240sq mm) or 30x6mm(180 sq mm) where as in copper strips the cross section will be much smaller for the same design and current requirement.

6.7 Types of Earth Electrodes

There are various types of conventional and non-conventional earth electrodes used in Radio installations depending on their power, operating frequency and the nature of the soil at site and are described as below:

6.7.1 GI Pipe Earth Electrode

The most prevalent conventional earth electrode system, as shown in fig 6.5, is a Galvanised Iron (GI) pipe earth electrode for general purpose earthing for power supply switchgears, audio equipments and RF equipments and systems in low & medium power Radio transmitter stations including production studios.

This conventional GI pipe earthing system, employing charcoal & salts are provided for various applications. BIS no; IS: 3043-1987 (Indian Standard code of practice for earthing) describes all the details and design requirements. Rule 51, 61 of the Indian Electricity Rule, 1956, defines the requirements and guidelines for a long lasting earthing system for various applications. Reasonably low value of Earth resistance is achieved easily if job is executed professionally. Its construction is simple & it does not require any special tools or arrangements. This is most widely used and is most cost effective.

Advantages; (a) Fabrication hardware & misc materials are easily available even in small towns; (b) A semiskilled worker can construct it under the guidance of a project technician, and c) If desired value of earth resistance is not achieved with one earth electrode, then two or more can be paralleled to achieve less than 1 ohm. However it has lot deficiencies and cannot meet stringent requirements of low value of earth resistance.

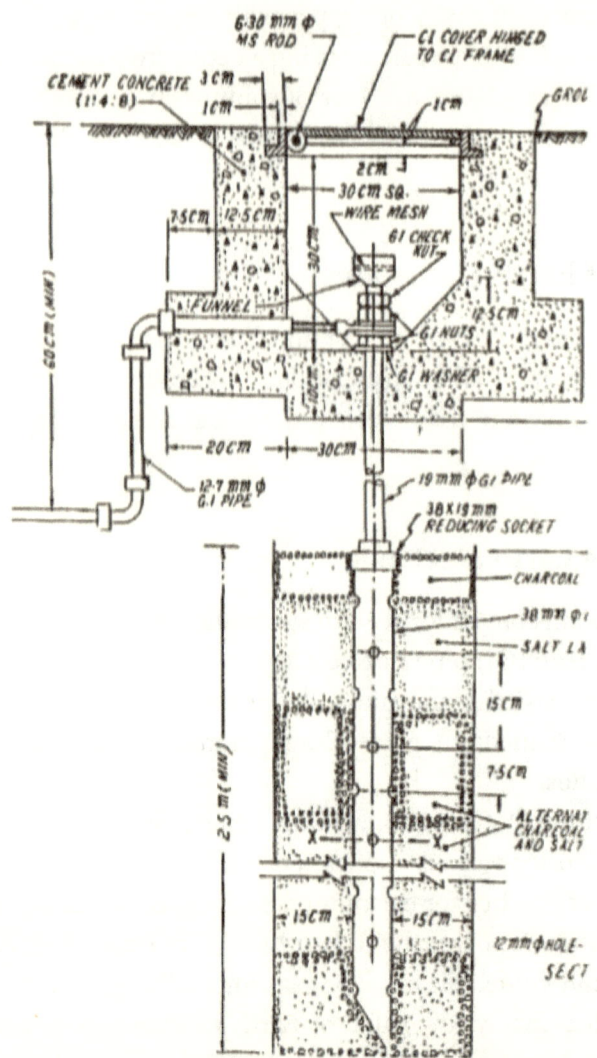

Fig 6.5: Typical Arrangement of GI Pipe Earth Electrode

6.7.2 Copper or GI Plate Earth Electrode

This earth electrode as shown in fig 6.6, is most widely used for RF earthing of low power MW as well SW transmitters, MW masts, Antenna tuning units, Guy anchors, RF coaxial feeder cables for VHF & UHF antenna and is also equally cost effective and can meets the design requirements if the soil conductivity is good.

Advantages; (a) It yields very good earth resistance if work is executed professionally within lowest budget; (b) Its construction also is simple as it does not require special tools or arrangements to construct, and (c) A semiskilled labourer can construct it under the guidance of a project technician.

Fabrication hardware & misc materials may not be easily available specially the copper plate in small towns but is available in cities as it needs copper plate along with good quality GI pipe. To prolong the performance normal but regular maintenance is required.

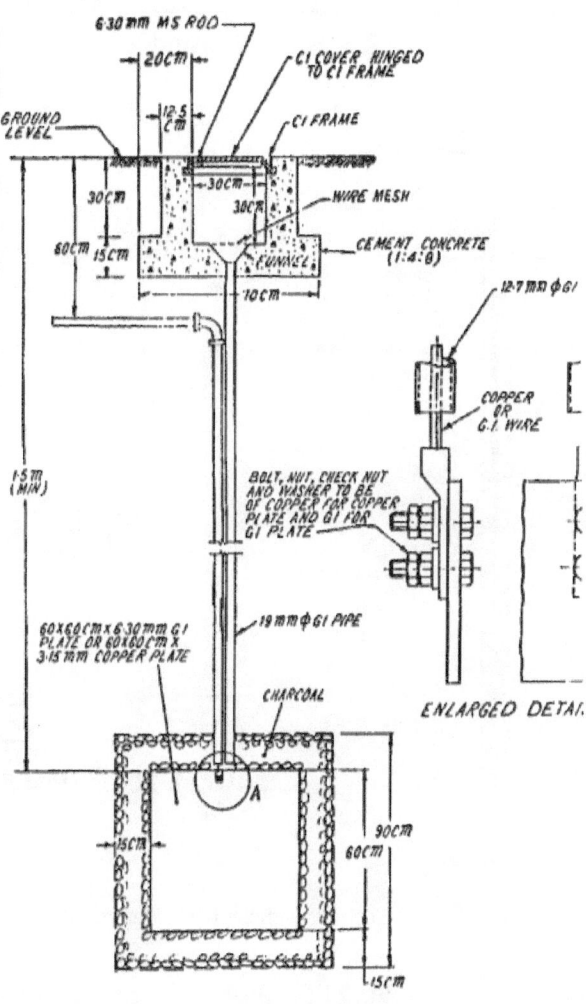

Fig 6.6: Typical Arrangement for Plate Earthing

6.7.3 GI Pipe with Copper Tube and Bentonite treatment

To make the soil around the electrode conductive and to maintain the moisture for consistent Earth resistance common salt and wood charcoal are used. As is known the common salt (sodium chloride) is a hygroscopic substance & it gets dissolved in water & losses its hygroscopic properties when it becomes water itself. The salt is also known to be a corrosive electrolyte which decays the pipe and the conductor used for earthing. Due to decay one does not get the consistent values of Earth resistance. The soft coke & charcoal also tend to become ash due to heavy heat generated by fault currents generated in the system especially at higher voltages of transmission distribution line & at substation and RF installations.

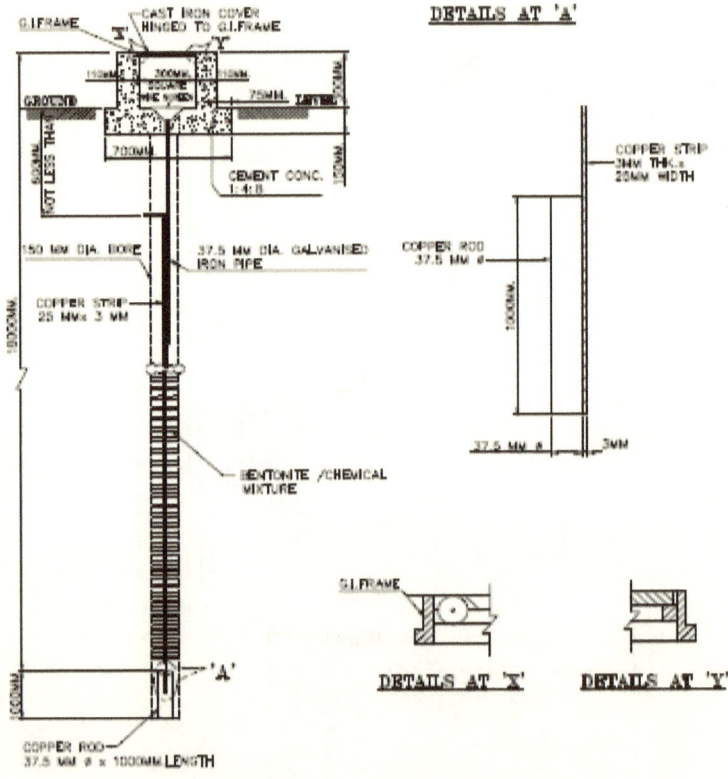

Fig 6.7: Earth Electrode with bentonite Treatment

Therefore due to above, the life of GI Pipe and Copper or GI Plate Earth Electrode is short. Also at places where the soil conductivity is poor or soil is rocky, sandy or there is lack of top or sub soil it is not feasible to obtain satisfactory and consistent Earth resistance. At such places the solution lies in providing the Earth electrode of GI pipe with an end copper tube surrounded by bentonite treated soil. This improves the conductivity of the soil for consistent and desired Earth resistance and as the corrosive salt is not there, of course it will prolong the life and performance of the Earth system also. Figure 6.7 shows such an Earth electrode system.

The earth electrode is constructed by drilling a bore of 150 mm diameter up to a depth of 18000mm in the ground which may be a rocky terrain. Then a 1000 mm long copper tube of 37.5 mm dia is attached to the end of a copper strip of size 25x3mm, top of which as shown in figure is attached to a perforated class 'B', GI pipe of 37.5mm dia & 1 mtr long. The Copper Strip is brazed with Copper Rod through full length of Rod. Then these are lowered into the bore and held in the centre with a forma. After putting the Copper Rod and Copper strip inside the bore, the bored hole is filled with the loamy/black cotton soil mixed with bentonite/Earthing Chemical & common salt. The soil and Bentonite are thoroughly mixed @ 35kg bentonite per mtr3 of soil and 10kg of common salt per mtr3. GI Pipe is provided with a funnel for watering and a masonry chamber over it including fixing of hinged C.I. cover. The Earth Pit Chamber is kept above the ground level. The soil around the Earth Pit Chamber is properly rammed. This type of Earth electrode easily yields resistance of less than 1Ω.

6.7.4 Chemical Earthing

This Earthing as shown in Fig 6.8 is a non-conventional type of earth electrode and now a day's used for its consistence earth resistance over prolonged use in all type of soil including rocky area and mountains.

The main reason for non consistent of earth resistance is the absence of sustained maintenance of the moisture in the soil. So the chemical earthing maintains the moisture of the soil with the help of a chemical compound called KIMMOIST (Grounding Back Fill Compound). It can absorb water 13 times of its weight. The pipe used for the chemical earthing are generally 2 to 3mtr long, therefore the earth bore need not be more than 250 to 300mm dia and 3mtr deep. The moisture is maintained at this small depth of 3mtr by using the KIMMOIST ground enhance material. The efficiency of the chemical earthing to maintain the moisture which is essential for low Ohmic values of earthing resistance is due to use of hygroscopic chemical, which

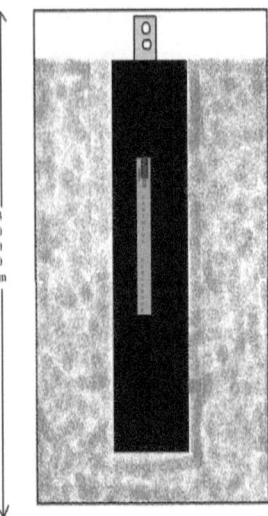

Fig 6.8: Typical view of a Chemical Earth Electrode

absorb the moisture but doesn't get dissolved in the water unlike salt. The heat generated due to electric faults developing heat of 1060 °C and above is resisted by CCM (Crystalline Conducive Material) which can withstand up to 2500°C. The CCM is filled in a pipe of size of 48 mm or 76 mm dia and sealed at both ends. It incorporates GI strip earth conductor of size 30x6mm or 40x6mm depending upon the individual design of user. The CCM filled in the pipe being a good conductor increases the fault current capacity of the system because the pipe is hollow but filled with CCM it behaves like almost a solid pipe. CCM resists cracking, warping, or shrinking even when temperature exceed 2500°C due to repeated faults which may happen in operations over the year.

Installation procedure

A chemical earth electrode requires:(a) About 50 kg of soil enrichment chemical mainly consisting of Aluminum Silicate; (b) GI pipe of 48 to 76 mm dia and 2 to 3 m in length; (c) A GI or Cu strip of size of 30x6 or 40x6mm depending upon the individual design of manufacturer as earth conductor; (d) Sealant for permanently sealing the top & bottom of the G I pipe, and (e) An auger or other means of boring 250 to 300 mm dia bore.

Chemical earthing is very compact and easy to install and locate near to the equipments may be 2-3 meters away depending on the situation. Earthing is installed in a bore of 8-10 inch of diameter of appropriate length depending upon size of electrode used. The electrode is installed in the bore and additional space surrounding the electrode is filled with the slurry of 35-70kgs of KIMMOIST.

Testing

The test is carried after one week of the installation of the pit. A week's time is necessary to allow curing and the chemical to mix with soil conditions. The Values obtained should be within the limit given as per IS: 3043-1987 i.e. less than 2 ohms. In practical the values obtained are between 0.5 to 1.5 ohms.

6.7.5 Multi-electrode Earth Pit

This type of earth as shown in fig 6.9 is a parallel equivalent of 3 pipe electrode earth pits used in high power hybrid and fully solid state MW S7HP transmitters manufactured by M/S Thomson France. In a big modern transmitter plant where spread of equipment is appreciable and transmitter is solid state, this type of construction is best solution. A RF grid mat is used in the transmitter hall for providing equipotential earthing below the cabinet, through multi-electrode earth pit.

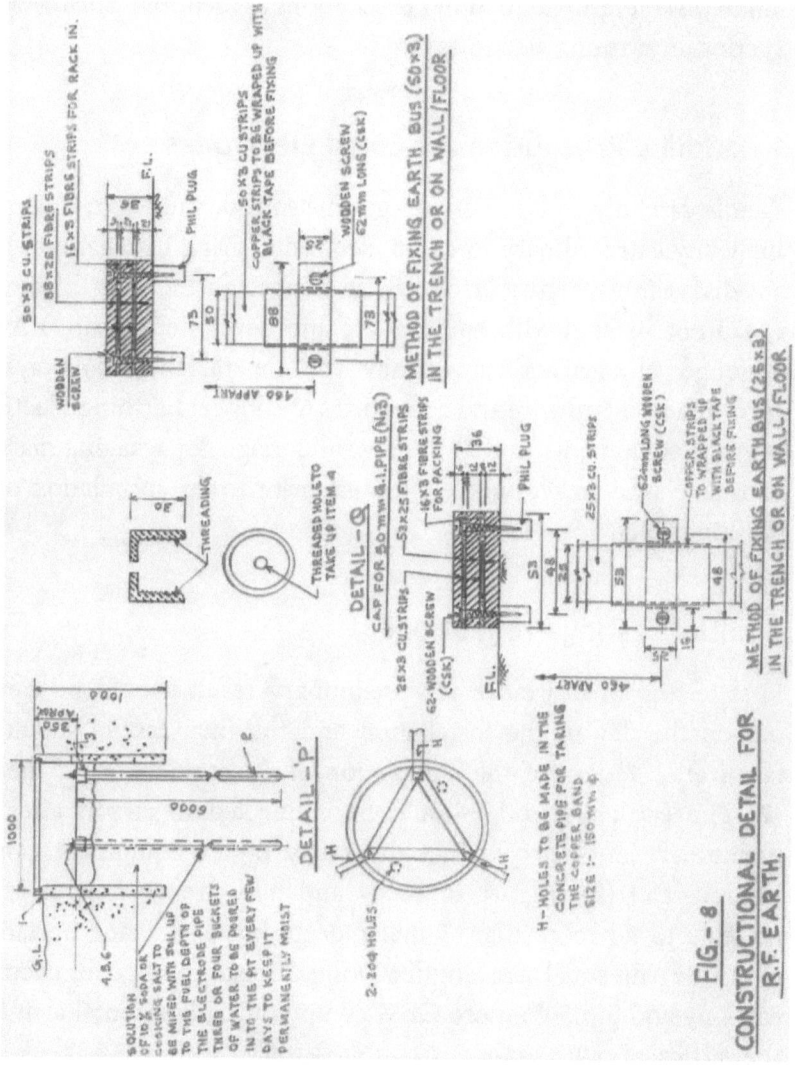

Fig 6.9: Multi-electrode Earth Pit System

This type of earthing pits gives very good results even in adverse soil conditions. This system however is limited to high power MW or SW solid state transmitters which are very sensitive to slight changes in SWR and are more prone to failure due to lightning and static discharges.

6.8 Initial Testing of Earthing Electrodes

As you are aware Earthing pits are kept as near to the Equipments as possible but beyond plinth protection of the building housing the equipments. In addition to this there are lot of other precautions as mentioned below while constructing an earthing electrode.

6.8.1 Handling Precautions of Earth Electrodes

(1) Handle carefully and do not plunge an electrode in the pit; 2) Do not apply any extraordinary force on electrode during installation;3) Do not disassemble, repair or otherwise tamper an Earthing Electrode; 4) Do not strike it with hammer etc and never step on it; 5) Avoid prolonged physical contact with any type of metal fence; 6) Always use very high standard wire and accessories to connect Earthing Electrode with the equipments, and 7) Remove all foreign & packaging material such as plastic cover, bubble wrap etc prior to the installation of an Earthing Electrode.

6.8.2 Initial Testing Precautions

(1) First testing of resistance for preliminary result should be carried out on the day of the installation and the final results should be taken after 7 days of the completion of the installation of the pit; (2) The earth electrode should be connected in circuit after the preliminary test only so that the safety of the equipment can be insured, and (3) In case of rocky and hilly areas where it is not possible to go more than 3 meter deep, install L-shape horizontal electrode with small arm length 600mm. In such cases go for chemical earthing and put 50% more GEM compound in horizontal arm than the vertical arm.

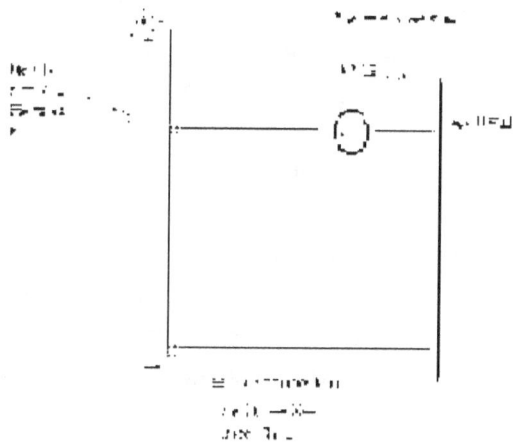

Fig 6.10: Voltmeter method for testing a new electrode

6.8.3 Method of Testing a New Earth Electrode

The method described below is the simplest test which can be done for a new electrode without any professional equipment, before connecting it to the equipment in the system.

1) Connect the Electrode to the Neutral of Power Supply Line by a very good secured connection either bolted, soldered or lugged. If bolted provide spring washers at both ends to ensure perfect ohmic contact; 2) Connect the Volt meter as shown in the Fig 6.10; and 3) When the Neutral is properly grounded and the earthing is successful the voltmeter should give nearly zero voltage between Earth and Neutral. Alternately a bulb also can be used and if it glows full, earthing is okay while connected to line and earth electrode.

6.9 Review of Earthing Systems

Herein a review of old traditional and present modern earthing systems shall be discussed.

6.9.1 Discard Pit Type Earthing?

The conventional GI or copper pit type earthing is considered outdated because of:

(1) The common salt (sodium chloride) used in pit type earthing is a hygroscopic substance & it gets dissolved in water & losses its hygroscopic properties. Also the salt is known to be a corrosive electrolyte which decays the pipe and the conductor used for earthing. Due to decay one does not get the consistent Ohmic values.

(2) Similarly soft coke & charcoal tend to become ash due to heavy heat generated by fault currents in the system especially at higher voltages, RF and due to lightning strikes as well static discharges.

(3) As you know the heat generated is proportional to $I^2 R t$. So with a fault current of 10,000 amps with an earthing resistance of 2 ohms in 0.1 seconds the heat generated shall be; $H = I^2 R t$

$$H = 10,000 \times 10,000 \times 2 \times 0.01 = 20,00,000 \text{ calories} = 1053 \,°C$$

This is the heat generated in one fault. Assuming 6 faults per year, in a period of four year-24 faults occur, each fault generating 1053 °C & above.

(4) Each fault of this magnitude will turn the soft coke/ charcoal into ash gradually in a period of 3-4 years. The earth system will deteriorate and give higher value of ohmic resistance thereby endangering the entire installation consequently people & property.

(5) If the ohmic values of earth resistance goes up from 2 to 3 ohms because of faulty earth systems, the heat generated will be 30,00,000 calories (i.e.1600°C). This shows that the pit type earthing is inefficient, not reliable & causes serious damage to life & property in a period of 3-4 years.

(6) To maintain moisture round the year, it needs a watering arrangement through a separate piping system. The water so poured in addition to the natural rain fall, will wash away the charcoal & the soft coke from its main position thereby it will further deteriorate the effectiveness of the earth system.

However for places having good conductivity, small earth electrodes in low power operations like broadcast recording studios dealing audio signals, this earthing system is more than adequate and should not be discarded. Like that for many more applications like low power MW & FM transmitter set up, earthing for Power supply equipments or D/G set, for speech input equipments of High Power transmitter set ups, and earth electrode system for Mast/ towers, its guy wires anchors and feeder hut etc, it's still adequate.

6.9.2 Switch to Chemical Electrode Earthing?

The above points conclude that pit type earthing is outdated and it is being replaced by chemical pipe earthing and preferred over for reasons given below:

(1) The pipe for the chemical earthing is generally smaller length of 2 or 3mtr. Therefore the earth bore need not be more than 3mtr deep and 250 to 300mm dia.
(2) At such a small depth of 3mtr the moisture is maintained by using ground enhance material, commonly known as KIMMOIST (Grounding Back Fill Compound).
(3) The efficiency of the chemical earthing to maintain the moisture which is essential for low ohmic values of earth resistance is due to use of hygroscopic chemical which absorb the moisture but doesn't get dissolved in the water unlike salt.
(4) The heat of 1060°C generated due to electric faults and above is resisted by CCM (Crystalline Conducive Material) as it can withstand heat upto 2500°C.
(5) The CCM is filled in a 48 or 76mm dia pipe and sealed at both ends. It incorporates the earth conductor of GI strip of suitable size of 30/40x6mm depending upon the design of each manufacturer.
(6) The CCM filled in the pipe being a good conductor increases the fault current capacity of the system because the pipe is hollow but when filled with CCM it behaves like almost solid pipe.
(7) The CCM resists cracking, warping, or shrinking, even when temperature exceed 2500°C due to severe repeated electric faults which may happen in operations over the year.

6.9.3 Comparison of chemical & Conventional Earth Electrode

The comparison shall be done on some of the most vital parameters like life, stability and consistency of earth resistance desired from an earthing system as well cost of maintenance incurred to maintain it.

(a) Life of Earthing Systems

Following paras will give some insight about the life of various type of earth electrodes so as to enable the user choose an earth electrode depending upon design, availability of finance and the life of the system for which it is needed. Also the user may like to initially go for an earth electrode pit with lesser cost and life and then change it with longer life system after initial heavy expenditure is over and finance is available for better system.

(1) Copper Plate or G.I Pipe Earth Electrode; the conventional copper plate pit type earthing has a very erratic behaviour and the ohmic values of earth resistance vary drastically on the seasonal factor: a) Summer Season, b) Winter Season, and c) Rainy Season

   Humidity factors in various seasons vary due to variation in ambient temp. The thumb rule for calculating humidity at a 20°c ambient temp and 1° variation in either side (+/-) will vary the humidity level in the ground by 16%. You can now understand why the ohmic values remain erratic during the life of the system.

   Contrary to the belief that winter, a wet climate, has a higher humidity level, is a myth and humidity is very low in winter because the ambient temp is in the vicinity of 4 to 6°c and as per thumb rule, the humidity level is reduced by 16% for every degree variation in temp. So in winter humidity is reduced by 16x (20-6) =224%. As such it needs watering in winter also through separate piping arrangement.

   The coal used in the earth pit burns and turns into ash under high level of fault currents and on the top of it the quality of coke used is very poor (Generally in Powder form). Also the salt used becomes water itself after few months of use and loses its hygroscopic properties. Therefore the life of pit type of earthing is not more than 3 to 4 years irrespective of the care and maintenance.

(2) Chemical Electrode earthing; following important factor are taken into the consideration:

   (a) CCM is a semi metal and the hollow pipe filled with CCM behave like a solid pipe and therefore the current carrying capacity increases substantially.

   (b) The CCM has a high melting temp of 2500°c and unlike coal it will not burn into ash under high temp of 1500°c under high fault currents occurring due to earth faults.

(c) Moisture level is maintained by KIMMOIST surrounding the electrode in a slurry form. The KIMMOIST has a property of absorbing water 15 times its weight and doesn't dissolve in the water. It remains moist and soft throughout the life. Even a rain fall of 2.5 cm makes the KIMMOIST moist and humid for next one year.

(b) Stability of the system
(1) Copper Plate or GI Pipe Earth Electrode; The ohmic values measured at different periods of the year vary drastically and sometimes reach failure level. There is no data available as the earthing installation is not done in the organized sector so far and only in the hand of unqualified, semi skilled traditional electrician without any electrical license whatsoever who use poor quality of copper, the poorest quality of coal (Powdered coal) and third grade quality of salt and complete the earth pit in a paltry sum of Rs. 6000/-(Including their profit).
(2) Chemical Electrode earthing; It is maintenance free earthing and experience over the years on periodically check-up have shown consistent results with respect to the ohmic values. Therefore the stability of this type of earthing is very good with no noticeable change. It gives safe and sure protection to the installation.

(c) Cost of maintenance
(1) Copper Plate or G.I Pipe Earth Electrode; In the pit type earthing some regular cost as under is required: a) the cost of labour for watering at periodical interval; b) Initial cost of laying pipe for watering grid from the source and the cost of maintenance of the water pipe going to the pit which gets choked due to scale formation.
(2) Chemical Electrode earthing; Cost on account of maintenance is negligible as pipe is sealed after filling the compound. The Feedback information answered in the questions generally asked by the project engineers and consultants that the chemical earthing need no maintenance and meets all the parameters of safe and reliable earthing System.

So it can be stated that Chemical Earth is almost a permanent solution for all earthing problems either due to poor soil resistivity or corrosive elements in the soil or environment or saline air near sea shores. The earth resistance results are repeatable irrespective of seasonal variations. This is onetime solution worth considering notwithstanding initial cost for sandy, rocky and mountainous regions.

(d) At a Glance Comparison

Table 6.2 below summarises the comparison between the chemical and conventional earth electrode.

Table 6.2: Comparison between Chemical and conventional Earth Electrode

S No	Chemical Earth Electrode	Conventional Earth Electrode
1	Employs highly secured strip on 48 or 76mm dia pipe instead of solitary 19 / 38 mm dia GI pipe.	It uses a copper or GI plate buried at the bottom of pit, prone to damage.
2	Pipe with flat strip, filled with graphite, becomes almost a solid to bear mechanical stresses developed during the fault.	Cu/GI plate placed amid alternate layers of charcoal and salt surrounded by soil is prone to damage.
3	Graphite has a burning temp of 2500^0c, so it can withstand high level of heat generated during faults.	Char coal becomes ash at 400^0c and after it no conductive material is left in pit for dissipation of heat.
4	Surface area for dissipation of heat generated is very large as the heat is dissipated from the total surface of the electrode.	Surface area for dissipation of heat is comparatively small in plate earthing being a length x breadth.
5	Bore filled with thick slurry of aluminum silicate and the entire electrode surrounded by it, retains moisture round the year.	Salt and charcoal looses moisture as salt being hygroscopic turns into water & coal turns into ash
6	Aluminum silicate absorb water 15 times its weight and does not dissolve in water, so doesn't dry-up with time.	Salt dissolve in the water and dries up gradually therefore leaving no utility after sometime.
7	Installation is easier as electrode is dug into ground up to appropriate length of 2 to 3 meter and dia of 200 to 300mm.	It is cumbersome to install plate Earthing. It involves additional water pipe with funnel for watering the pit.
8	It's maintenance-free as no need to pour water at regular interval-except in sandy soil. It however needs pre installation watering of bore to moist the soil surrounding the electrode.	Needs regular maintenance and watering to retain moisture in soil surrounding the electrode. But this creates mud due to coal ash and it loses its utility.
9	Initial ohmic value of resistance remains same for many years as there is virtually no corrosion at all.	Due to salt it corrodes in 3/4 years and resistance increases. Even resistance values vary with in summer, winter and rains.
10	Long life of up to 15 years	Short life of upto 4 year in spite of maintenance.

6.10 A typical Earth schematic layout of a High Power MW Transmitter

A transmitter set up have 3 independent earthing Systems viz. earthing for; transmitter plant and its ancillaries, Power supply Equipments and Audio Equipment System. In addition, there are separate earthing system consisting of various type e.g.; (a) for feeder line 2 no of 8 SWG copper wire buried below it; (b) for Feeder Hut, 4 single pipe electrode earth, one on its four corners, interconnected to inside shielding and a ring all around it; (c) earth radials for tower, and (d) Pits in the aerial field one for each Guy wire for static earthing.

Fig 6.11 shows earth schematic for a high power MW broadcasting transmitter setup in normal conditions. Earth system of each type is run independently to each other on insulated supports inside the building and is not connected to each other. However all the earth systems are connected externally to have an earth connection for safety and protection of man and machine under worst conditions of failure of earth pits? The earth electrode for transmitter and its ancillaries including Power Supply system is with multi-electrode pits to have lowest possible earth resistance value for the full life of the equipments.

Each system have 2 earth pits at a distance of 7.0 mtr from each other and are interconnected with a copper band of 50 x 3mm. Copper bands are drawn through 150 mm dia cement pipes upto building earth entry pit beyond plinth protection. Inside the building these strips are laid in trenches or imbedded in floor in the form of a grid before final floor is laid while constructing the building, as per design if equipotential surface is to be provided (normally in case of SW transmitters). When it is embedded in the floor 2 pig tails of suitable sizes are provided at each equipment location for extending the earth connection to them.

6.11 Measurement of earth Resistance and Soil Resistivity

Broadly speaking, "earth resistance" is the resistance of the soil to the passage of electric current. It is earth's abundance and availability that make it an indispensible component of a properly functioning earth system.

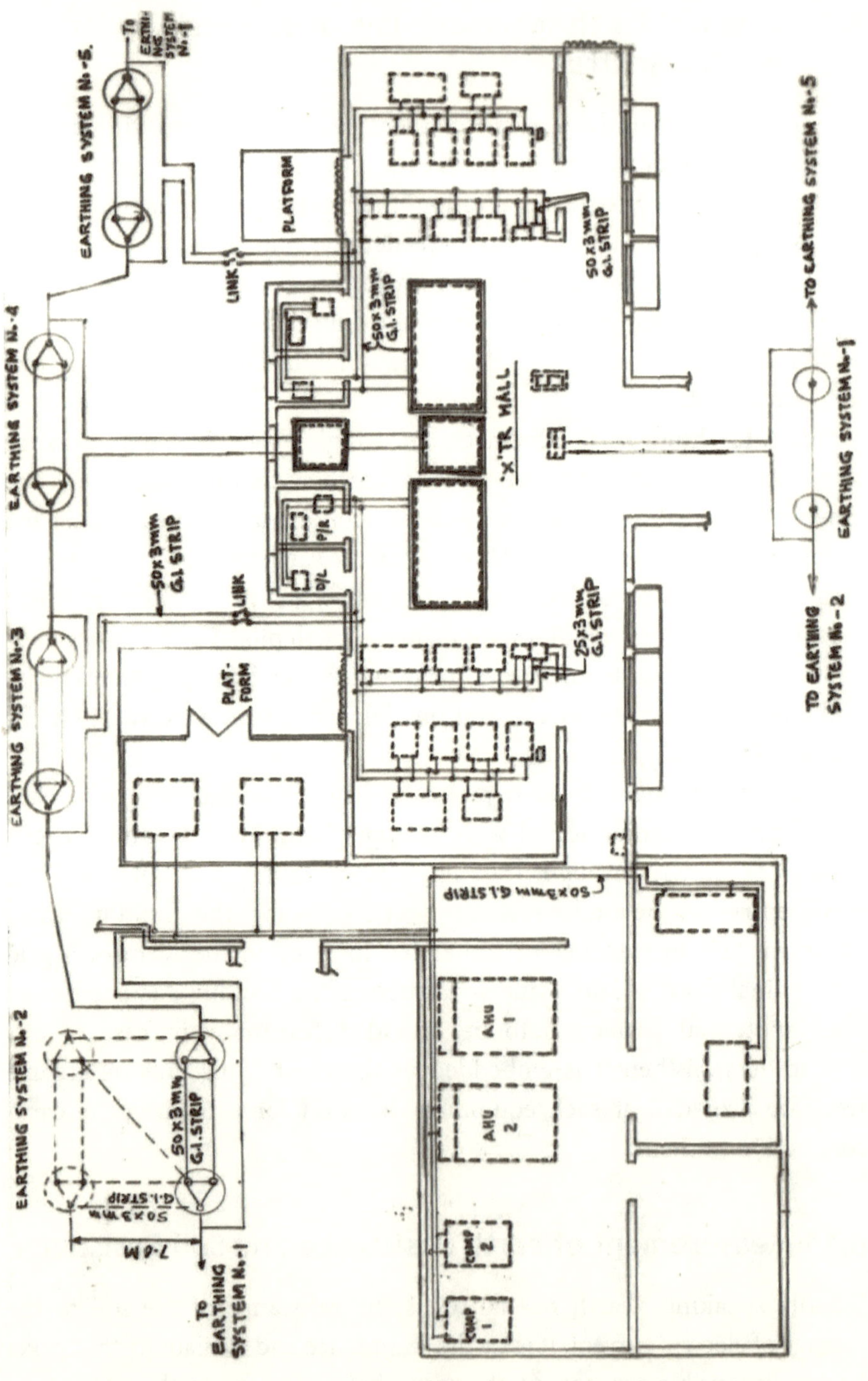

Fig 6.11: A Typical Earth schematic layout for a High Power MW Transmitter

Earth resistance is measured in two ways for two important fields of use: (1) Determining effectiveness of "ground" grids and connections that are used with electrical systems to protect personnel and equipment; and (2) Prospecting for good (low resistance) "ground" locations, or obtaining measured resistance values that can give specific information about what lies some distance below the earth's surface (such as depth to bed rock).

This section describes the principles and methods for measuring the resistance of earth electrode used in broadcasting facilities and the soil resistivity while measuring the earth resistance.

6.11.1 Principles of Ground Resistance Measurement

The object of an earth electrode is to provide a low resistance path to fault currents that may cause injury to the operating staff or damage or the equipments. The currents will dissipate safely when properly conducted to earth via earth electrode. As shown in Fig 6.12, there are three components of the earth resistance;

(1) Ohmic Resistance of the electrode materials and connections to them,
(2) Contact resistance between the electrode and the soil surrounding it, and
(3) Resistance of the surrounding earth

The resistance of the electrode materials is purposely made small so its contribution to the total resistance is negligible. Generally, copper materials are used throughout. Ground rods usually are copper-coated steel for strength, although galvanized steel ground rods are also found in applications where corrosion is a problem.

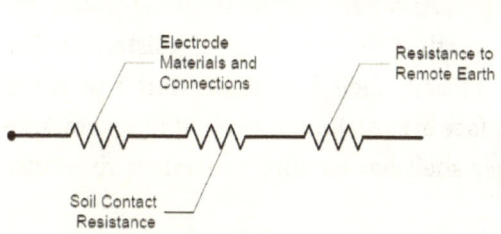

Fig 6.12: Equivalent Resistance of an Earth electrode System

The contact resistance between the electrode and soil is negligible if the electrode materials are clean and unpainted when installed and the earth is packed firmly.

Normally, the resistance of the surrounding soil will be the largest of the 3 components. An earth electrode buried in the earth radiates current in all directions and eventually dissipates some distance away depending on the soil's resistance to the current flow, as dictated by its resistivity.

An earth electrode system consists of all interconnected buried metallic components including ground rods, ground grids, buried metal plates, radial ground systems and buried horizontal wires, water well casings and buried metallic water lines, concrete encased electrodes (Ufer grounds), and building structural steel.

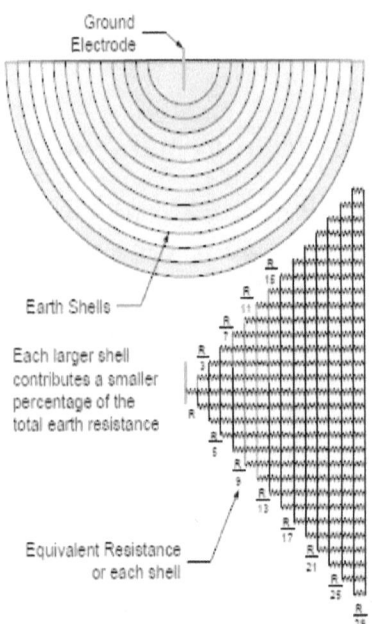

Fig 6.13: Concentric earth shells around an earth Electrode

An earth electrode in ground can be thought of as being surrounding by shells of earth, each of the same thickness as shown in fig 6.13. The shell closest to the electrode has the smallest surface area and offers the highest resistance. The next shell has larger area, so lower resistance, and so on. A distance eventually will be reached where the additional earth shells do not add significantly to the resistance. Earth electrode resistance is measured to remote earth, which is the earth outside the electrode's influence. A larger electrode system requires greater distance before its influence decreases to a negligible level.

Another way of thinking about the earth shells is as parallel resistances. The closest shell has some unit resistance. The next larger shell has more surface area so it is equivalent to several unit resistances in parallel. Each larger shell has smaller equivalent resistance due to more parallel resistances.

The resistance of the surrounding earth depends on the soil resistivity. Soil resistivity is measured in ohm-meters (ohm-m) or ohm-centimeters (ohm-cm) and is the resistance between two opposite faces of a 1 meter or 1 centimeter cube of the soil material. The soil resistivity depends on the type of soil, salt concentration and its moisture content and temperature. Frozen and very dry soils are a good insulators (have high resistivity) and are ineffective as earth electrodes.

If the earth impedance is greater than 1 Ω then it is treated that the reactive component of the earthing impedance is negligible and the earth impedance is mainly contributed by the earth resistance. The reactive component is generally taken into account when ohmic value of the earthing under test is less than 0.5 Ω.

6.11.2 Earth Resistance Measuring Equipments

Three typical test sets are shown in Fig 6.14. A typical test set has four terminals–two current terminals marked C1 and C2 and two potential terminals marked P1 and P2. Cheap test sets may have only three terminals; these should be avoided because of their limited usefulness and questionable accuracy and preciseness. Also, it has been found in the field that some expensive digital test sets are worthless.

One current terminal (C1) is connected to the earth electrode under test and the other (C2) to a probe driven in the earth some distance away. The test set injects a current into the earth between the two current terminals. One potential terminal (P1) also is connected to the earth electrode but the other potential terminal (P2) is connected to a separate probe driven in the earth between the electrode and the current probe (C2).

The potential probes detect the voltage due to the current injected in the earth by the current terminals. The test set measures both the current and the voltage and internally calculates and then displays the resistance;

$$R = V / I \text{ ohms}$$

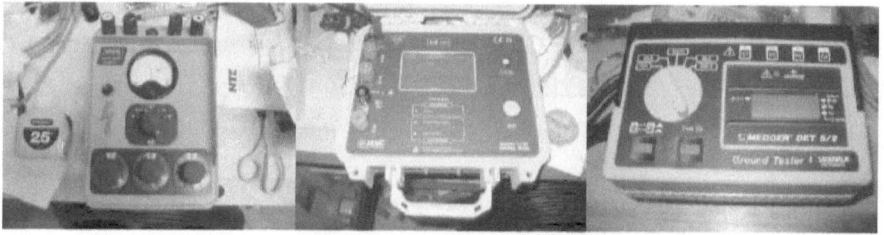

Fig 6.14: Typical earth resistance test sets
(L-R: Biddle "Megger" 240241, AEMC 4630, and Biddle DET5/2)

Some test sets use a resistive bridge circuit that requires manual adjustment of the resistance to balance the bridge as indicated by a null meter. When the meter has a null reading, the value is taken from the resistance controls. Most modern test sets require no manual adjustment.

The resistance indicated by the test set is the resistance between the earth electrode and the potential probe. If the current and potential probes both are far enough away from the electrode, that is, outside the influence of the electrode, the reading corresponds to the resistance of the electrode to remote earth.

When measuring the resistance of an earth electrode, C1 and P1 are connected together either at the test set or at the electrode. If connected at the test set, say with a jumper, the resistance of the wire from the test set to the earth electrode adds to the reading. The value of this resistance can be measured easily with the test set:

(1) Short C1 and P1 together with a short jumper; (2) Short P2 and C2 together with a short jumper; (3) Connect the test lead between P1 and P2 and measure the resistance; and (4) Subtract the reading from all electrode resistance measurements subsequently taken.

To avoid having to make these additional measurements and calculations, always use a short, low-resistance conductor between the test set and the electrode when C1 and P1 are shorted at the test set (Fig 6.15-left), or always connect C1 and P1 to the electrode with separate leads (Fig 6.15-right). If the two terminals are connected together at the electrode, the length and size of the conductor does not affect the measurement. The sizes of wires connected to the other two terminals, P2 and C2, are not important from an electrical point of view but should be large enough to provide the needed physical strength.

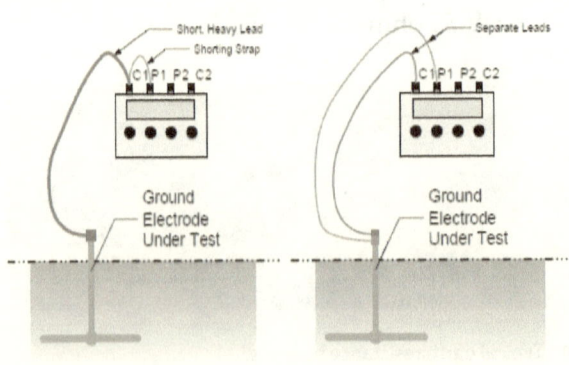

Fig 6.15: Test Equipments to Earth Electrode connections

Test sets with 4 terminals also may be used to measure the soil resistivity (test sets with three terminals cannot be used to measure soil resistivity). Each terminal is connected to an independent probe resulting in two potential and two current probes. Soil resistivity measurements are covered later.

6.11.3 Methods of Earth Resistance Measurement

Resistance of an earthing system is usually determined with alternating current of power frequency to avoid possible polarization effects when using direct current. When the power frequency current is injected into the earthing system under test, care has to be taken to avoid the interference in measurement system due to the power supply leakage currents. So generally the current signal at frequency close to the power frequency such as either 48Hz or 52Hz but not exactly 50Hz, to avoid the power frequency signal interference which can distort the measurement results, is injected. Commercially available earth testers inject switched DC signal at 96Hz to 128Hz to avoid the interference with the fundamental and third harmonic component of the power frequency signals of 50Hz or 60Hz.

Methods* for measurement of earth resistance as described by various authors are: (1) fall of potential method, (2) two-point method, (3) three-point method, (4) ratio method, (5) staged fault test, (6) single clamp earth resistance testing, (7) double clamp earth resistance testing, and (8) attached rod technique. These methods are briefly described below. Three point method and ratio method are derived from the analytical calculations related to the fall of potential theory. Two point method and clamp method of measuring earth resistance should be limited to small earthing systems. It should be noted that the measured ohmic value is called resistance. However, when the measured ohmic value of the resistance is generally less than 0.5Ω, there is a reactive component that should be taken into account if the earthing system is of a relatively large extent.

[*Impulse Measurements in Earthing System-Deepak Lathi – Nov 2012 School of Engineering Cardiff University]

(1) Fall-of-Potential Measuring method

This method involves passing a current in the earth electrode whose resistance is to be measured and recording the influence of this current in terms of potential between the earth electrode under test and another electrode called potential electrode. Prior to the conduction of the fall of potential test, the earth electrode may be disconnected if possible from the system to which it is providing protection. However, the fall of potential method of measuring earth impedance can also be carried out with the earth electrode connected to the power system provided that the earth leakage current from the power system is measured appropriately.

The Fall-of-Potential method (also called 3-Terminal method) is the most common way to measure earth electrode system resistance, but it requires special procedures when used to measure large electrode systems. For small electrodes, such as one or a few ground rods or a small loop, the procedure is as below;

Procedure

First connect the test set terminals C1 and P1 to the earth electrode under test, connect the test set C2 terminal to current probe located some distance from the earth electrode and finally connect the test set P2 terminal to a potential probe located a variable distance between. The two probes normally are located in a straight line. At each potential probe location, the resistance is recorded. The results of these measurements are then plotted to graphically determine the electrode resistance.

(a) Measuring Small Electrodes

(1) Connect C1 and P1 terminals on the test set, as shown in Fig 6.16, to the earth electrode.

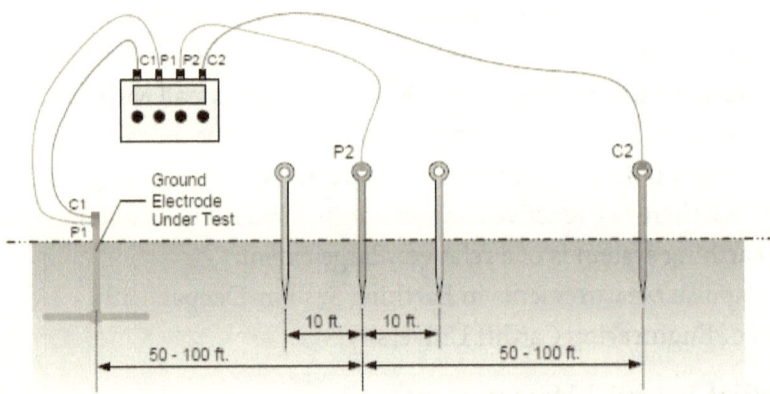

Fig 6.16: Resistance Measurements of Small Electrode system

(2) Drive a probe into the earth 100 to 200 feet from the center of the electrode and connect it to terminal C2. This probe should be driven to a depth of 6 – 12 inches.

(3) Drive another probe into the earth midway between the electrodes and probe C2 and connect it to terminal P2. This probe should be driven to a depth 6 – 12 inches.

(4) Record the resistance measurement.
(5) Move the potential probe 10 feet farther away from the electrode and make a second measurement.
(6) Move the potential probe 10 feet closer to the electrode and make a third measurement.
(7) If the three measurements agree with each other within a few percent of their average, then the average of the three measurements may be used as the electrode resistance.
(8) If the three measurements disagree by more than a few percent from their average, then additional measurement procedures are required as done in case of Measuring Large Electrodes.

(b) Measuring Large Electrodes

A quick set of measurements as shown in Fig 6.17, can be made. This method eliminates many tedious measurements but may not yield good accuracy unless the current and potential probes are outside the electrical influence of the electrode system:

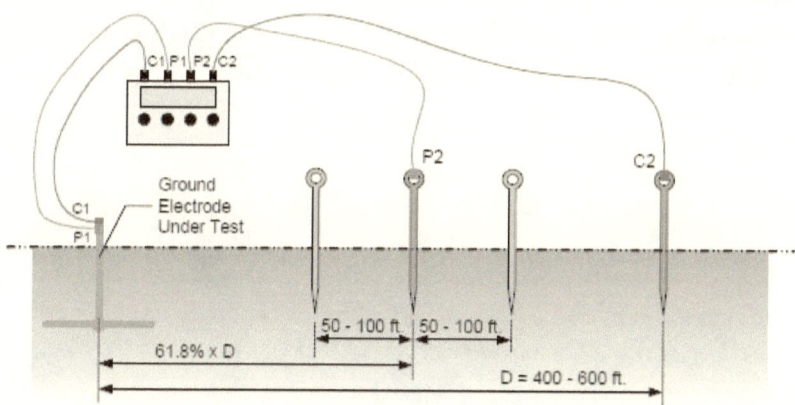

Fig 6.17: Quick Resistance Measurements of Large Electrode system

(1) Place the current probe 400 – 600 feet from the electrode;
(2) Place the potential probe 61.8% of the distance from the electrode to the current probe;
(3) Measure the resistance;
(4) Move the current probe farther away from its present position by 50–100 feet;

(5) Repeat steps 2 and 3;
(6) Move the current probe closer by 50 – 100 feet;
(7) Repeat steps 2 and 3; and
(8) Average the three readings

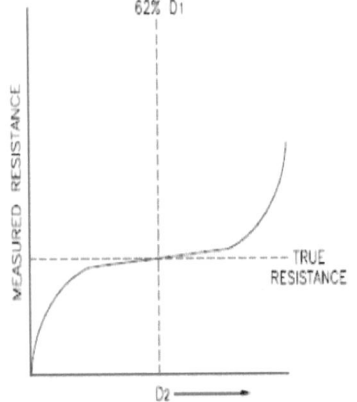

Fig 6.18: Plot for FOP Ground resistance

A value of resistance is obtained at each step. This resistance is plotted as a function of distance, and the value in ohms at which this plotted curve appears to level out is taken as the true resistance value of the earthing system under test as shown in fig 6.18.

As described in beginning of this para, the earth electrode can be thought of being surrounded by concentric shells of earth. The current probe also is surrounded by earth shells but with a smaller influence. In order to obtain a flat portion of the curve, it is necessary for the current electrode to be placed outside the influence of the earth electrode to be tested so the influential shells do not overlap as shown in Fig 6.19.

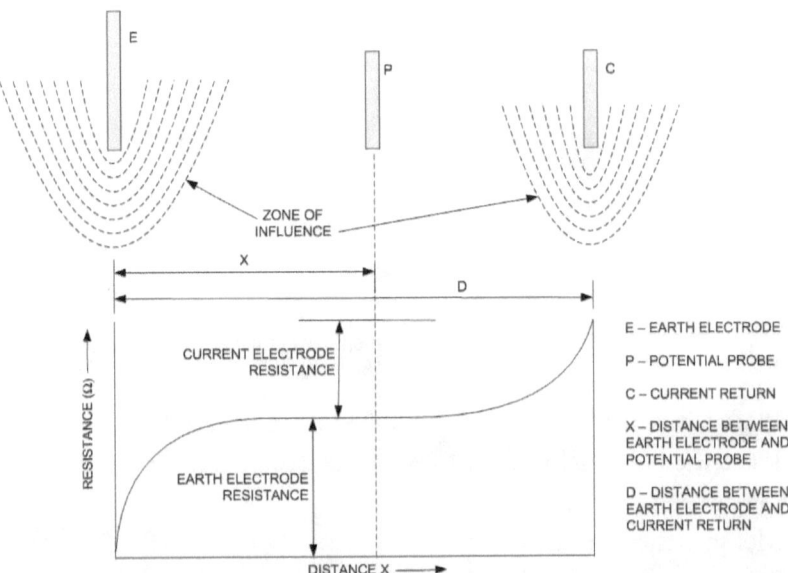

Fig 6.19: Overlapping and Non-overlapping Shells of Earth

Fall-of-Potential measurements are based on the distance of the current and potential probes from the center of the electrode under test. Farther a probe is placed from the electrode, smaller the influence. The best distance for the current probe is at least 10 to 20 times the largest dimension of the electrode (see Table 6.3).

Table 6.3: Current Probe Distance from Electrode

Maximum Electrode Dimension (Feet)	Distance to Current Probe from Electrode Center (Feet)
15	300
30	450
60	600

However, terrain or obstructions such as buildings and paved paths, can limit the space available for measuring large systems. Nevertheless, acceptable accuracy can be obtained by using smaller distances also.

It can be shown that the actual electrode resistance is measured when the potential probe is located 61.8% of the distance between the center of the electrode and the current probe. For example, if the current probe is located 400 ft from the electrode center, then the resistance can be measured with the potential probe located 61.8% x 400=247 ft from the electrode center. The 61.8% measurement point assumes the current and potential probes are located in a straight line and the soil is homogeneous (same type of soil surrounding the electrode area and to a depth equal to 10 times the largest electrode dimension). While the latter condition is almost never known with certainty, the 61.8% measurement point still provides suitable accuracy for most measurements if other cautionary measures, as described below, are taken.

As mentioned above, the electrode center location seldom is known. In such case, make at least 3 sets of measurements, each with the current probe a different distance from the electrode, preferably in different directions. However, when space is not available or obstructions prevent measurements in different directions, suitable measurements usually can be made by moving the current probe in a line away from or closer to the electrode. For example, the measurement may be made with the current probe located 200, 300 and 400 ft along a line from the electrode.

Each set of measurements involves placing the current probe and then moving the potential probe in 10 feet increments toward or away from the electrode, depending on the starting point. The starting point is not critical but should be 20 to 30 feet from the electrode connection point, in which case the potential probe is moved in 10 feet increments toward the current probe, or 20 to 30 ft from the current probe, in which case the potential probe is moved in 10 ft increments back toward the electrode.

The spacing between successive potential probe locations is not particularly critical, and does not have to be 10 ft, as long as the measurements are taken at equal intervals along a line between the electrode connection and the current probe. Larger spacing means quicker measurements at the expense of fewer data points; smaller spacing means more data points at the expense of slower measurements.

Based on the above, the basic steps for accurately measuring earth electrodes of the type are as follows. If obstructions do not permit using the probe spacing specified, make as many measurements as possible with the current probe as far and in different directions from the electrode under test as possible:

(1) Place the current probe (C2) 200 feet from the electrode under test
(2) Place the potential probe (P2) 20 feet from the point at which the test set is connected to the earth electrode
(3) Measure and record the resistance
(4) Move the potential probe (P2) farther away by 10 feet
(5) Repeat steps 3 and 4 until the potential probe (P2) is within 20 or 30 feet of the current probe (C2)
(6) Move the current probe (C2) 300 feet from the electrode under test
(7) Repeat steps 2 through 5
(8) Move the current probe (C2) 400 feet from the electrode under test 9. Repeat steps 2 through 5

Once all measurements have been made, the data is plotted with the distance from the electrode on the horizontal scale and the measured resistance on the vertical scale. The curves for each data set should be smooth with no significant peaks or valleys. If there are departures from a, smooth curve, these data points can be re-measured, ignored or replaced artificially by interpolating between two good data points. If the current probe is outside the electrode's influence, the curves will rise as the potential probe is moved away from the electrode, increase with a slight positive (increasing) slope and then level off just beyond the mid-point between the electrode and current probe. As the potential probe approaches the current probe, the slope will increase sharply. If the curves do not level off in the middle but have a small slope, the probes were only partially influenced by the electrode, and the resistance can be read from the curve at a point that is 61.8% of the distance to the current probe.

If the curves have a steep slope, the electrode influenced the measurements and reading the resistance at the 61.8% point usually gives a resistance that is higher than actual. In this case, the Intersecting Curves and Slope procedures (refer Principles and Practice of Earth Electrode Measurements: Author: Whitham D. Reeve –Reeve Engineers – for details) may be used to manipulate the data to yield a more accurate resistance value.

(2) Two-point method

In this method as shown in Fig 6.20, total resistance of the unknown and an auxiliary earth are measured. This method is generally used to measure the resistance of a single rod driven earth which has metallic water pipes in close vicinity and which can be used as an auxiliary earth. The earth resistance of metallic water pipes without insulating joints is assumed to be of the order of 1 Ω which is low in relation to the driven earth resistance which is usually of the order of 25 Ω.

The resistance of the auxiliary earth (metallic water pipes) is assumed to be negligible in comparison with the resistance of the unknown earth, and the measured value of the resistance is taken as the resistance of the unknown earth. This method is not suitable for low resistance earths. This method can be used for driven earths where a rough estimate of earth resistance is required.

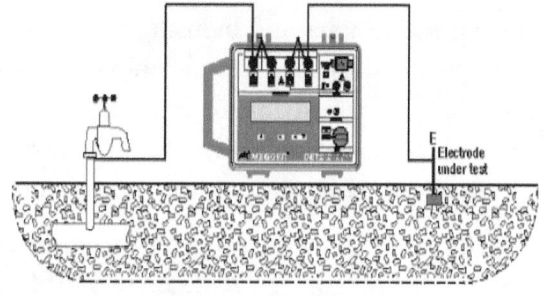

Fig 6.20: Two-Point Method

(3) Three point method**

This method involves the use of two test electrodes with the resistance of the test electrodes taken as R2 and R3. The resistance of the earth electrode under test is taken as R1. The resistance between each pair of electrodes is measured and designated as follows;

R_{12} – Resistance between earth electrode and test electrode 1
R_{13} – Resistance between earth electrode and test electrode 2
R_{23} – Resistance between test electrode 1 and test electrode 2

We know that; $R_{12} = R_1+R_2$, $R_{23} = R_2+R_3$, and $R_{13}= R_1+R_3$

Solving the above simultaneous equations, we get: $R_1 = (R_{12}+R_{13} - R_{23}) / 2$

Thus, by measuring the series resistances of each pair of earth electrodes and substituting the values in equation of R1, the value of earth resistance can be calculated. If the magnitude of the resistance of two test electrodes is comparatively higher than the earth electrode, then this method will not give accurate results. The spacing of the test electrodes for driven earths should be more than 5 mtr. This method is not suitable for large area earthing systems.

(4) Ratio method**

In this method, the resistance of the electrode under test is compared with a known resistance, usually by using the same electrode configuration as in the Fall of Potential Method. As this method is a comparison method, the ohmic readings are independent of the test current magnitude. The test current magnitude can be kept high enough (few tens of milli-amps) to give adequate sensitivity.

(5) Staged fault test**

The most representative measurement method of the earth impedance of an installation is the staged fault test. This test produces realistic fault current magnitudes and by using the remote voltage reference the rise of earth potential and earth impedance can be calculated. The staged fault test is seldom performed due to economic penalties and the system operational constraints. Staged high current tests may be required for those cases where specific information such as the integrity of the earthing system is desired.

[**IEEE Guide for measuring earth resistivity, earth impedance and earth surface potentials of a ground system, Part I – Normal Measurements", IEEE Std. 81-1983]

(6) Single clamp earth resistance testing method*** (Megger)

In this method, the auxiliary test probes for injecting current and measuring the voltages are not needed when measuring the earth resistance. Earth resistance can be measured without disconnecting the connections between the earthed body and the metal work of the electrical plant. This means that there is no need to turn off the equipment power or disconnect the earth rod. The clamp-on methodology is based on Ohm's Law (R=V/I). Fig 6.21 shows the typical set-up of the clamp-on earth tester.

A known voltage is inductively applied to a complete circuit with the help of the source coil inside the clamp of the earth tester inducing the voltage. The resulting current flow in the earthing circuit due to the induced voltage is measured by the current coil installed in the same clamp of the earth tester. The resistance of the circuit can then be calculated by taking the ratio of the induced voltage and the circulated current in the earthing circuit.

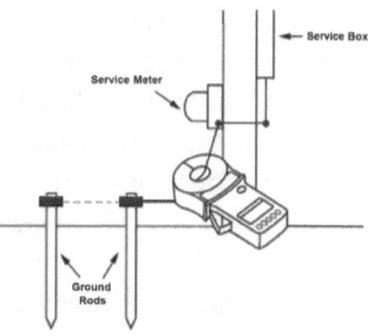

Fig 6.21: Clamp on earth resistance testing

It has to be ensured that the earthing system under test is included in the current circulation loop. The clamp-on earth tester measures the resistance of the path traversed by the induced current. All elements of the loop are measured in series. This method assumes that only the resistance of the earthing system under test contributes significantly.

According to the manufacturer (Megger), this method has the following disadvantages;

(a) If the frequency of AC current injected into the earth by the tester happens to be of the same as that of disturbance current in the earth, then the accuracy of the readings are seriously affected.

(b) The mutual inductance between the voltage and current loops of the clamp tester has some effect on the measurement accuracy.

(c) The clamp-on method is only effective in situations with multiple earthing electrodes are in parallel and a closed circuit is available for the current circulation. It cannot be used on isolated grounds, as there is no return path, thus making it in-applicable for installation checks or commissioning new sites.

(d) Measurement of low earth resistance of the order of 0.5 ohms is difficult with this method.

(e) The internal diameter of the clamp is usually small of the order of 25 mm which prevents the measurement of earth rods greater in size.

*** ["*Getting Down to Earth, A Practical Guide to Earth Resistance Testing*", by Megger, 1-866-254-0962, www.megger.com]

(7) Double clamp earth resistance online tester****

This method as shown in Fig 6.22 is slightly different to single clamp measurement method in the sense that it uses two separate clamps instead of one to avoid the effect of mutual coupling between the two coils installed in the same clamp. This system can be used without isolating the earth conductor or turning off the power. The tester uses a variable low frequency AC voltage generator, two separate clamps for voltage and current.

The frequency of injected current can be automatically adjusted so as to eliminate the effects of earth disturbance current frequency. As shown in fig 6.22, 'V' is the voltage applied to the voltage clamp by the signal generator. By electromagnetic induction, voltage 'v' will be induced in the earth loop. This voltage is given by the equation; $v = V/N_v$ Where N_v is the number of turns on the voltage clamp.

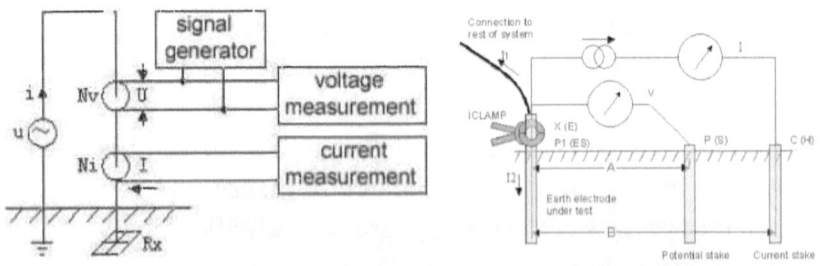

Fig 6.22: Double clamp online earth tester

Fig 6.23: Attached Rod Technique measurement (Megger)

The induced voltage 'v' produces an electric current 'i' in the earth loop. This current is given by the equation;

$$i = v/R_x$$

where Rx is the Resistance of the earth path. The current measured by the current clamp can be expressed as $I = i/N_i$, Where N_i is the number of turns on the current coil.

We know $R_x = v/i$, therefore $R_x = (V/N_v) / (I \times N_i)$ or $V/(N_v \times I \times N_i)$

If the value of N_v and N_i are set as 1, then the earth resistance can be worked out as $R_x = V/I$

In order to improve the anti-interference ability of the tester, the frequency of the AC voltage produced is kept variable and typically can be chosen automatically from 4 values as 94, 105,111 and 128 Hz. These frequencies avoid the interference from the fundamental power frequencies of 50Hz or 60Hz and their third harmonics.

[****Liwei Li, Jiyan Zou, Hui Sun, "Research on the New Clamp-on Ground Resistance On-line Tester Based on AC Variable Frequency", Proceedings of the 6th World Congress on Intelligent Control and Automation, Vol. 2, pp. 5286-5289, Dalian, China, June 21 – 23, 2006.]

(8) Attached Rod Technique by Megger***

The Attached Rod Technique (ART), as shown in Fig 6.23, provides some of the advantages of clamp-on testing (not having to disconnect the ground electrode) while remaining true to the theory and methodology of fall-of-potential testing. A fall-of-potential measurement can be made without disconnecting the ground electrode if additional measurements were made with an earth leakage clamp meter (milliamp meter). In case of an interconnected earthing system if the fall of potential test is carried out, it will measure the effective resistance of the interconnected earthing systems. If the earthing resistance of the individual earthing system is of interest, then it is necessary to know the current carried by the individual earthing system during the test conditions. This is accomplished with an additional clamp-on meter used in the ART

It involves three steps: (a) The first step is to measure the effective resistance (R) of the entire system using a typical fall-of-potential configuration with the applied voltage V, (b) The second step is to measure the total current (I) being injected into the system and (c) The third step is to measure the amount of current (I1) flowing to the rest of the interconnected system other than the earthing under test. The difference of the current (I) and the current (I1) gives the current (Ie) passing through the earthing system under test. Using this current, the resistance of the earthing (Re) system under test can be calculated as; Re = R x (I / Ie)

6.11.4 Soil Resistivity Measurements

Soil resistivity is a measure of how well the soil passes electrical current. It's reverse of conductivity. Soil with high resistivity does not pass current well. Resistivity is defined in terms of the electrical resistance of a cube of soil. The resistance of this cube, as measured across its faces, is proportional to the resistivity and inversely proportional to the length of one side of the cube.

The resistivity of the soil in which the earth electrode is buried is the limiting factor in attaining low electrode resistance. Soil with low resistivity is better for constructing low resistance earth electrodes than soil with high resistivity. Soil resistivity depends on soil type, salt concentration, moisture content and temperature.

From an operational point of view, soil resistivity measurements allow for initial earth electrode design based on the standard formula and in prediction of the effects of changes to an existing electrode.

Resistivity measurements provide an average reading. The method almost always used and described here assumes the soil is homogeneous (all the same or similar kind). Soil resistivity is measured by injecting a current into the earth, measuring the voltage drop, and then calculating the resistance. Four-terminal test sets do the resistance calculations automatically. If the four probes are placed equidistance from each other and in a straight line as shown in Fig 6.24, then resistivity (ρ) can be calculated;

$\rho = 2 \pi a R$ = 1.195 x a (ft) x R ohm-m or ρ = 6.283 x a (mtr) x R ohm-m

Where: ρ = Soil resistivity, ohm-m
a = Distance between the four equally spaced probes, ft or mtr
R = Measured resistance, ohms (Also R= ρ x l /a ohms)

Most soil resistivity tables are in units of ohm-m or ohm-cm but most measurements are in ft or mtr. The above formulas are valid only if the probes are buried not more than 1/20 of the separation distance, a. For example, if the probes are spaced 5 ft (1.5 m), they cannot be driven more than about 3 inch (0.075 m). Every effort should be made to drive all four probes to the same depth.

As stated above, this is average resistivity to a depth equal to the probe separation distance. For example, if the probes are spaced 10 ft, the measurement provides the average resistivity to a depth of about 10 ft.

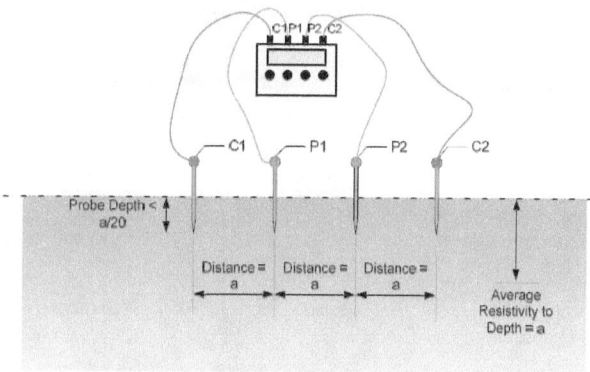

Fig 6.24: Soil Resistivity Measurement Setup

Generally, soil resistivity measurements are taken at several probe spacing's starting at 5 or 10 ft spacing and then increasing in 5 or 10 ft increments. The measurements may indicate that soil resistivity decreases sharply as the depth increases. If so, the earth electrodes should be buried deep to take advantage of the low resistivity.

Fig 6.25 shows a plot of Soil Resistivity Measurements taken at a typical site for having an idea as to how the soil restively varies with the depth of probes. The resistivities of various types of soil are given in Table 6.4. A wide range of soil types may be found in any given area or the soil types may vary widely with depth, or both. The season and weather conditions will greatly affect the soil resistivity. For example, high resistivity rock can be overlain by clays that have low resistivity in the summer but have high resistivities in the winter when frozen.

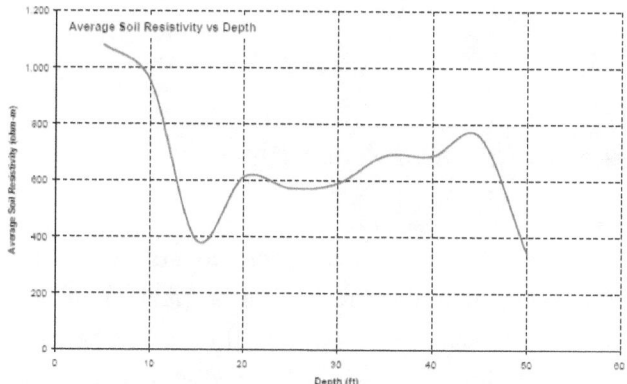

Fig 6.25: Soil Resistivity Measurements at a typical site

Table 6.4: Resistivity of various types of Soils (ohm-m)

Soil Type	Median	Min	Max	Soil Type	Median	Min	Max
Topsoil, loam	26	1	50	Clayey gravel, poorly graded gravel, sand-clay mixture	300	200	400
Inorganic Clay of High Plasticity	33	10	55	Rocky soil, steep hills	750	500	1,000
Fills – ashes, cinders, brine wastes	38	6	70	Well graded gravel, gravel-sand mixtures	800	600	1000
Gravelly clays, sandy clays, silty clays, lean clays	43	25	60	Granites, basalts, etc.	1000	—	—
Slates, shales	55	10	100	Sandstone	1010	20	2000
Silty or clayey fine sands with slight plasticity	55	30	80	Poorly graded gravel, gravel-sand mixtures	1750	1000	2500
Clayey sands, poorly graded sand-clay mixtures	125	50	200	Gravel, sand, stones, little clay or loam	2585	590	4580
Fine sandy or silty clays, lean clays	190	80	300	Sandy, dry, flat (typical coastal country)	2750	500	5,000
Decomposed gneisses	275	50	500	Surface limestone	5050	100	10000
Silty sands, poorly graded sand-silt mixtures	300	100	500	City, industrial areas	5500	1000	10000
				Note: To convert to ohm-cm, multiply by 100			

6.12 General points and Precautions

(a) Precautions
(1) The resistances of parallel electrodes, do not combine like parallel resistors because of mutual resistance between them. So resistances of interconnected parallel electrodes always will be higher than the parallel combination of the equivalent individual resistances.

(2) Never disconnect operating equipments from the earth electrode before switching off to ensure your safety.
(3) In a new setup, it will be possible to test individual grounding system components as they can be isolated from each other. Otherwise, in a working setup during periodical checks, measure the overall system.
(4) Usually, in broadcast transmitters set up facility earth systems consist of several components and the facility operates normally with all the components interconnected. Therefore, if your goal is to measure only the broadcast facility earth electrode system, make sure all components, except the electric utility, are bonded as they would be in the final operating configuration and measure them as a complete system.
(5) If you can safely disconnect an individual electrode component from the grounding system, then you can correctly measure that component.
(6) Frozen ground is an insulator. Readings taken during the summer when the active soil layer is thawed are meaningless for winter. Where the winter frost depth exceeds the depth of earth electrodes, the electrodes are insulated from remote earth and the resistance of the electrode system will be very high. To reduce potential differences for safety and operation, all metallic components of the broadcast facility must be bonded together.

(b) Test Instruments
(1) Battery operated test sets with an analog null meter are the best, most consistent and are the most reliable.
(2) The hand-crank test set works fine but are relatively difficult and awkward to use.
(3) Some digital test sets most often do not yield usable data. This is particularly the case if the earth electrode system being measured is near to a power plant or substation.
(4) Always test your test set before taking it into the field. The manual tells you how to do this.
(5) Take an extra set of batteries, if the test set uses rechargeable batteries, make sure they are not worn out and have enough charge for the measurements to be made.

(6) Occasionally check calibration especially if the test set has not been taken care of in the field. The manual tells you how to do this (you will need some resistors).

(7) For electrode measurements, always short P1 and C1 terminals together or use separate test leads, one from P1 to the earth electrode and one from C1 to the earth electrode. The latter provides the best accuracy.

(c) Test Leads – General Information

(1) Use a heavy-gauge (14 – 16 AWG) test lead wire.

(2) The lead connected to the current probe (C2) should have a length of 300 – 400 ft or more.

(3) The lead for the potential probe (P2) should have a length of 300-400 ft or more (same length as C2 lead).

(4) The lead connected from C1/P1 to the grounding electrode system should be as short as possible, preferably less than 10 ft long. If this lead is too long, the readings will not be reliable. The resistance of this lead directly affects the measured resistance. One way to eliminate this measurement error is to use a separate lead from C1 terminal to grounding electrode system and a separate lead from the P1 terminal to the grounding electrode.

(d) Test Lead Kit

(1) Have approximately 600 ft – 800 ft of test lead wire (300 ft – 400 ft for P2 and 300 ft – 400 ft for C2).

(2) Make up two test leads, each 100 ft long, with a large (2 or 3 inch) alligator clip at one end and 1/4 inch fork terminal lug at the other. One will connect to the C2 terminal and the other to the P2 terminal on the test set. Use rubber insulating boots over the alligator clips

(3) Use the remaining test lead wire to make 100 ft sections with large insulated alligator clips at each end.

(4) Make up one test lead, 10 ft long, with large alligator clip at one end and 1/4 inch Fork terminal lug at the other. This will be used to connect P1/C1 terminal on the test set to the grounding electrode system. Also make up extra test leads if it is necessary to locate the test set some distance from the connection point at the earth electrode (one test lead for C1 and one for P1).

(5) Put test leads on 12 or 14 inch "Cord wheels." These are available from hardware stores and are designed for coiling extension cords but work well with test leads.
(6) Put spare terminal lugs and alligator clips in the kit along with a crimping tool, screwdriver and any other tools you may need to repair the test leads in the field. Be sure to include a set of good gloves.
(7) The kit should have safety glasses, gloves, and a short-handle, 2kg, double-head hammer, and test probes.

(e) Test Leads – Connections
(1) When stringing out the test leads, make sure the insulation is not damaged. Use gloves to handle the wire, especially if the leads are strung below or above power lines.
(2) Where test lead sections are connected together to make a longer lead, be sure to insulate the alligator clips from the earth and foliage. Use 8 inch long, 1-1/2 or 2 inch flexible non-metallic conduit or other insulating material as insulator sleeves (these are slipped over the alligator clips to keep them from touching the ground).
(3) Make sure to clean the point of connection to the grounding electrode system. Use ScotchBrite or wire brush.
(4) Farther the current probe (C2) is from the grounding electrode, the better. The object is to get the P2 probe out of the influence of the grounding electrode system when P2 is 61.8% of the distance to C2. If the largest dimension of the electrode system is x ft, then C2 probe should be 10x ft to 20x ft from the electrode system.

(f) Test Probes
(1) Test probes should be 18–20 inch long and made from strong galvanized steel. They should have a loop at one end and sharpened to a point at the other.
(2) It is a good idea to carry a couple longer probes made from a cut ground rod, 3 to 4 ft long, for use where the shorter probes do not make good enough contact with the earth. Weld a T-bar near the top of these longer rods so you can twist and pull them out of the ground.
(3) The probes normally only have to be driven to a depth of 6–12 inch. The deeper the probe the harder it is to pull out. Wear safety glasses when driving the probes.

(4) Remove probes by twisting and pulling on the loop at the top of the probe; do not hit the side of the probe with a hammer or kick it to loosen it because all you will do is bend the probe and ruin it.
(5) If you have to place the probes in a parking area or roadway with very hard ground, concrete or pavement, use metal plates in place of the pointed probes. Lay the plates on the ground and wet them and the area around them thoroughly. An alternative is to drill through the pavement or concrete with a hammer drill.
(6) Make sure the alligator clip makes a good connection where it is connected to the probe. Use 3M ScotchBrite or wire brush to clean the probes where you connect the alligator clips.
(7) P2 should be placed in a straight line between the connection point and C2. Sometimes it will be necessary to place C2 in one direction but move in another direction with P2, although some test set manufacturers recommend against this procedure. If you do this, move P2 in a straight line away from the connection point and away from the electrode system.
(8) Avoid placing the probes in line with underground facilities (electric, telephone, metallic water and sewer pipes, etc.) but sometimes you have no choice.

(g) Readings
(1) Start your readings with the P2 probe 10–20 ft from the point of connection.
(2) Take a reading every 10 ft to within 20–30 ft of C2. You will notice the readings are low when you are close to the point of connection and slowly climb. If C2 is far enough away, the readings will level off in the middle. As you approach C2, the readings climb rapidly.
(3) If C2 is far from the influence of the electrode system, the readings will level off near 61.8% point. When P2 is closer to the connection point, the readings will be lower and when closer to C2, the readings will be higher. Readings that decrease as you move away from the connection point indicates underground metallic facilities are interfering with the measurements or that the P2 probe is over the grounding electrode. An occasional dip in the readings is not unusual, but if several readings follow this pattern, you have to move C2 and start over.

(4) Always take at least three separate sets of readings. Usually, you can take two sets of readings with the C2 probe at two different distances along the same direction; say 300 and 250 ft, although 400 and 300 ft are preferable. Take 3rd set of readings in another direction if possible. Depending on topography and obstructions, may have to take all 3 readings in the same direction (say 400, 300 and 200 or 300, 250 and 200 ft).

(5) If you are able to move the test set analog meter needle to either side of null (with the resistance knobs), the probes are deep enough. If the needle moves only to one side (usually the "– "side), one or both probes are not deep enough, you have a bad connection, or the alligator clips of lead are in contact with the earth.

Check all connections and try driving the probes a little deeper or use the longer probes. You also can try pouring water around the probe. In any case, you must be able to move the needle to both sides of null with the resistance knobs; otherwise, the reading is not reliable.

(6) With analog instruments, the null sometimes will be slow (move slowly as you adjust the resistance knobs), which indicates underground metallic facilities or inadequate probe contact. For the latter situation, drive the probes deeper or use longer probes.

(7) With analog instruments, the meter sometimes will oscillate around null. You will have to adjust the resistance knobs so the meter moves an equal distance left and right of null (center) and then take a reading; this takes practice and a little practice.

(8) Digital instruments include fault indicators that indicate excessively high probe resistance or bad connections. If you see this indication, check all connections and, if necessary, drive the probes deeper.

(h) Data Collation

(1) The actual electrode system resistance is measured when P2 is 61.8% of the distance C2 is from the center of the electrode system. However, the actual center of the electrode system seldom is known except in small systems, so the location of this 61.8% point is never known accurately. A small system would be a few ground rods that are not connected to anything else other than each other.

(2) If measuring a small system, put C2 outside the influence (10x–20x as mentioned above) and then put P2 at a point 61.8% of this distance. The reading you take at this point will be close to the actual electrode

system resistance. Additional readings are not necessary, but they should be taken to ensure that the data is valid.

(3) If measuring large Systems, use "Intersecting Curves" or "Slope" method (refer to main source).

References

1. Nanometrics Equipment Grounding Recommendations Nanometrics Inc. Seismological Instruments Nanometrics Systems Engineering November 3, 2003
2. Principles and Practice of Earth Electrode Measurements: Author: Whitham D. Reeve –Reeve Engineers –USA
3. Chemical Earthing: Alfredkim Systems & Solutions Pvt Ltd, 14/3, Bolton Compound Mathura Road, Faridabad – 121003 (Haryana) – India (*www.alfredkim.net*)
4. A practical guide to earth resistance testing – Application notes of M/S Megger website; *www.megger.com*. 2621 Van Buren Avenue, Norristown PA 19403-2329, 1-866-254-0962,
5. Designing for A Low Resistance Earth Interface (Grounding) Roy B. Carpenter, Jr. Joseph A. Lanzoni An LEC Publication Revised 2007, 6687 Arapahoe Road, Boulder Co, 80303-1453 USA.
6. Impulse Measurements in Earthing Systems by Deepak Lathi A Thesis for the degree of Doctor of Philosophy November 2012 School of Engineering Cardiff University
7. IEEE Guide for measuring earth resistivity, earth impedance and earth surface potentials of a ground system, Part I – Normal Measurements", IEEE Std. 81-1983
8. Liwei Li, Jiyan Zou, Hui Sun, "Research on the New Clamp-on Ground Resistance On-line Tester Based on AC Variable Frequency", Proceedings of the 6th World Congress on Intelligent Control and Automation, Vol. 2, pp. 5286-5289, Dalian, China, June 21 – 23, 2006.]

CHAPTER 7

RADIO FREQUENCY INTERFERENCE AND SHIELDING

7.1 The EM environment

An equipment in the electromagnetic (EM) environment is exposed to electromagnetic interference (EMI) originated by physical phenomena or generated by various other equipments. EMI is somewhat arbitrarily defined to cover the frequency spectrum from about 10 Hz to 100 GHz. For radiated emissions a lower frequency limit of 10 kHz is often used, although EMI can exist in many equipment and systems even below this frequency. Except for electrostatic discharge (ESD) there rarely exists a pure DC EMI problem.

The EM environment will be variable from place to place, and between various locations. Estimation of EMI environment in any situation is required before adequate protection methods, which will enable equipment to operate without error in all environments, can be selected. E.g., if solenoid control valve operation of a dummy load of high power transmitter is to be initiated automatically by a micro computer on the basis of impedance, then the equipment must be capable of continuous operation in high electromagnetic field of the transmitter complex.

7.2 Electromagnetic Compatibility and Interferences

Performance of an electric / electronic equipments and circuits in particular analog circuit is often affected adversely by high frequency signals from nearby electrical / radio frequency activity. And equipment containing analog circuitry may also adversely affect the systems external to it. So Electromagnetic Compatibility (EMC) based on the IEC-60050 standard definition is:

"**EMC** is the ability of a device, or a system to function satisfactorily in its EM environment without introducing intolerable electromagnetic disturbances to other gadgets in that environment". OR

"**EMC** is the ability to either equipment or a system to function as designed without degradation or malfunction in their intended operational electromagnetic environment. Further, the equipment or system should not adversely affect the operation of, or be affected by any other equipment or system".

The other terms in EM environment are defined as;

Interference can be defined as the undesirable effects of noise. If noise voltage causes unsatisfactory operation of a circuit, it is interference. Usually noise cannot be eliminated but only reduced in magnitude until it no longer causes interference.

Susceptibility is the characteristic of electronic equipment that results in undesirable responses when subjected to electromagnetic energy. The susceptibility level of a circuit or device is the noise environment in which the equipment can operate satisfactorily.

Electrostatic discharge (ESD) is a phenomenon that is becoming an increasingly important concern with today's integrated circuit technology. The basic phenomenon is the build-up of static charge on a person's body or furniture which discharge to the product when the person or furniture touches the product. In dry atmosphere and especially where carpets are used in a computer room, the operator can be charged to high voltage. When the discharge occurs, a relatively large current momentarily passes through the product. These currents can cause IC memories to clear, machines to reset etc. Such discharges can play havoc with the control circuit of transmitters and other devices. If a computer unit is touched by such a charged operator a discharge spark can occur and result in malfunctioning. In unfavourable condition, the static discharge can approach 25 kV in magnitude, but normally not more than 6 kV by contact and 8 kV in air. So such charges must be discharged to ground before they damage equipments.

Harmonic distortion is a phenomenon when non-linear loads, e.g. static power converters, arc discharge devices, change the sinusoidal nature of the AC power thereby resulting in the flow of harmonic currents in the AC system. The degree to which harmonics can be tolerated is determined by the susceptibility of the load (or power source) to them. The least susceptible type of equipment is that in which the main function is heating. The most susceptible type of equipment is that whose design or construction assumes a (nearly) perfect sinusoidal fundamental input e.g. communication or data processing equipment. Other types of electronic equipment can be affected by transmission of AC supply harmonic through the equipment power supply or by magnetic coupling of harmonics into equipment components. Computers and similar equipment such as programmable controllers frequently require AC sources that have no more than 5% harmonic voltage distortion factor with the largest single harmonic being no more than 3% of the fundamental voltage.

EMC therefore has two aspects; (1) It describes the ability of electrical and electronic systems to operate without interfering with other systems. (2) It also describes the ability of such systems to operate as intended within a specified EM environment.

So, complete EMC assurance mean that the equipment should neither produce spurious signals, nor should it be vulnerable to out-of-band external signals. It is the latter class of EMC problem to which analog equipment most often falls prey. And, it is the handling of these spurious signals that are studied and emphasized in this chapter.

The externally produced electrical activity may generate noise, and is referred to either as EMI, or RFI. In this chapter, we will refer to EMI in terms of both electromagnetic and radio frequency interference. One of the more challenging tasks of the analog circuit designer is the control of equipment against undesired operation due to EMI. It is important to note that, EMI and or RFI is always detrimental once it enters equipment, as it can and will degrade its operation, quite often considerably.

This section is oriented heavily towards minimizing undesirable analog circuit operation due to EMI/RFI. Misbehaviour of this sort is also known as EMI or RFI susceptibility, indicating a tendency towards anomalous equipment behaviour when exposed to EMI/RFI. There is of course a complementary EMC issue, namely with regard to spurious emissions. However, since analog circuits typically involve fewer pulsed, high speed, high current signal edges that give rise to such spurious signals (compared to high speed logic; for example), this aspect of EMC isn't as heavily treated here. Nevertheless, the reader should bear in mind that it can be important, particularly if the analog circuitry is part of a mixed-signal environment along with high speed logic.

7.3 EMI / RF Interferences Mechanisms

To understand and properly control EMI/RFI, it is helpful to first segregate it into manageable portions. Thus it is useful to remember that when EMI / RFI problems do occur, it can be fundamentally broken down into a Source, a Path, and a Receiver. A system designer, have under his direct control the receiver part of this mechanism, and perhaps some portion of the path. But seldom will the designer have control over the actual source.

7.3.1 EMI Noise Sources

Depending on the different environments, a wide variety of interference sources can be encountered. Power convertors, switch gears, contactors, relays, welders, radio and television transmitters and mobile radios, are among the most conspicuous EMI sources. Transient disturbances, which occur most frequently, usually for short random periods of time and mostly result from interferences caused by lightning, earth-faults or switching of heating furnaces or welding machines and inductive loads/ circuits.

These disturbances can have a frequency range from 50 Hz up to a few hundred MHz with time duration including transients ranging from less than 10 nanoseconds to a few seconds. Some common sources of externally generated noise;

(a) All broadcasts transmitters' in particular Radio and TV,
(b) All communications system including Mobile Radio
(c) All type of walkie –Talkie including Cellular Telephones,
(d) Vehicular Ignition and household gadgets.
(e) Arc welding machines and Lightning strikes,
(f) Utility Power Lines and substations,
(g) Electric Motors and their drives,
(h) Inverters, Computers, UPS,
(i) Garage Door Openers or other automatic remote devices, and
(j) Telemetry Equipments

Since no control is possible over these sources of EMI, the only best management tool to control these, one can exercise over them is to recognize and understand the possible paths by which they couple to the equipment.

7.3.2 EMI Noise Coupling Paths

The system boundaries for penetration of interference may be power feed lines, input signal lines, output signal lines or equipment enclosure. With only a few exceptions, EMI begins as a desirable signal current flowing along an induced path. The signal current becomes a source of interference when it is diverted to one or more unintentional paths that lead, ultimately to a victim. Some circuit elements may generate new voltages, currents, or fields. Typical transition points include the generation of voltages by ground currents flowing through the distributed impedance of ground, the generation of fields by currents flowing along conductors, and the leakage of currents to nearby circuit elements through stray capacitances. It is important to identify the transition points along the coupling paths to a victim because these points make the mostly effective locations for EMI fixes.

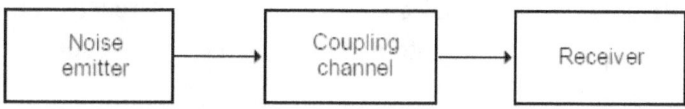

Fig 7.1: Typical EMI Noise Path

As seen in Fig 7.1, in a typical EMI noise path, three elements are necessary to produce a noise problem, i.e. there must be a noise source, a receiver circuit that is susceptible to the noise, and finally a coupling channel to transmit the noise from the source to the receiver. It follows that there are three ways to break the noise path: the noise can be suppressed at the source, the receiver can be made insensitive to the noise, or the transmission through the coupling channel can be minimised. In some cases, noise suppression techniques must be applied to two or to all three parts of the noise path.

So EMI / RFI noise sources can couple from anywhere. There are countless ways in which undesired noise can couple into an analog circuit to ruin its functioning, reliability and accuracy.

The EMI coupling paths are actually very few in terms of basic number. Three main general paths are:

(a) Interference due to conduction (common-impedance)
(b) Interference due to capacitive or inductive coupling (near-field interference)
(c) Interference due to Electromagnetic radiation (far-field interference)

7.3.3 Noise Coupling Mechanisms

EMI /RFI energy may enter wherever there is an impedance mismatch or discontinuity in a system. In general this occurs at the interface where cables carrying sensitive analog signals are connected to printed circuit boards (PCB), and through power supply leads. Improperly connected cables or poor power supply filtering schemes are often perfect conduits for entering of interference.

Conducted noise may also be encountered when two or more currents share a common path (impedance). This common path is often a high impedance "ground" connection. If two circuits share this path, noise currents from one will produce noise voltages in the other. Steps may be taken to identify potential sources of this interference. Some of the general ways noise can enter into a circuit from external sources are given below;

(a) Impedance mismatches and discontinuities
(b) Common-mode impedance mismatches → Differential Signals
(c) Capacitive Coupled (Electric Field Interference) → dV/dt →Mutual Capacitance →Noise Current (Example: 1V/ns produces 1mA/pF)
(d) Inductively Coupled (Magnetic Field) di/dt →Mutual Inductance →Noise Voltage (Example: 1mA/ns produces 1mV/nH)

There is a capacitance between any two conductors separated by a dielectric (air and vacuum are dielectrics, as well as all solid or liquid insulators). If there is a change of voltage on one conductor there will be change of charge on the other, and a displacement current will flow in the dielectric. Where either the capacitance or the dV/dT is high, noise is easily coupled. A 1-V/ns rate-of-change gives rise to displacement currents of 1 mA/pF.

If changing magnetic flux from current flowing in one circuit couples into another circuit, it will induce an emf in the second circuit. Such mutual inductance can be a troublesome source of noise coupling from circuits with high values of dI/dT. A mutual inductance of 1 nH and a changing current of 1 A/ns will induce an emf of 1 V.

7.3.4 Reducing Common-Impedance Noise

Steps to be taken to eliminate or reduce noise due to the conduction path sharing of impedances, or common-impedance noise are as below.

Common-impedance noise
(a) Decouple op amp power leads at LF and HF, (b) Reduce common-impedance, and (c) Eliminate shared paths.

Techniques
(a) Bypass Low impedance electrolytic (LF) and local low inductance (HF), (b) Use ground and power planes, and Optimize system design.

Power supply rails feeding several circuits are good common-impedance examples. Real world power sources may exhibit low output impedance, or may not—especially over frequency. Furthermore, PCB traces used to distribute power are both inductive and resistive, and may also form a ground loop. The use of power and ground planes also reduces the power distribution impedance. These dedicated conductor layers in a PCB are continuous (ideally, that is) and as such, offer the lowest practical resistance and inductance.

In some applications where low-level signals encounter high levels of common-impedance noise it will not be possible to prevent interference and the system architecture may need to be changed. Possible changes include:

(a) Transmit signals in differential form, (b) Amplify signals to higher levels for improved S/N (signal to Noise ratio), (c) Convert signals into currents for transmission, and (d) Convert signals directly into digital form.

7.3.5 Noise Induced by Near-Field Interference

Crosstalk is the second most common form of interference. In the vicinity of the noise source, i.e., near-field, interference is not transmitted as an electromagnetic wave, and the term crosstalk may apply to either inductively or capacitive coupled signals.

(a) Reducing Capacitance –Coupled Noise

This noise may be reduced by reducing the coupling capacitance (increase conductor separation), but is most easily cured by shielding. A conductive and grounded shield (known as a Faraday shield) between the signal source and the affected node will eliminate this noise, by routing the displacement current directly to ground.

With the use of such shields, it is important to note that it is always essential that a Faraday shield be grounded. A floating or open-circuit shield almost invariably increases capacitive-coupled noise.

Methods to eliminate capacitance-coupled interference are summarized below;

a) Reduce Level of High dV/dt Noise Sources, b) Use Proper Grounding Schemes for Cable Shields, c) Reduce Stray Capacitance by; (1) Equalize Input Lead Lengths, (2) Keep Traces Short, (3) Use Signal-Ground Signal-Routing Schemes, and d) Use Grounded Conductive Faraday Shields to Protect against Electric Fields.

(b) Reducing Magnetically – Coupled Noise

Methods to eliminate interference caused by magnetic fields are summarized below;

a) Careful routing of Wiring, b) Use Conductive Screens for HF Magnetic Shields, c) Use High Permeability Shields for LF Magnetic Fields (mu-Metal), d) Reduce Loop Area of Receiver by adopting; (1) Twisted Pair Wiring, (2) Physical Wire Placement, (3) Orientation of Circuit to Interference, and e) Reduce Noise Sources by using; (1) Twisted Pair Wiring, and (2) Driven Shields

To illustrate the effect of magnetically-coupled noise, consider a circuit with a closed-loop area of A cm2 operating in a magnetic field with an rms flux density value of B gauss.

The noise voltage Vn induced in this circuit can be expressed by the following equation:

$$V_n = 2 \pi f B A \cos\theta \times 10^{-8} \text{ V}$$

In the equation, f is the frequency of the magnetic field, and θ is the angle of the magnetic field B to the circuit with loop area A. Magnetic field coupling can be reduced by reducing the circuit loop area, the magnetic field intensity, or the angle of incidence. Reducing circuit loop area requires arranging the circuit conductors closer together. Twisting the conductors together reduces the loop net area. This has the effect of cancelling magnetic field pickup, because the sum of positive and negative incremental loop areas is ideally equal to zero. Reducing the magnetic field directly may be difficult. However, since magnetic field intensity is inversely proportional to the cube of the distance from the source, physically moving the affected circuit away from the magnetic field has a great effect in reducing the induced noise voltage. Finally, if the circuit is placed perpendicular to the magnetic field, pickup is minimized. If the circuit's conductors are in parallel to the magnetic field the induced noise is maximized because the angle of incidence is zero.

There are also techniques that can be used to reduce the amount of magnetic-field interference, at its source. In the previous paragraph, the conductors of the receiver circuit were twisted together, to cancel the induced magnetic field along the wires. The same principle can be used on the source wiring. If the source of the magnetic field is large currents flowing through nearby conductors, these wires can be twisted together to reduce the net magnetic field.

Shields and cans are not as effective against magnetic fields as against electric fields, but can be useful on occasion. At low frequencies magnetic shields using high-permeability material such as Mu-metal can provide modest attenuation of magnetic fields. At high frequencies simple conductive shields are quite effective provided that the thickness of the shield is greater than the skin depth of the conductor used (at the frequency involved). *Note*—copper skin depth is $6.6/\sqrt{f}$ cm; here is f in Hz.

7.4 Filters with Passive Components

Passive components, such as resistors, capacitors, and inductors, are powerful tools for reducing externally induced interference when used properly by arranging them in various form of Low Pass (LP) filters.

Simple RC networks make efficient and inexpensive one-pole, low-pass filters. Incoming noise is converted to heat and dissipated in the resistor. But note that a fixed resistor does produce thermal noise of its own. Also, when used in the input circuit of an op amp or in-amp, such resistor(s) can generate input-bias-current induced offset voltage. While matching the two resistors will minimize the dc offset, the noise will remain. Table 7.1 below summarizes some popular low-pass filters for minimizing EMI.

Table 7.1: Using Passive Components within Filters to Combat EMI

S. No	LP Filter Type	Advantage	Disadvantage
1	RC-Section(R-C)	-Simple -Inexpensive	-Resistor Thermal Noise, $-I_B \times R_{drop}$ —— offset, – Single-Pole Cut off
2	LC-Section(Bifilar) (L-C)	-Very Low Noise at LF Very, -Low IR Drop, -Inexpensive, -Two-Pole Cutoff	-Medium Complexity -Nonlinear Core Effects Possible
3	Π-Section(C-L-C)	-Very Low Noise at LF -Very Low IR Drop -Pre-packaged Filters -Multiple-Pole Cutoff	-Most Complex -Non linear Core Effects Possible -Expensive

In applications where signal and return conductors are not well-coupled magnetically, a common-mode (CM) choke can be used to increase their mutual inductance. Note that these comments apply mostly to in-amps, which naturally receive a balanced input signal (whereas op amps are inherently unbalanced inputs—unless one constructs an in-amp with them). A CM choke can be simply constructed by winding several turns of the differential signal conductors together through a high-permeability (>2000) ferrite bead. The magnetic properties of the ferrite allow differential -mode currents to pass unimpeded while suppressing CM currents.

Capacitors can also be used before and after the choke, to provide additional CM and differential-mode filtering, respectively. Such a CM choke is cheap and produces very low thermal noise and bias current-induced offsets, due to the wire's low dc resistance. However, there is a field around the core. A metallic shield surrounding the core may be necessary to prevent coupling with other circuits. Also, note that high-current levels should be avoided in the core as they may saturate the ferrite.

The third method for passive filtering takes the form of packaged π-networks (C-L-C). These packaged filters are completely self-contained and include feed through capacitors at the input and the output as well as a shield to prevent the inductor's magnetic field from radiating noise. These more expensive networks offer high levels of attenuation and wide operating frequency ranges, but the filters must be selected so that for the operating current levels involved the ferrite doesn't saturate.

7.4.1 Reducing System Susceptibility to EMI

The general examples discussed above and the techniques illustrated earlier in this section outline the procedures that can be used to reduce or eliminate EMI/RFI. Considered on a system basis, a summary of possible measures is given below;

(a) Always assume that Interference exists?
(b) Use Conducting Enclosures against Electric and HF Magnetic fields
(c) Implement Cable Shields Effectively
(d) Use Feedthrough Capacitors and Packaged PI Filter

7.5 A Review of EM Field

In the region around an electric lead that carries an alternating current, an electromagnetic field is set up. The field changes in strength and direction in phase with the alternating current. The field propagates away from the lead as electromagnetic waves with the speed of light.

All macroscopic electromagnetic phenomena may be expressed mathematically (Maxwell's equations). The equations describe the distributed-parameter nature of electromagnetic fields, i.e. the electromagnetic quantities, distributed throughout space, e.g. a set of partial differential equations being functions of spatial parameters x, y, z in three dimensional space as well as time. From a mathematical standpoint, these equations are difficult, although they are quite easy to describe in conceptual terms. Where appropriate one uses approximations, and the governing equations become ordinary differential equations where the variables are functions of only one parameter i.e. time.

An approximation of the field intensity and radiated power from antenna in free space:

The power density S at a point due to the power P_T radiated by an isotropic radiator is given as follows:

$$S = P_T / 4\pi r^2 \text{ and } S = E^2 / 120\pi$$

Where S = power density (W/m2); r = distance (mtr); P_T = transmitted power (Watt).

The field intensity of an isotropic radiator in free space is;

$$E = 5.5 / r \cdot \sqrt{P_T}$$

For a half-wave dipole in the direction of maximum radiation;

$$S = 1.64 \times P_T / 4\pi r^2 \text{ and } E = 7.01 / r \, (\sqrt{P_T})$$

Where 1.64 = maximum gain of a half-wave dipole.

The efficiency of an antenna is the ratio of the radiation resistance to the total resistance of the system. The total resistance includes radiation resistance, resistance in conductors and dielectric, including the resistance of loading coils if used and the resistance of the earthing system.

The use of transceivers of which the antenna is too close to electronic equipment is a matter of great importance. A separation distance of 2 m between the antenna and the equipment is highly recommended. In addition, operation at reduced power ratings will materially reduce the influence of radiated interference resulting from the use of portable transceivers.

The ratio of the electric field E to the magnetic field H is the wave impedance. In the far field this ratio E/H equals the characteristic impedance of the medium e.g. $E/H = Z_o \approx 377$ for air or free space. In the near field the ratio is determined by the characteristics of the source and distance from the source to where the field is observed. If the source has high current and low voltage ($E/H < 377\Omega$) the near field is predominantly magnetic. Conversely, if the source has low current and high voltage ($E/H > 377\,\Omega$) the near field is predominantly electric.

For a rod or straight wire antenna, the source impedance is high. The wave impedance near the antenna – predominantly an electric field is also high. As distance is increased, the electric field loses some of its intensity as it generates a complementary magnetic field. In the near field, the electric field attenuates at a rate of $1/r^3$ whereas the magnetic field attenuates at a rate $1/r^2$. Thus the wave impedance from a straight wire antenna decreases with distance and asymptotically approaches the impedance of the free space in the far field.

For a predominately magnetic field – such as produced by a loop antenna – the wave impedance near the antenna is low. As the distance from the source increases, the magnetic field attenuates at a rate $1/r^3$ and the electric field attenuates at a rate of $1/r^2$. The wave impedance therefore increases with distance and approaches that of free source at a distance of $\lambda/2\pi$. In the far field, both the electric and magnetic fields attenuate at a rate of $1/r$.

At frequencies less than 1 MHz, most coupling within electronic equipment is due to the near field, since the near field at these frequencies extends out to 50 metres or more.

At 30 kHz, the near field extends out to 1.59 km. Therefore, interference problems within any given equipment should be assumed to be the near field problems unless it is clear that they are far field problems. In the near field the electric and magnetic fields must be considered separately, since the ratio of the two is not constant. In the far field, however, they combine to form a plane wave having an impedance of 377 Ω. Therefore, when plane waves are discussed, they are assumed to be in the far field. When individual electric and magnetic fields are discussed they are assumed to be in the near field.

7.6 Shielding-General

The term 'shield' usually refers to a metallic enclosure that completely encloses an electronic product or portion of that product. The perfect shield is a barrier to transmission of electromagnetic fields. The effectiveness of a shield is the ratio of the magnitude of electric (magnetic) field that is incident on the barrier to the magnitude of the electric (magnetic) field that is transmitted through the barrier. An effectiveness of 100 dB means that the incident EM field has been reduced by a factor 1, 00,000 as it exits the shield.

As you know (as per Ampere's Law) Magnetic fields are formed around electrical conductors. Coupling between a magnetic field and an adjacent electrical conductor will occur unless the adjacent conductor has a shield preferably of high permeability ferrous materials (iron, mumetal etc.) or is physically separated. A shield with a high magnetic permeability is the best solution for magnetic field shielding, if grounding techniques cannot be properly employed. A copper-braided shield that would serve well as a shield against electric field, the coupling is however, not as effective for magnetic field shielding.

Electric fields are much easier to guard as compared to magnetic fields. EMI resulting from electric field coupling becomes a great concern as frequency increases. Thus cables operating at frequencies of 1 MHz or more are prone to emit this type of EMI, resulting in 'crosstalk' when it occurs between adjacent cables.

For both magnetic and electric field shielding, effectiveness depends upon achieving the highest degree of EMI loss through absorption and reflection. There are several ways to characterise the effectiveness of a shield. First the sum of absorption and reflection loss will yield the total loss i.e. total EMI reduction for a shield. A second means of measuring shielding effectiveness is to measure the fraction of the electric or magnetic field that reaches the other side of the barrier (shield). Total shielding effectiveness (SE) may be given by the following equation;

$$SE = A + R + B$$

A = Absorption loss R = Reflection loss B = Secondary reflection loss

Absorption loss is generally the greatest contributor to shielding effectiveness because of the large amount of EMI that shields can conduct away. The magnitude of absorption loss varies directly with the thickness of shielding barrier, the electrical conductivity of the barrier or the magnetic permeability of the shielding material. Thicker a shield as well higher it's electrical conductivity or magnetic permeability, the better the shielding effectiveness. Most shielding materials with high electrical conductivity often have a low magnetic permeability. The electrical conductivity of copper is higher than of any other commercial metal–five times more electrically conductive than steel, for example. But the magnetic permeability of iron is 1000 times than that of copper. A high electrical conductivity is the most important quality that a shield for high-frequency cable should have.

Magnetic fields are more difficult to shield, since the reflection loss may approach zero for certain combinations of material and frequency. With decreasing frequency, the magnetic field reflection and absorption losses of nonmagnetic materials such as aluminium, decrease.

Consequently, it is difficult to shield against magnetic fields using non magnetic materials. At high frequencies the shielding efficiency is good due to both reflection and absorption losses, so that the choice of materials becomes less important. The use of non-magnetic shields around conductors provides nil magnetic shielding. Conducting material can provide magnetic shielding, i.e. the incident magnetic field induces currents in the conductor producing an opposite field to cancel the incident field in the region enclosed by the shield.

Table 7.2: Shielding effectiveness of magnetic and non-magnetic materials

S.No	Material	Frequency (KHz)	Absorption loss[1] (all fields)	Reflection loss Magnetic field[2]	Reflection loss Electric field	Reflection loss Plane wave
1	Magnetic $\mu_r=1000$	<1	0-30dB	0-10dB	>90dB	>90dB
2		1-10	30-90dB	0-30dB	>90dB	>90dB
3		10-100	>90dB	10-30dB	>90dB	60-90dB
4		>100	>90dB	10-60dB	60-90dB	30-90dB
5	Non Magnetic $\mu_r=1$	<1	0-10dB	10-30dB	>90dB	>90dB
6		1-10	0-10dB	30-60dB	>90dB	>90dB
7		10-100	10-30dB	30-60dB	>90dB	>90dB
8		>100	30-90dB	60-90dB	>90dB	>90dB

1) Absorption loss for 0,8mm thick shield.
2) Magnetic field reflection loss for a source distance of 1m. (Shielding is less if distance is less than 1m and more if distance is greater than 1m). Rating of shielding: 0-10dB=Bad; 10-30dB=Poor; 30-60dB=Average; 60-90dB=Good; >90dB= Excellent

Table 7.2 summarises the Shielding Effectiveness (SE) of the magnetic and non-magnetic materials at various frequencies for magnetic and non-magnetic materials.

7.6.1 Shielding Effectiveness

Applying the concepts of shielding effectively requires an understanding of the source of the interference, the environment surrounding the source, and the distance between the source and point of observation (the receiver). If the circuit is operating close to the source (in the near, or induction-field), then the field characteristics are determined by the source. If the circuit is remotely located (in the far, or radiation-field), then the field characteristics are determined by the transmission medium.

A circuit operates in near-field if its distance from the source of the interference is less than the wavelength (λ) of the interference divided by 2π or $\lambda/2\pi$. If the distance between the circuit and the source of the interference is larger than this quantity, then the circuit operates in the far field. For instance, the interference caused by a 1-ns pulse edge has an upper bandwidth of approximately 350 MHz. The wavelength of a 350-MHz signal is approximately 32 inches (the speed of light is approximately 12"/ns). Dividing the wavelength by 2π yields a distance of approximately 5 inches, the boundary between near – and far-field. If a circuit is within 5 inches of a 350-MHz interference source, then the circuit operates in the near-field of the interference. If the distance is greater than 5 inches, the circuit operates in the far-field of the interference.

Regardless of the type of interference, as you know there is characteristic impedance associated with it. The characteristic or wave impedance of a field is determined by the ratio of its electric (E) field to its magnetic (H) field. In the far field, the ratio of the electric field to the magnetic field is the characteristic (wave impedance) of free space, given by $Z_o = E/H = 377$ Ω. In the near field, the wave-impedance is determined by the nature of the interference and its distance from the source. If the interference source is high-current and low-voltage (for example, a loop antenna or a power-line transformer), the field is predominately magnetic and exhibits wave impedance of less than 377 Ω. If the interference source is low-current and high-voltage (for example, a rod antenna or a high-speed digital switching circuit), the field is predominately electric and exhibits a wave impedance of more than 377 Ω.

Conductive enclosures can be used to shield sensitive circuits from the effects of these external fields. These materials present an impedance mismatch to the incident interference, because the impedance of the shield is lower than the wave impedance of the incident field. The effectiveness of the conductive shield depends on two things: First is the loss due to the reflection of the incident wave off the shielding material. Second is the loss due to the absorption of the transmitted wave within the shielding material.

The amount of reflection loss depends upon the type of interference and its wave impedance. The amount of absorption loss, however, is independent of the type of interference. It is the same for near – and far-field radiation, as well as for electric or magnetic fields. Reflection loss at the interface between two media depends on the difference in the characteristic impedances of the two media. For electric fields, reflection loss depends on the frequency of the interference and the shielding material. This loss can be expressed in dB, and is given by:

$$Re \text{ (dB)} = 322 + 10\log_{10}(\sigma_r / \mu_r \times f^3 \times r^2)$$

Where σ_r = relative conductivity of the shielding material, in Siemens per meter; μr = relative permeability of the shielding material, in Henries per meter; f = frequency of the interference, and r = distance from source of the interference, in meters

For magnetic fields, the loss depends also on the shielding material and the frequency of the interference. Reflection loss for magnetic fields is given by;

$$Rm \text{ (dB)} = 14.6 + 10\log_{10}(f \cdot r^2 \cdot \sigma_r / \mu_r),$$ and, for plane waves ($r > \lambda/2\pi$), the reflection loss is given by;

$$Rpw \text{ (dB)} = 168 + 10\log_{10}(\sigma_r / \mu_r \cdot f)$$

Absorption is the second loss mechanism in shielding materials. Wave attenuation due to absorption is given by;

$$A \text{ (dB)} = 3.43 \, t \sqrt{(\sigma_r \cdot \mu_r \cdot f)}$$

Where t = thickness of the shield material, in inches. This expression is valid for plane waves, electric and magnetic fields. Since the intensity of a transmitted field decreases exponentially relative to the thickness of the shielding material, the absorption loss in a shield one skin-depth (δ) thick is 9 dB. Since absorption loss is proportional to thickness and inversely proportional to skin depth, increasing the thickness of the shielding material improves shielding effectiveness at high frequencies.

Reflection loss for plane waves in the far field decreases with increasing frequency because the shield impedance, Zs, increases with frequency. Absorption loss, on the other hand, increases with frequency because skin depth decreases. For electric fields and plane waves, the primary shielding mechanism is reflection loss, and at high frequencies, the mechanism is absorption loss.

Thus for high-frequency interference signals, lightweight, easily worked high conductivity materials such as copper or aluminium can provide adequate shielding. At low frequencies however, both reflection and absorption loss to magnetic fields is low. It is thus very difficult to shield circuits from low-frequency magnetic fields. In these applications, high-permeability materials that exhibit low-reluctance provide the best protection. These low-reluctance materials provide a magnetic shunt path that diverts the magnetic field away from the protected circuit. Examples of high-permeability materials are steel and mu-metal.

So to summarize the characteristics of metallic materials commonly used for shielded purposes: Use high conductivity metals for HF interference and high permeability metals for LF interference.

A properly shielded enclosure is very effective at preventing external interference from disrupting its contents as well as confining any internally-generated interference. However, in the real world, openings in the shield are often required to accommodate adjustment knobs, switches, connectors, or to provide ventilation. Unfortunately, these openings may compromise shielding effectiveness by providing paths for high-frequency interference to enter the instrument.

The longest dimension (not the total area) of an opening is used to evaluate the ability of external fields to enter the enclosure, because the openings behave as slot antennas. Formula below can be used to calculate the shielding effectiveness, or the susceptibility to EMI leakage or penetration, of an opening in an enclosure:

$$\text{Shielding Effectiveness (dB)} = 20 \log_{10} (\lambda / 2.L)$$

Where λ = wavelength of the interference and L = maximum dimension of the opening

Maximum radiation of EMI through an opening occurs when the longest dimension of the opening is equal to one half-wavelength ($\lambda/2$) of the interference frequency (0-dB shielding effectiveness). A rule-of-thumb is to keep the longest dimension less than 1/20th of wavelength of the interference signal, as this provides 20-dB shielding effectiveness. Furthermore, a few small openings on each side of an enclosure are preferred over many openings on one side. This is because the openings on different sides radiate energy in different directions, and as a result, shielding effectiveness is not compromised. If openings and seams cannot be avoided, then conductive gaskets, screens, and paints alone or in combination should be used judiciously to limit the longest dimension of any opening to less than 1/20 wavelength. Any cables, wires, connectors, indicators, or control shafts penetrating the enclosure should have circumferential metallic shields physically bonded to the enclosure entry point. In those applications where unshielded cables/wires are used, then filters are recommended at the shield entry point.

7.6.2 General Points on Cables and Shields

The improper use of cables and their shields can be a significant contributor to both radiated and conducted interference. Rather than developing an entire treatise on these issues here, the interested reader may consult detailed studies on this matter. As shown in Fig 7.2, proper cable/enclosure shielding confines sensitive circuitry and signals entirely within the shield, with no compromise to shielding effectiveness.

As seen from this diagram, enclosures and shield must be grounded properly; otherwise they can act as an antenna, thereby making radiated & conducted interference problem worse (rather than better).

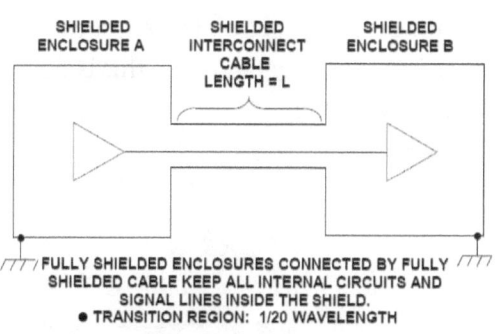

Fig 7.2: Shielded Interconnect Cables either Electrically Long or Short

Depending on the type of interference (pickup/radiated, low/high frequency), proper cable shielding is implemented differently and is very dependent on the length of the cable. The first step is to determine whether the length of the cable is electrically short or electrically long at the frequency of concern. A cable is considered electrically short if the length of the cable is less than 1/20 wavelength of the highest frequency of the interference. Otherwise it is considered to be electrically long.

Example, at 1MHz, an electrically short cable is any cable length less than 15 meters, where the primary coupling mechanism for these low frequency electric fields is capacitive. As such, for any cable length less than 15 meters, the amplitude of interference will be same over the entire length of cable.

In applications where the length of the cable is electrically long, or protection against HF interference is required, then the preferred method is to connect the cable shield to low-impedance points, at both ends. As will be seen shortly, this can be a direct connection at the driving end, and a capacitive connection at the receiver. If cable is left ungrounded, unterminated transmission lines effects can cause reflections and standing waves along the cable. At frequencies of 10 MHz and above, circumferential (360°) shield bonds and metal connectors are required to maintain low-impedance connections to ground.

In summary, for protection against low-frequency (<1 MHz), electric-field interference, grounding the shield at one end is acceptable. For high-frequency interference (>1 MHz), the preferred method is grounding the shield at both ends, using 360° circumferential bonds between the shield and the connector, and maintaining metal-to-metal continuity between the connectors and the enclosure.

However in practice, there is a caveat involved with directly grounding the shield at both ends. When this is done, it creates a low frequency ground loop, shown in Fig 7.3.

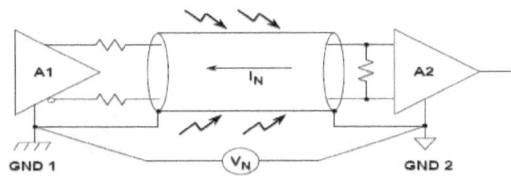

Fig 7.3: Ground Loops in Shielded Twisted Pair Cable Can Cause Errors

Whenever two systems A1 and A2 are remote from each other, there is usually a difference in the ground potentials at each system, i.e., V_N. The frequency of this potential difference is generally the line frequency (50 or 60 Hz) and in multiples thereof. But if the shield is directly grounded at both ends as shown noise current I_N flows in the shield. In a perfectly balanced system, the common-mode rejection of the system is infinite, and this current flow produces no differential error at the receiver A2. However, perfect balance is never achieved in the driver, its impedance, the cable, or the receiver, so a certain portion of the shield current will appear as a differential noise signal at the input of A2.

As noted above, cable shields are subject to both low and high frequency interference. Good design practice requires that the shield be grounded at both ends if the cable is electrically long to the interfering frequency, as is usually the case with RF interference.

Fig 7.4 below shows a system of Hybrid Grounding of a remote passive RTD sensor connected to a bridge and conditioning circuit by a shielded cable. The proper grounding method is shown in the upper part where the shield is grounded at the receiving end.

However, safety considerations may require that the remote end of the shield also be grounded. If this is the case, the receiving end can be grounded with a low inductance ceramic capacitor (0.01 µF to 0.1 µF), still providing high frequency grounding. The capacitor acts as a ground to RF signals on the shield but

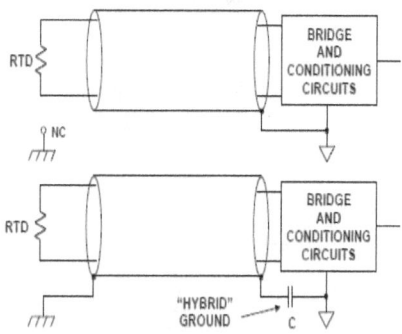

Fig 7.4: Hybrid grounding of shielded cable with Passive Sensor

blocks low frequency line currents to flow in the shield. This technique is often referred to as Hybrid Ground. A case of an active remote sensor and/or other electronics is shown Fig7.5. In both of the two situations, a hybrid ground is also appropriate, either for the balanced (upper) or the single-ended (lower) driver case. In both instances the capacitor "C" breaks the low frequency ground loop, providing effective RF grounding of the shielded cable at the A2 receiving end at the right side of the diagram. There are also some more subtle points that should be made with regard to the source termination resistances used, Rs. In both the balanced as well as the single-ended drive cases, the driving signal seen on the balanced line originates from a net impedance of Rs, which is split between the two twisted pair of legs as twice Rs/2. In the upper case of fully differential drive, this is straight forward with an Rs/2 valued resistor connected in series with complementary outputs from A1.

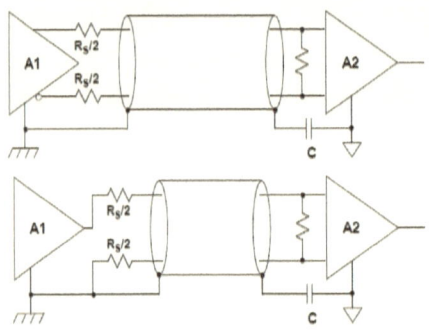

Fig 7.5: Impedance –Balanced drive of balanced shielded cable aides Noise –Immunity with either

In the bottom case of the single-ended driver, note that there are still two Rs/2 resistors used, and one each in series with both legs. Here the grounded dummy return leg resistor provides an impedance-balanced ground connection drive to the differential line aiding in overall system noise immunity. Note that this implementation is useful for those applications with a balanced receiver at A2, as shown.

Balanced or Single –Ended source Signals: Coaxial cables are different from shielded twisted pair cables in that the signals return current path is through the shield. For this reason, the ideal situation is to ground the shield at the driving end and allow the shield to float at the differential receiver (A2) as shown in the upper portion of Fig 7.6. For this technique to work; however, the receiver must be a differential type with good high frequency CM rejection.

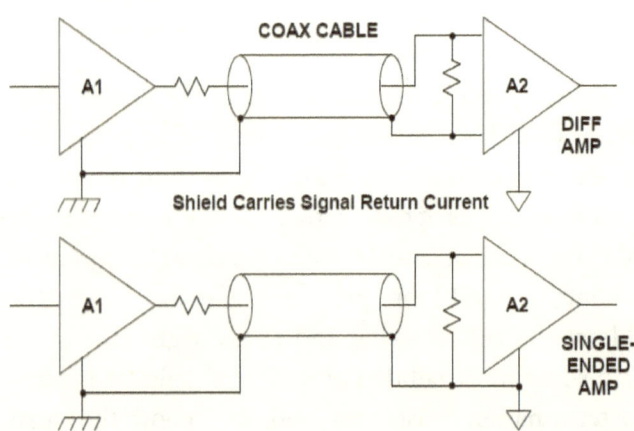

Fig 7.6: Coaxial Cables can use either Balanced or Single–Ended Receivers

However the receiver may be a single-ended type, such as typical of a standard single op amp type circuit. This is true for the example of Fig7.6, so there is no choice but to ground coaxial cable shield at both ends for this case.

7.7 Components design and selection for minimising EMC

The primary objective of EMC should be to minimise the amount of noise generated by the equipment, as the noise may interfere with other equipment. It is better to control as much noise at the source as possible, since that approach can avoid an interference problem for countless number of receiver circuits. By selecting the proper noise reduction method and components, EMI can in many instances be reduced considerably.

7.7.1 Capacitors

Capacitors are generally effective noise de-couplers means as they divert (by-pass) the noise. Use of parallel capacitors in low-impedance circuits is usually insufficient. They are most effective with high-impedance loads. Whenever a parallel suppression component is used, the impedance levels of not only the element but also the parallel path should be computed or estimated at the desired frequency.

7.7.2 Inductors

Unlike capacitors which divert noise currents, the inductors are placed in series with wire to block noise currents. This will be effective if impedance of inductor at frequency of noise current is larger than original series impedance seen looking into the wire. Series inductors are most effective in low-impedance circuits.

7.7.3 Electromechanical devices

A number of electric products such as typewriters, printers and servo/robotic devices use small electromechanical devices such as DC motors, stepper motors, AC motors and solenoids to translate electrical energy into mechanical motion. These devices can create significant EMC problems. Brushes of DC motors generate high-frequency spectra due to arcing across them as well as they provide paths for common-mode currents through their frames. The spectral content tends to create radiated emission problems in the radiated emission regulatory limit frequency range between 200MHz and 1GHz, depending on the motor type. In order to suppress this arcing resistors or capacitors may be placed across the commutator segments.

For AC motors, the rotor and stator consist of closely spaced inductors, the problem of large parasitic capacitance between the rotor and stator exits. If high-frequency noise is present on the AC-wave form feeding these motors then it is likely that this noise will be coupled to the chassis or to the AC power cord, where its potential for radiated or conducted emissions will usually be enhanced.

7.7.4 Ferrite components

Ferrite materials are basically non conductive ceramic materials, consisting mainly of iron oxide that is blended with other metallic oxides, calcined, then sintered, resulting in a polycrystalline, spinel structured ceramic. The material differs from other ferromagnetic materials such as iron in that they have low eddy-current losses at frequencies up to hundreds of MHz. Thus they can be used to provide selective attenuation of high-frequency signals that we may wish to suppress from the standpoint of EMC and not affect the more important lower-frequency components of the functional signal.

7.7.5 EMI gaskets

EMI gaskets are employed for either temporary or semi permanent sealing applications between joints or structures in order to reinstate loss of shielding integrity at seams and joints where other than permanent fastenings methods are permitted e.g.:(a) Securing access doors to enclosures, cabinets, or equipment, (b) Mounting covers plates or removal panels for equipment maintenance, alignment, or other purposes

7.7.6 Cabling and connectors

Cables and connectors should be designed to achieve a system's specified levels of emission suppression and resistance to outside EMI. Cable assemblies should also deliver the required undistorted signals while achieving proper mechanical performance. It is necessary to give special attention to the cable-connector interface and the environment in which the cable assembly must perform.

Cable shielding is an effective means of controlling / limiting radiated EMI. Conducted EMI is difficult to overcome without the use of filters, but experts suggest trying proper shield grounding techniques before filters are introduced. Radiated EMI, however, can be controlled with shielding. Cable shields must provide protection against both magnetic fields and electric fields. Each field requires a different shielding mechanism.

Twisted pairs can achieve some magnetic field shielding because twisting provides equal and opposite induced voltage. A short pair pitch (no of twists per unit length) will provide a higher degree of magnetic field rejection.

Note: First step in designing a system to meet EMC requirements is to separate power cables and data cables.

This will limit the opportunity for magnetic field interference of non-power cables by the magnetic field generated by the load cables. If sufficient separation is not possible, the non power cable shield must provide shielding against both magnetic fields and electric fields. Power cables should have a magnetic field shield. This can be done by locating the cables near a part of the cabinet frame which can act as a high permeability shield and ground plane. The four most common possibilities of shielding are; (a) foil laminates, (b) braided shields, (c) optimized braids, and (d) combination shields.

The effectiveness of these approaches can be adjusted by changing the coverage of a braided shield of the amount of the overlap in foil laminate shield. Varying the coverage (solid tube equals 100 percent coverage) can lead to greater shielding effectiveness. Likewise, increasing the thickness, overlap or electrical conductivity will improve the performance of the shield.

Because braided shields introduce apertures, 100 percent coverage is theoretically impossible. Therefore these shields represent a departure from the solid 'tube' approach to shielding, and are thus less effective at achieving reflection loss than are foil laminates in some cases. But, because they generally contain a large mass of metal per unit of cable length, characterizing a low impedance path, they can accomplish more absorption loss. Braided shields are most effective at shielding against radiation frequency interference.

Optical coverage indicates the amount of the surface of an insulated conductor that is covered by shielding, as viewed with the eye perpendicularly. Greater optical coverage does not necessarily mean greater shielding effectiveness with respect to reflective and absorptive loss. But, braid angle and individual wire diameter have a significant effect on shield effectiveness.

There are several approaches for terminating a cable shield at the connector. Proper consideration must be given to whether the shields are to be grounded through connector's backside at one or both ends and that the cable shield at back shell offer equivalent shielding quality as both separate components and an assembled part.

7.8 Installation

The exposure of an item of equipment to interference can be related to the electrical environment in which it is situated. The degree of interference is related to the characteristics of the source, the nature of coupling impedances, the sensitivity of electronic equipment and the quality of the earthing and protective measures utilised at the installation site.

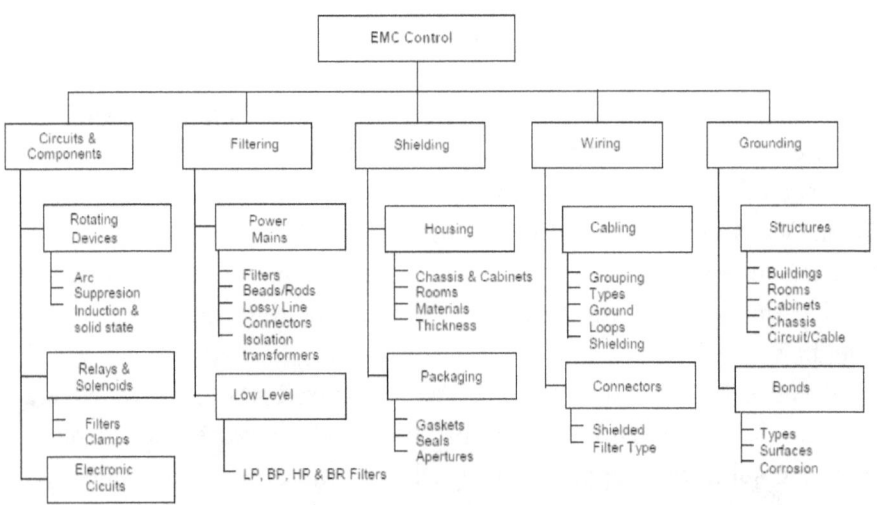

Fig 7.7: An Overview of available options to achieve EMC control

In the installation of electrical and electronic systems, a number of options are available as to how to earth the signal circuits, to choose cable shields and earthing of the shield, each of which can contribute to the reduction of the interference. In addition, the treatment of signal lines and power cables with respect to the cable routing and cable separation, the use of filters and enclosures, bounding practices, etc. are ways by which coupling of interference to sensitive circuits can be reduced. Fig 7.7 gives an overview of available options to achieve EMC control.

7.8.1 Circuits and components

The sensitivity of a circuit to EMI is dependent upon the input impedance. The higher the input impedance, the more sensitive the circuit is to EMI. The effect of an unsymmetry is also greater for high impedance inputs. By using minimum bandwidth for the signal inputs, the equipment's response to EMI is reduced. Since now-a-days standard integrated circuits having very large bandwidth are commonly used, filters or other components should be used to reduce the bandwidth. When possibilities of common-mode voltages occur in extremely sensitive equipment, unsymmetry in cable routing and the equipment's impedances to ground must be avoided. Good symmetry can be achieved by using a balanced isolation transformer or especially symmetric equipment.

Relay coils and its contacts which have direct connection to semiconductors and/or secondary contacts in the electronic cabinet must be EMI suppressed. A relay without such protection must not be connected to supply voltage circuits for the electronics. All other relay coils and operating coils should be suppressed if possible.

7.8.2 Filtering

Construction and application of EMI filters in each case should be considered in the actual situation. The equipment manufacturer's specifications should be followed where these are available.

Power supply for electronic equipment should normally be filtered. This is applicable to both DC and AC voltages.

Electronic equipment representing heavy load should be fed via separate lines from the main switch board. Power supply filters should have separate shields (boxes). Filters should be used at the signal input of sensitive equipment and the noise source if needed.

The distance between input and output terminals on the filter should be maximised. If a filter is used in a symmetric circuit, the filter should be balanced. Filters should be used at all analog inputs. Filters should be used at all digital inputs if the filter itself does not introduce functional degradation.

The use of galvanic isolators should be considered for the following cases;

(a) Isolation of sensitive measuring and control equipment from noisy AC power supply.
(b) Isolation of noise generating equipment from noise sensitive equipment when both are using same AC power supply source.
(c) For minimising differential-mode noise (noise across winding) resulting from common-mode noise (noise between winding and ground).
(d) For maximum common-mode noise voltages.
(e) Separate measuring equipment and cables galvanically from central, electronic signal processing equipment. This is to prevent that unintentional supply of high voltages on signal and control cables should damage the signal processing equipment.

Isolation can be done in many different ways, depending on the application. For Signals as well control circuits both opto-coupler and isolation transformers can be used. For power supply system isolation transformers can be used. Isolation transformers should have a ground screen between the windings to reduce the capacitive coupling. The capacitance between the windings should be much less than 1pF. Any isolation device has to withstand the highest noise voltage likely to occur in the system.

7.8.3 Screens and shields

Electric/electronic equipments which are susceptible to EMI or might be EMI emitting should generally have a metallic shield. From an EMI viewpoint, it would be favourable to use screened cables everywhere, with the screen grounded everywhere. For all EMI susceptible and EMI emitting cables on board, conductive screens are recommended. Many commonly used cables in wiring the electronic equipments whether on land or air or in sea of radio transmitters etc having copper braided screens are satisfactory for reduction of capacitively coupled EMI voltage, shield factor 60 dB provided satisfactory grounding of the screens.

Commonly used cables will usually have a modest shield factor for magnetic fields at frequencies up to 1 kHz; the best achievable is down to 2 dB. If better shielding effect at low frequencies is desired, a screen made of magnetic material should be used. At higher frequencies the shield factor increases at two – and multi point grounding, e.g. 60 dB at 1 MHz. Cables should be as short as possible, and routed as close as possible to the main ground system when routed outside cable trays. This will also have a certain shielding effect.

In order that the screening of the conductor and the equipment to which it is connected shall be completely effective against high frequencies, all outer cable screens must be in contact, all around, with the screening enclosure of the equipment.

Preferred low frequency shield grounding for both shielded twisted pair and coaxial cable are shown in Fig 7.8. The shield grounding in A through C are grounded at the amplifier or structure, but not at both ends. When the signal circuit is grounded at both ends, the amount of noise reduction possible is limited by difference in ground potential and the susceptibility of the ground loop to magnetic fields.

The preferred shield ground configurations for cases where the signal circuit is grounded at both ends are shown in D and E. A transformer may be used to break the ground loop. In case E the shield is grounded at both ends to force some ground-loop current to flow through the lower impedance shield, rather than the centre conductor. In case of circuit D the shielded twisted pair is also grounded at both ends to shunt some of ground loop current from the signal conductors. If additional noise immunity is required, the ground loop must be broken e.g. using transformers, optical couplers or a differential amplifier. Note that an unshielded twist pair, unless it is balanced, provides very little protection against capacitive pickup, but is very good for protection against low frequency signals. The effectiveness of twisting increases with the number of twists per length.

In exceptional cases in which signals of frequencies from 20 Hz to about 10 MHz must be transmitted via a coaxial cable in an unsymmetrical system e.g. video lines, analog signals, it is necessary to prevent the flow of an interference current e.g. with net frequency in the cable outer conductor, the return conductor of the signal circuit. Since in this case, on account of the large frequency range, it is not possible to depart from coupling the cable screen to ground at each end, it is necessary to use double screen circuit. In this case the inner conduits and the inner screen form a normal coaxial system which must be driven as such. The circuits connected together have, however, only a connection to ground at one end (preferably at the transmitter end or according to the manufacturer's specifications). The outer screen must however, be connected all around at both ends to the particular equipment housings.

The cable screen can only be effective against magnetic fields if it is electrically connected at both ends, so that a current can flow in the screen.

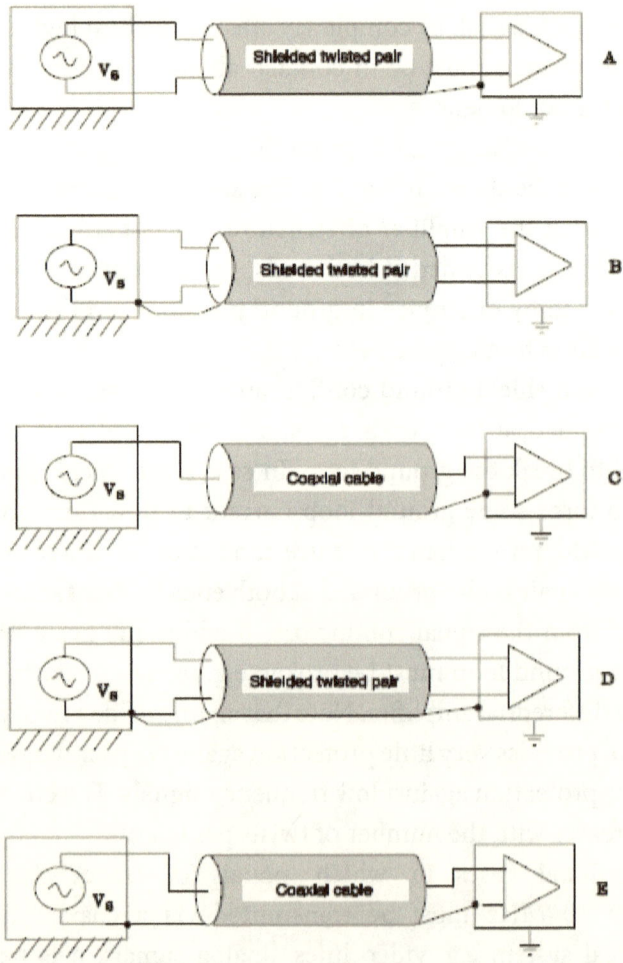

Fig 7.8: Preferred grounding of cable screens

Single ended connected cable screens are only effective against electrical fields and only when the screened length is not greater than $\lambda/10$ of the highest frequency in the EMC zone be considered. In this case it must be taken into account that the equipment with working frequencies below e.g. 100 kHz however can be interfered by higher frequencies. It is equally possible that these equipment form interference sources for frequency range being far away from their working frequencies.

In order to prevent the electrostatic charging ESD of objects, grounding of metallic enclosures is highly recommended. Insulating material such as floor coverings, seating etc. should have a limited conductivity to avoid charging-up. As a target value, specific resistance of the insulating material should be taken as 10^7 ohm-cm. Rotating and movable parts (machine parts, propeller, shafts, swing-out equipment etc.) are to be conductivity connected to ground e.g. by means of grounding brushes, slide contacts, conducting grease etc.

Efforts to reduce EMI inside and outside the radio room of a ship etc have two purposes;

(a) Prevent other equipment on board from disturbing transmitting from and receiving to the radio room.
(b) Prevent the radio station from disturbing other equipment on board.

The radio room must be shielded. Normally steel bulkhead and steel deck in radio room will form a natural part of the shield. All joints in the shield should be continuously welded where practically possible. If the whole shield or parts of it must be joined in another way e.g. by bolted connections, special efforts (e.g. EMI gaskets) should be executed to make the shield as electrically 'tight' as possible. All necessary apertures (door, window, ventilation openings etc.) in the shield should be made as electrically non-penetrating as possible. This involve e.g.:

(a) EMI gaskets around doors (door, or door panel of same material as the rest of the shield).
(b) Multi-layer screened covers or honeycomb aperture covers at ventilation openings.
(c) Conductive glass in windows not facing free air.

7.8.4 Wiring

Cables may be grouped in different classes, according to EMI generation and EMI susceptibility. There are cables of different functions in each class because the number of actual functions is greater than the suitable number of interference classes. Most of the cables can be categorized in 5 different classes as shown in Table 7.3 below;

Table 7.3: Class / Category of different type of Cables

S. No	Type of cable/ Functions	EMI Generating or not	Class
1	Power cables, control cables in circuit using mechanical contacts & relay coils	EMI generating, Not EMI susceptible 24-600V, DC,50,60,400Hz High Power up to 10KV	A
2	Telephone cables, Signal Cables, Synchro circuits (50-400 Hz)	Slightly EMI generating, Slightly EMI susceptible. 0.5 to 50 V, Low Frequency	B
3	Video, Data Transmission, Analog measuring values after converter.	EMI generating, EMI susceptible, 0.1 – 5 Volt, 50 Ω, pulse,0.1 – 50 V, DC	C
4	Receiving Antennas, Microphone & Audio cables, Analog measuring values	Highly EMI susceptible, 10 uV-100mV, 50 – 200 Ω, DC, AF, HF	D
5	Radio Transmitter, TV Transmitter, Radar, Sonar, Rectifiers, Thyristors control	Highly EMI generating	E

The classification is schematic, and in some instances it will be a matter of judgement whether a cable belongs to e.g. group A, B or in two groups.

Note: As a general rule, all different cable classes A through E should have separate routing, and the distance should be as large as possible. However, the benefit from separation is not linearly dependent upon the separation distance and the first tens of millimetres are the most significant.

Table 7.4 shows the minimum distances between cables of the different classes when routed on cable trays or directly onto the chassis of a vehicle, steel hull of a ship, or in trenches in heavy duty floors etc.

If there are only screened cables of reasonably good quality e.g. copper or iron braided screen with outer non-metallic sheath, the distances may be reduced, except for cables in class E. To allow distance reduction between class E and other classes, cables of class E must be routed in tubes, conduits or boxes having a minimum wall thickness of 1.5 mm steel with good electrical connection to the chassis plate at least in both ends.

Radio Frequency Interference and Shielding

Table 7.4: Minimum distances in mtr between cables of different classes when routed on cable trays or directly onto the chassis plate etc

Cable class	EMI generating Not susceptible	Slightly EMI generating & Susceptible	EMI generating & susceptible	Highly EMI susceptible	Highly EMI generating
	A	B	C	D	E
Case 1: unscreened cables paralleled over more than 2 meters					
A	0	0.25	0.25	0.50	0.25
B	0.25	0	0.25	0.25	0.25
C	0.25	0.25	0	0.25	0.50
D	0.50	0.25	0.25	0	0.50
E	0.25	0.25	0.50	0.50	0
Case 2: unscreened cables at 90 degree[1] crossing angle					
A	0	0	0.15	0.30	0.15
B	0	0	0.15	0.15	0.15
C	0.15	0.15	0	0	0.30
D	0.30	0.15	0	0	0.30
E	0.15	0.15	0.30	0.30	0
Case 3: grounded & screened Cables					
A	0	0	0.10	0.10	0.25[2]
B	0	0	0.10	0.10	0.25[2]
C	0.10	0.10	0	0	0.50[2]
D	0.10	0.10	0	0	0.50[2]
E	0.25[2]	0.25[2]	0.50	0.50	0

Legends:

(1) If the crossing angle is less than 90 degree then the distance should be increased towards the values in case 1.
(2) No separation requirements if Steel tubes or conduits of at least 1.5 mm wall thickness are used around class E cables.

Special cables that may be both highly EMI generating and highly EMI susceptible (e.g. combined cables from antenna to transmitter/ receiver etc), i.e. cables that may alternate between class E and D, should be routed separately with distances as shown in Table 7.4, case 1. Alternately to avoid requirement of separation such cables may be routed in separate steel conduits having a wall thickness of at least 1.5 mm. Using only screened cables in an installation, separation in meters should be as per table7.4, case 3.

(a) Installation of cable trays

Cables should be mounted either directly to the conducting chassis plates or on trays made of at least 1.5 mm perforated steel plate. The plate is for mechanical reasons usually made with a bent edge. To reduce EMI this edge should be higher than the height of the cable bundle. The cables should as far as practically possible, be installed in a single layer.

The plate sections should be welded together and to the fasteners which in turn are welded to the chassis plate. Alternatively one might use at least 1 screw and lock washer per 0.2 m joint. The distance between cable trays for different cable classes is specified in Table7.4. At feed-through points the distance between the trays might be reduced over a short distance.

Single core cables for AC current rating in excess of 250 A, and single core cables for DC with high ripple content, should not be mounted directly to the chassis plate or other magnetic material, but at a distance of at least 50 mm. If not, large losses and additional voltage drop will occur due to magnetic hysteresis.

(b) Conductors in cables

In a multi-conductor cable the different circuits should normally have the same function and the same power/voltage level. Deviation from this rule might be accepted if a cable is running between parts of a single self-contained system.

Twisted pairs should have a pitching of at least 10 turns /meter. Analog and digital signals should have separate cables. Signal conductor and return should be adjacent conductors in the same cable, and the difference in power/voltage level between the circuits in one cable should normally not exceed one order of magnitude.

Using circuits with twisted pair cables and symmetric terminations appreciably greater difference in power/voltage level can be tolerated. Such circuits could also be used for low level signals as an alternative to coaxial cable, especially for low frequencies. In special cases it might be necessary to use coaxial cables, double screen cables, cable routing in tubes, conduits, etc. This must be considered in each individual case and in agreement with the equipment supplier.

7.8.5 Grounding

Grounding is one of the primary mean to minimise unwanted noise and pick-up. Proper use of grounding and shielding in conjunction can solve a large percentage of all noise problems.

In the most general sense a ground can be defined as an equipotential point or plane which serves as a reference voltage for a circuit or system. It may or may not be at earth potential. If ground is connected to the earth through a low impedance path, it can then be called an 'earth ground'.

Safety grounds are always at earth potential, whereas signal grounds are usually but not necessarily at earth potential. In many cases, a safety ground is required at a point which is unsuitable for a signal ground, and this may complicate the problem.

In the context of EMC it is imperative to think of 'ground' as a path for the current to flow instead of an equipotential surface. Currents with frequency components from DC to100 MHz typically pass through 'ground'. At frequencies in the MHz range resistance of the conductor, even including skin effect, is negligible compared with the impedance due to the ground conductor inductance.

There are two basic objectives involved in designing good grounding systems;

(a) To minimise the noise voltage generated by currents from two or more circuits flowing through common ground impedance.
(b) To avoid creating ground loops which are susceptible to magnetic fields and differences in ground potential?

Note; An improper Grounding can become a primary means of noise coupling however.

When grounding, the reference conductors, the frequency range of the signals to be transmitted, the mode of transmission as well as the electromagnetic environment are to be taken into account. For frequencies f<100 kHz only the point of symmetry in a symmetrical transmission can be grounded. In an unsymmetrical system the reference conductor is only to be grounded at one point. One reference conductor can be used for several signal conductors. If several reference conductors are used, these are to be grounded at only one point, the common ground point.

For frequencies f > 100 kHz and for pulse techniques a reference conductor system grounded at the common point is no longer applicable.

As a general rule, equipment housings must be grounded. For equipment whose dimensions are smaller than $\lambda/10$ for the highest considered frequency, it is normally sufficient to ground the housing at one point. If the housing dimensions exceed $\lambda/10$ then the housing is to be grounded along the longest edge at several points at separations not greater than $\lambda/10$ in order to reduce the antenna effect of the housing. For separations of less than 0.3 m between ground points, in general no improvement is to be expected. The highest frequency considered is dependent on the electromagnetic environment in which the equipment operates.

In the radiation field of an antenna, metal parts can act as secondary radiators. If these metal parts have connection with ground or with each other which varies strongly with time (loose contacts) or is corroded (semiconductor effect), then these varying contact resistance can cause new frequencies (harmonics, interference spectra) to arise, which by means of the antenna effect of the metal parts can be radiated and considerably disturb radio receiving. Movable rods, links, ladders, turn-buckles, cables, doors, hatch covers and tools etc. are therefore to be connected to ground or isolated.

(a) Main ground system

Metallic floorings and superstructure, including details in these which are welded together, are presumed to make a satisfactory main ground system. If metallic parts of the flooring or superstructure are bolted together, they are presumed to be part of the main ground system, provided measures have been taken to secure good and permanent electrical conducting contact at the bolted joints.

In non-metallic parts of the plant/equipment, e.g. a plastic superstructure, the main ground should be formed by interconnected copper bus-bars of at least 50 mm^2 along all cable routings.

Aluminium superstructures which are provided with insulated material between aluminium and steel in order to prevent galvanic action are to be grounded to the flooring. For this purpose, corrosion-resistant metal wires or bands are to be used. Provisions are to be made for preventing galvanic action at the terminals of these connections by using 'Cupal terminals' when copper wires or bands are connected to the aluminium constructions.

(b) Signal reference system
The signal reference system consists of the electrical conducting material (copper bars) of a common reference between radio transmitter, communicating electronic measuring and/or control equipment. This reference can be either connected to the main ground system or be floating. The signal reference system is constructed as a 'star' network emanating from the functionally central instrument in the communicating electrical / electronic systems.

(c) Ground network configurations
When a galvanic conducting part of an installation e.g. a signal reference system or a cable screen is connected to the main ground system in one point only, this is called single point grounding. When the conducting part of an installation is connected to the main ground system in several geographically separated points e.g. at each instrument on a control system, this is called a multi point grounding.

(d) Grounding rules
The signal reference system should be grounded at the functionally central instrument, the rest of the system being insulated from ground. As mentioned above, the reference system might also be floating. If the reference within an electronic system must be grounded in several points e.g. at many instruments geographically separated, the reference must be split by means of galvanic separation to avoid ground loops via the signal reference system. The connection between the reference system and the main ground system should be as short as possible and not in common with any other grounding except at one point at the main ground system.

Signal cable screen between equipment operating at or being susceptible to frequencies having a wave length $\lambda > 20L$, where L is the cable length, should have single point grounding. (Wave length $\lambda = 300/f$ (m); where f=frequency (MHz)). Below 100 kHz single point grounding is generally recommended.

The grounding should be made at the cable end where the circuits connected provide the lowest impedance to ground. This is usually in the functionally central instrument, where the reference system is also grounded. Exemption from this is cable screens from thermocouples which might have a ground point at the sensor end.

Signal cable screen between equipment operating at high, or low frequencies, and being susceptible to frequencies having a wave length $\lambda < 20L$, should be grounded at least at both ends. Above 10MHz, multi point grounding is generally recommended. In a transition region from 100 kHz to 10MHz, both single point and multi-point grounding may be combined for complex systems.

Power cable screen (DC, 50Hz, 60 Hz, 400 Hz, etc.) should be grounded to main ground whenever possible, at least at both ends. This is safety grounding, which is also favourable from an EMI point of view. Above cabinet or deck the cables are especially exposed to radio signals. If the highest radio frequency is 22 MHz, the power cable screen above deck should be grounded at least at every 2.5 m (approx. 0.2λ).

Single core cables for AC and special DC-cables with high ripple content (e.g. for thyristor equipment) are to be grounded at one end only.

All metallic racks, cabinets, cases, etc. surrounding electric/electronic equipment must be grounded. Large units should have several ground points distributed around the unit.

To prevent electrostatic charging of insulated mounted metallic parts in the vicinity of antennas or cable routing, these should also be grounded.

The equipment supplier's specification regarding the grounding network should be considered according to the guidelines given above. In case where the two specifications do not agree, one should carefully examine the total grounding system to avoid inferior overall solutions.

(e) Shielding, procedures and choice of materials

Cabinets, racks, etc. should be all seamless welded to give continuous, homogenous joint of the separate parts. For any opening in the shield (cabinet, rack, etc.), the largest diameter should be less than $\lambda/40$ of the EMI that might disturb the equipment.

For equipment working with signals having fast rise times (less than 10 nsec.): $L < 25 t$; where L = largest acceptable aperture (mm); t = rise time of signals (pulse) – nanoseconds

When using EMI gaskets on doors, covers, etc. in cabinets, racks and boxes, one should ascertain that there is a sufficient pressure on the gasket to achieve good contact to both surfaces. The surfaces for the gasket must be electrically conducting, i.e. surface treatment like painting, plastic coating etc. must be removed. Possible natural deformation or compression of the gasket must be taken care of (at doors, covers, etc.)

Doors, hinged covers, etc. should preferably have a continuous (long) hinge to ascertain better electrical contact (higher pressure, evenly distributed). Especially this is important where EMI gaskets are used.

(f) Grounding, procedures and choice of materials

The technique for making the ground connection and type of materials used in the straps, conductors, etc. and the joints are of great importance to have a lasting, good connection. The ground connection can be done either by direct bonding to the main ground or by a dedicated ground lead. The latter is the case when the equipment is mounted on vibration isolators. The mating surfaces should be clean and free from oil and oxides. Special attention should be paid to corrosion at the joints.

The dedicated ground conductors should be solid, flat, metallic conductors or a woven braid configuration where many conductors are effectively in parallel. The conductor must be of a sufficient cross section. Ground straps should be protected against corrosion. Ground straps should be as short as possible. The ground strap, connecting two points together should be insulated if possible, to prevent undesired metallic contact between the ground strap and other items. If components such as instruments signal lamps, etc., are mounted on hinged doors in a rack, the door should be connected to the rack via a separate flexible ground strap.

The same applies to doors in radio communication rooms or other rooms where the bulkhead and decks serve as shielding against EMI. At such doors care should be taken to ascertain a good electrical connection all around the door by means of EMI gaskets.

Welded or brazed bonds are preferred over all other types. Bolted connection or shrinkage is preferred over soldering whenever practical. Especially when mechanical bolt connection or shrinkage is used in joints between ground point and ground conductor, it is important that all mating surfaces are cleaned to base metal (made electrical conducting). Be aware of that anodised aluminium has a poorly conducting surface. Ensure good ground contact by use of washer and serrated lock washer or equivalent.

When using cable trays, cable 'ladders', etc. that are not welded together, good and lasting electrical connection between the different sections of the cable tray, cable 'ladder', etc., should be taken care of by using bonding straps or by screws and lock washers. Make all cable trays electrically continuous.

Any ground connection should provide a low resistance path to ground, AC and/or DC. Depending on environment, structural details, etc., there will practically be one or more metals suitable as ground connector. Due to corrosion possibilities it is important that metals having large electrochemical potential difference do not make contact with each other, e.g. copper and aluminium. Generally speaking, metals having low electromotive force (EMF) are corroding more rapidly than those having higher EMF. When different metals are combined, it is desirable to use metals from the same group. In cases where contact between metals having large EMF difference are unavoidable, metals having the lowest EMF are used in easily changeable parts such as bolts, lock washers, nuts, etc., such that more solid and 'unchangeable' parts having higher EMF.

All radio room equipment must have its individual ground connection directly to the main ground system. Use of one or more common ground buses which in turn are connected to the main ground system should be avoided because this can lead to noise voltages via common impedance (common mode coupling effects).

The radio transmitters' cabinet should be grounded at several places. If special efforts are made to assure good electrical contact through the bolted connection fastening the cabinet to the vehicle chassis in case of a mobile or to a ship's hull (shield around radio room) in case of marine, such grounding will often be sufficient. Dedicated ground straps are necessary if the transmitter is mounted on isolating vibration dampers as in case of mobile transmitters.

Power and instrumentation cables to /from equipment that are not situated in the radio room must not be routed through the radio room (i.e. not inside the shield around the radio room). Cables, including metallic cable screen that go to/from the radio room may carry considerable amounts of EMI. Especially when the radio transmitter is operating, cable screens on cables in the radio room can carry considerable currents in the radio frequency range (about 2-30MHz). Cables screens should therefore not go through the wall (shield) in the radio room, but be terminated in the shield around the radio room. As a general practice all cables to/from the radio room should have noise reducing filters at the point where they leave the radio room. Special efforts should be made to reduce coupling of EMI via the power supply system to/from the radio room. E.g. isolation transformers with grounded metal screen between primary and secondary windings could be used for AC supplies. DC power supply could e.g. go via a converter/rectifier system. (This system must be relatively free from EMI).

The commonly used frequencies for the radio station are between 2-30MHz. In this range (actually 100 kHz-10MHz) the electrical and magnetic coupling between circuits are determined by several variables. (But normally the electric field is predominant). The correct grounding method will be multi-point ground or a combination of single and multi-point ground of cable screens.

Screened power cables in and near by the radio room and transmitter antennas should normally be grounded at least at both ends. All cables passing through transmitter antennas area within a certain distance should be shielded. The precautions taken must comply with governmental regulations.

The EMC checklist in table 7.5 below summarises the commonly used noise reduction techniques.

Table 7.5: EMC check list commonly used noise reduction techniques

EMC Check List for noise reduction technique		
Items	Y/N	Comments
A. Suppressing noise at source:		
Enclose noise source in a shielded enclosure		
Filter all leads leaving a noisy environment		
Limit pulse rise times		
Relay coils should be provided with some form of surge damping		
Twist noisy leads together		
Shield and twist noisy leads		
Ground both ends of shields to suppress radiated interference		(shield does not need to be insulated)
B. Eliminating noise coupling:		
Twist low-level signal leads		

Contd...

EMC Check List for noise reduction technique

Items	Y/N	Comments
Place low-level leads near chassis (especially if circuit impedance is high)		
Twist & shield signal leads (coaxial cable may be used at high fre.)		
Shielded cables used to protect low-level signal leads should be grounded at one end only		At high frequencies use Coaxial cable shield grounded at both ends.
When low-level signal leads and noisy leads are in the same connector, separate them and place the ground leads between them		
Carry shield on signal leads through connector on separate pin		
Avoid questionable or accidental grounds		
For very sensitive applications, operate source and load balanced to ground		
Avoid common ground leads between high and low level equipment		
Keep hardware grounds separate from circuit grounds		
Keep ground leads as short as possible		
Use conductive coating in place of non conductive coatings for protection of metallic surfaces		
Separate noisy and quiet leads		
Ground circuits at one point only (except at high frequencies)		
Place sensitive equipment in shielded enclosures		
Filter or decouple any leads entering enclosures containing sensitive equipment		
Keep the length of sensitive leads as short as possible		
Keep the length of leads extending beyond cable shields as short as possible		
Use low-impedance power distribution lines		
Avoid ground loops by using devices		To break ground loops: Use Isolation or neutralising transformers; Optical couplers; Differential amplifiers; Guarded amplifiers; Balanced circuits

EMC Check List for noise reduction technique		
Items	Y/N	Comments
Use steel cabinets of 1-2mm thickness		Avoid using plastic enclosures
Aluminium superstructures to be grounded by CUPAL-foils every 10m		
When coaxial cables are used the cable screen may be used as a return and ground		It may be advantageous to use a filter to prevent ground loops
Avoid coaxial cables with BNC connectors		Use twinax or triax having an outer screen that does not carry signal
Avoid using RS 232 for data communication as it is not balanced i.e. conductors do not have the same impedance with respect to ground and to all other conductors.		RS485 is better. TTY current loop 20mA is balanced, sender and receiver are separated (low capacity)
Make proper connection and clean the surface		Use stainless steel bolts, zinc spray, lock washers, etc.
Check if copper straps used for grounding is at least of 25mm^2 cross –section.		Use at least one strap per metre of the cabinet and preferably at both bottom and top
C. Reducing noise at receiver:		
Use only necessary bandwidth		
Use frequency selective filters when applicable		
Provide proper power supply decoupling		
Bypass electrolytic capacitors with small high-frequency capacitors		
Separate signal, noisy, and hardware grounds		
Use shielded enclosures		

7.9 Testing for EMC

The type of test required should be determined on the basis of interference to which the equipment may be exposed when it is installed, taking into consideration the arrangement of the circuit i.e. the manner of earthing the circuit and shields, the quality of shielding applied and the environment in which the system as a whole is required to work.

It is impossible to simulate all conditions that may be encountered in the field. But a very good indication as to susceptibility of an equipment can be obtained from the application of a few standard tests. The tests should be considered to be basic tests which cover a sufficiently wide range of interferences to test industrial and field process measurement and control equipment.

Interference susceptibility tests are essentially equipment withstand-tests designed to demonstrate the capability of equipment to function correctly when installed in its working environment. Interference tests should be carried out with the system 'live' i.e. with the functional signals present, which in practice may be simulated. When considering the severity of interference tests to be applied the intention is to simulate, as closely as possible, the conditions which can actually exist in normal applications. It is therefore appropriate that high test values should be chosen but not extreme values. For comparison of equipments it is necessary to simulate a test signal that is relatively uniform and repeatable. A typical EMC test setup is shown below in Fig 7.9.

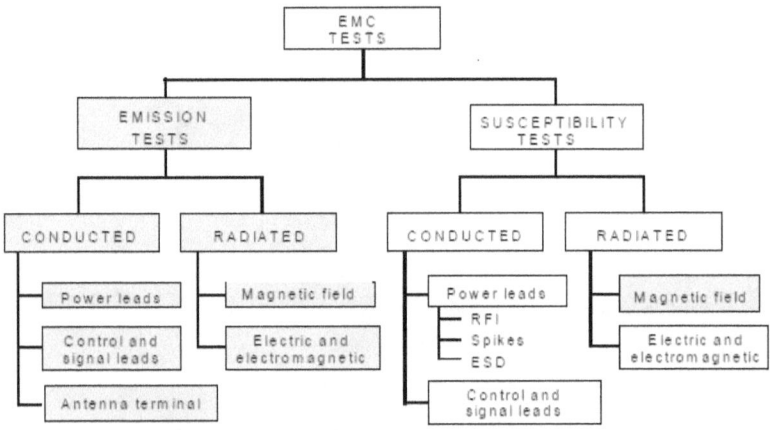

Fig 7.9: A typical EMC test setup

7.9.1 Conducted interference

Products can be susceptible to a wide range of interference signals that enter it via AC power cord. An obvious example is lightning induced transients. Thunderstorms frequently strike radio towers, power lines and transformer stations. Circuit breakers are intended to momentarily clear any faults and reclose after a short while.

The intent of the conducted emission limits is to restrict the noise current passing out through the product's AC power cord. The reason for this is that the noise currents will be placed on the common power net of the installation. The common power net of an installation is an array of interconnected wires in the installation walls, and as such represents a large antenna. Noise currents so generated on this power net will therefore radiate quite effectively, which can produce interference.

7.9.2 Radiated interference

The purpose of these tests is to insure that the product will operate properly when it is installed in the vicinity of high power transmitters. The common types are AM and FM transmitters and surveillance radars. Manufacturers test their products to these types of emitters by subjecting the product to typical waveform and signal level representing the worst-case exposure of the product and determining whether the product will perform satisfactorily.

If the product cannot perform satisfactorily in such installation, this deficiency should be determined prior to its marketing so that fixes can be applied to prevent a large number of customer's complaints and service calls.

7.10 Rules and Regulations of EMC

The requirements imposed by the government agencies are legal requirements and generally cannot be waived. These requirements are imposed in order to control the interference produced by the products. However, compliance with these EMC requirements does not guarantee that the product will not cause interference when used in the field. It only allows the country imposing the requirement to control the amount of electromagnetic 'pollution' that the product generates. On the other hand EMC requirements that the manufacturers impose on their products are intended to result in customer satisfaction. They are imposed for the purpose of insuring a reliable, quality product.

From a classification point of view, the safety of the any installation can be verified by well planned functional testing of equipment and systems, possibly if required backed up by selected measurements of radiation levels and EMI currents flowing in cables and cable screens. This may however lead to situations where the problems are revealed close to or during the field testing or trials.

Implementation of formal and complete EMC management procedures based on susceptibility and emission levels for all equipment backed up by equipment tests over a wide frequency range, provides a systematic method of predicting problem areas and planning for the optimum solutions. This method may be an expensive approach to solving problems, but the expenses must be balanced against the benefits. So all user departments must have a Type Approval schemes for instruments and equipments to be introduced in the projects for getting conducted Susceptibility EMI testing. These tests requirements should be similar to those required by other classification bodies / societies. The test levels should provide a trade-off between costs and documentation of equipment properties i.e. one is able to shortlist equipment which is definitely not suitable for installation in a particular project.

7.10.1 EEC and the EMC Directive

The EMC Directive 89/336/EEC was enforced on January 1st, 1992 and is mandatory to the member countries. Hence the national legislation of these countries must be adjusted in line with the EMC-directive. The EEC/EMC-directive comprises equipment and systems. A system shall be tested and approved as a whole system. An installation can however, seldom be tested as such. Therefore one must assume that the requirements of the directive have been met if the individual parts of the installation have been EMC approved (-which in practice may turn difficult since the EMC performance depends on a number of measures which need to be taken care of during the installation e.g. cable routing, grounding, screening). The distinction between system and installation can in some instances be hard to define. A computer system comprising a PC and some peripheral units should be regarded as a system, if the units are of the same make and destined to be integrated. On the other hand a computer system, configured with units of different makes by the user, should be regarded as an installation.

In order to release a product in the EEC countries, a proof must be produced justifying that the EMC-directive is complied with. This shall be stated by the manufacturer. Documentation must be produced to verify such compliance. As a proof of compliance with the EEC regulations, the product in question shall be marked with a sticker; 'CE' (Communauté Européenne). To demonstrate that a piece of equipment complies with a harmonised standard, the manufacturer (or Agent) shall provide a declaration of manufacture. Verification of compliance according to the standard requirements should be by means of a test protocol issued by an impartial testing laboratory, or in case the manufacturer has the necessary resources himself. Compliance without complete fulfilment of the harmonised standard requirements shall be documented in a technical construction file. Radio equipment for broadcasting shall be provided by an EEC-type approval certificate issued by a notified body. Note that the product requirements therewith may apply several directives according to the so-called 'new approach', which needs to be complied with by all e.g. EMC directive 89/336/EEC, 91/263/EEC and 89/392/EEC. Note that the latter directive refers to the 73/23/EEC (LVD) regarding safety of personnel.

7.10.2 Parts of EU Directive 89/336/EEC

The data in the following tables gives an overview of the testing requirements of EU and DNVC respectively. Note that in some cases the tests are based on different test set-ups and a straight comparison may be awkward.

The following documents have been used:

(a) EN 50081-1 January 1992 Electromagnetic compatibility – Generic emission standard Part 1: Residential, commercial and light industry
(b) EN 50081-2 August 1993 Electromagnetic compatibility – Generic emission standard Part 1: Industrial environment
(c) EN 50082-1 January 1992 Electromagnetic compatibility – Generic immunity standard Part 1: Residential, commercial and light industry (To be replaced in the near future)
(d) EN 50082-2 March 1995 Electromagnetic compatibility – Generic immunity standard Part 2: Industrial environment

7.10.3 Indian Standard for EMC

BIS (Bureau of Indian Standards) a govt of India body under the aegis of Ministry of Consumers Affairs is responsible for maintaining and framing standards for all type of Electrical / Electronic equipments including broadcasting and other engineering tools & implements to be manufactured or used in India. All standards before implementing are studied critically, examined by a committee of Experts in the field for any difficulties which one may experience in ones business or profession, if it is finally adopted as National Standards.

Indian standards are largely mirror image and based on the international standards prepared and issued by IEC (International Electrotechnical Commission. BIS have an understanding with IEC, for using their standards as base for adapting or formulating a new standard.BIS draft documents are identical with corresponding IEC Standard. Table 7.5 shows a list of BIS standards on EMC and their equivalent IEC standards. In these adopted standards, reference appears to certain International Standards for which Indian Standard also exist;

Table 7.5: list of BIS and their equivalent IEC standards on EMC

S.No	IEC standard	BIS standard
1	IEC-61000-1-1(1992) Electromagnetic compatibility (EMC) – Part 1: General – Section 1: Application and interpretation of fundamental definitions and terms Equipment connected thereto	IS 14700(Part 1/Sec 1):2000 Electromagnetic compatibility (EMC) Part 1: General Sec 1: Application and interpretation of fundamental Definitions and terms
2	IEC 61000-3-2(2005) Electromagnetic compatibility (EMC) – Part 3-2: Limits – Limits for harmonic current emissions (equipment input current <= 16 A per phase)	IS14700(Part3/Sec2):2008-Electromagnetic compatibility Part 3 Limits Sec 2 Limits for harmonic current emissions (equipment input Current =< 16A per phase) (First Revision)
3	IEC 61000-3-3(2005) Electromagnetic compatibility (EMC) – Part 3-3: Limits – Limitation of voltage changes, voltage fluctuations and flicker in public low-voltage supply systems, for equipment with rated current ≤ 16 A per phase and not subject to conditional connection	IS 14700(Part 3/Sec 3):2008 Electromagnetic compatibility Part 3: Limits Sec-3 Limitations of voltage fluctuations and flicker in low voltage Supply systems for equipment with rated current =< 16A Phase and not subject to conditional connection (First Revision)
4	IEC 61000-4-1(2006) Electromagnetic compatibility (EMC) – Part 4-1: Testing and measurement techniques – Overview of IEC 61000-4 series	IS 14700(Part 4/Sec 1):2008 Electromagnetic compatibility (EMC) Part 4 Testing and measurement techniques Section 1: Overview of Immunity tests Basic EMC publication (First Revision)

S.No	IEC standard	BIS standard
5	IEC 61000-4-2(2001) Electromagnetic compatibility (EMC) – Part 4-2: Testing and measurement techniques – Electrostatic discharge immunity test	IS 14700(Part 4/Sec 2):2008 Electromagnetic compatibility (EMC) Part 4: Testing and measurement techniques Section 2: Electrostatic Discharge immunity test (First Revision)
6	IEC 61000-4-3(2006) Electromagnetic compatibility (EMC) – Part 4-3: Testing and measurement techniques – Radiated, radio-frequency, electromagnetic field immunity test	IS14700(Part4/Sec3):2008 Electromagnetic compatibility (EMC) Part 4: Testing & measurement techniques Sec3Radiated, RF Electromagnetic field immunity test (First Revision)
7	IEC 61000-4-4(2004) Electromagnetic compatibility (EMC) – Part 4-4: Testing and measurement techniques – Electrical fast transient/burst immunity test	IS 14700(Part 4/Sec 4):2008 Electromagnetic compatibility(EMC) Part4 Testing and Measurement techniques – Sec 4 Electrical fast transient/burst Immunity test (First Revision)
8	IEC 61000-4-5:2005 Electromagnetic compatibility (EMC) – Part 4-5: Testing and measurement techniques – Surge immunity test	IS 14700(Part 4/Sec 5):2012 Electromagnetic Compatibility (EMC) – Part 4 Testing and measurement techniques: Sec 5 Surge immunity test
9	IEC 61000-4-8(2001) Electromagnetic compatibility (EMC) – Part 4-8: Testing and measurement techniques – Power frequency magnetic field immunity test	IS 14700(Part 4/Sec 8):2008 Electromagnetic compatibility (EMC) Part 4: Testing and Measurement techniques Section 8: Power frequency magnetic field immunity test (First Revision)
10	IEC 61000-4-9(2001) Electromagnetic compatibility (EMC) – Part 4-9: Testing and measurement techniques – Pulse magnetic field immunity test	IS 14700(Part 4/Sec 9):2008 Electromagnetic compatibility (EMC) Part 4: Testing and measurement techniques Section 9: Pulse magnetic field immunity test (First Revision)
11	IEC 61000-4-11(2004) Electromagnetic compatibility (EMC) – Part 4-11: Testing and measurement techniques – Voltage dips, short interruptions and voltage variations immunity tests	IS 14700(Part 4/Sec 11):2008 Electromagnetic compatibility (EMC) Part 4: Testing and measurement techniques Section 11: Voltage dips, short interruptions and voltage variations immunity tests
12	IEC 61000-4-12(2006) Electromagnetic compatibility (EMC) – Part 4-12: Testing and measurement techniques – Ring wave immunity test	IS 14700(Part 4/Sec 12):2008 Electromagnetic compatibility (EMC) Part 4: Testing and measurement techniques Section 12: Ring wave immunity test (First Revision)
13	IEC 61000-4-15(2003) Electromagnetic compatibility (EMC) – Part 4: Testing and measurement techniques – Section 15: Flicker meter – Functional and design specifications	IS 14700(Part 4/Sec 15):2008 Electromagnetic compatibility (EMC) Part 4: Testing and measurement techniques Section 15: Flicker meter functional and design specification (first Revision)
14	IEC 61000-4-16(2002) Electromagnetic compatibility (EMC) – Part 4-16: Testing and measurement techniques – Test for immunity to conducted, common mode disturbances in the frequency range 0 Hz to 150 kHz	IS 14700(Part 4/Sec 16):2008 Electromagnetic compatibility (EMC) Part 4: Testing and measurement techniques Section 16: Test for immunity to be conducted common mode disturbances in the Frequency range 0Hz to 150 kHz (First Revision)

Contd...

S.No	IEC standard	BIS standard
15	IEC 61000-4-24(97) Electromagnetic compatibility (EMC) – Part 4: Testing and measurement techniques – Section 24: Test methods for protective devices for HEMP conducted disturbance – Basic EMC Publication	IS 14700(Part 4/Sec 24):2007 Electromagnetic compatibility (EMC) Part 4: Testing and measurement techniques Section 24: Test method for protective devices for hemp conducted
16	IEC 61000-6-1(2005) Electromagnetic compatibility (EMC) – Part 6-1: Generic standards – Immunity for residential, commercial and light-industrial environments	IS 14700(Part 6/Sec 1):2008 Electromagnetic compatibility (EMC)-Part 6 Generic standards Sec 1 Immunity for residential, commercial and light-industrial environments
17	IEC 61000-6-2(2005) Electromagnetic compatibility (EMC) – Part 6-2: Generic standards – Immunity for industrial environments	IS 14700(Part 6/Sec 2):2008 Electromagnetic compatibility (EMC)-Part 6 Generic Standards Sec 2 Immunity for industrial environments
18	IEC 61000-6-3(1996) Electromagnetic compatibility (EMC) – Part 6: Generic standards – Section 3: Emission standard for residential, commercial and light-industrial environments	IS 14700(Part 6/Sec 3):2002 Electromagnetic compatibility (EMC) Part 6: Generic standards Section 3: Emission standards for residential, commercial and light industrial environment

7.11 EMC Management

Implementation of EMC management control is a systematic approach to help in decision making, describe remedial measures and make possible an estimate of the time and cost expenditure. It will normally be easier to remedy those problems that do surface in the design in contrast to problems arising at the final testing after installation in the field under actual working conditions. The management control procedures serve to bring in a systematic manner the possible interference in and between the systems, to investigate them qualitatively and quantitatively and to form the basis for working out the remedial measures for EMC.

It consists of setting up a general list as well as a review plan to take in all the equipments of the system that can have an effect on the EMC of the system: In the data list are to be included:

(a) EMC requirement as to emission of disturbance.
(b) EMC specification values for equipment already developed. Obtain data from the manufacturers. If there are no measured values available, then estimate values and values from experience may be put in.

(c) Consequence classes for immunity margin.
(d) Data concerning the transmitting and receiving equipments
(e) Data on the electromagnetic environment that the equipment is at times or continuously put into, e.g. data on useful emissions, on emission of disturbance, on immunity to disturbance and on the field strengths of strong transmitters in the neighboring systems as well as the requirements to take into consideration the thresholds of disturbance of sensitive sensors (antenna, receiver input) in these neighboring systems.
(f) Equipment installation location, antenna sites.
(g) Scaled drawings of the entire system or part of the system, from which data can be taken on the spatial arrangement and the placing of the equipment, antenna and cabling.
(h) Information concerning the cable installation e.g. type, routing, paths, lengths, and other characteristics of the cable installation which are relevant to EMC to be seen.
(i) Information about screening, grounding, earthing & further measures for potential equalization in the system.

All the equipments of the system are to be classified into the following classes as shown in Table 7.6, according to the consequence of disturbance. This classification serves to set the immunity margins to assess the quantum of disturbances in the processing of the EMC analysis.

Table 7.6: Immunity Margins and consequences of disturbances

class	Margin	Consequence of Disturbance
0	0 dB	No harmful effect
1	6 dB	Equipment for which disturbance of the function can lead to an additional load on the operators or a limitation on the system efficiency
2	10 dB	Equipment for which disturbance of the function can lead to wounding, damage to system or a restriction on efficiency of the system.
3	20 dB	Equipment for which disturbance of the function can lead to loss of life, loss of system, or unjustifiable restriction on the system efficiency

The allowable level for emission of disturbance depends essentially on the decoupling of equipment units, lines and cables from the system's receiving antennas and from each other. The inherent coupling attenuation of various systems depends primarily on the design and dimensions of the equipment location. The coupling attenuation of an equipment location is essentially characterised by;

(a) The distance of equipment units, lines and cables from the antennas and between each other,
(b) Metal surfaces for decoupling, and
(c) The existing screening against antennas and other equipment units, at the place of installation.

As shown in Table 7.7 systems are classified by means of these characteristics into decoupling levels. The decoupling is given by a typical value. This value may be corrected according to the actual situation;

Table 7.7: Decoupling levels and typical characteristics (not including distance related decoupling)

Equipment location characteristics	Decoupling (±10dB)	Typical EMC features
No screening and no decoupling metal surfaces	Level 0 0 dB	Lines & cables have large antenna heights and loop widths; therefore little decoupling from each other & from the antennas
No substantial screening, but decoupling metal surfaces	Level 1 15 dB	Lines and cables in contact with or closely above the metal surfaces have minimum antenna heights and loop widths; therefore decoupling from each other and from the antennas by at an average level
Screening as well as decoupling metal surfaces	Level 2 30 dB	Antenna heights and loop widths of lines and cables same as for level 1; average antenna decoupling of equipment to screened areas
Closed screening or multiple screening e.g. by nested screens	Level 3 50 dB	Very high tightness against electromagnetic fields; average antenna decoupling of equipment within closed screens or double screening

The level of equipment emission of disturbances are based on a measuring distance $r = 1$ mtr assuming reflection-free propagation of electric and magnetic fields. For other distances the respective coupling values between source and sink of disturbance must be calculated or corrected according to the nomograph Fig 7.10.

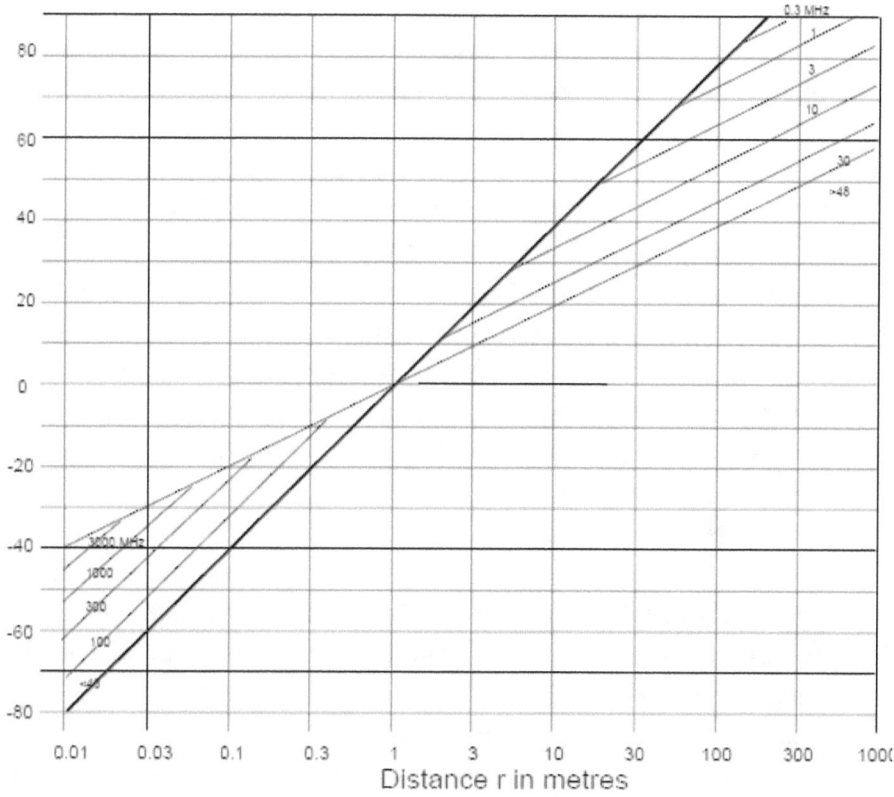

Fig 7.10: Coupling attenuation as function of distance r (in meters) and Frequency (MHz)

The field strength present at the location of equipment installation and the interfering currents flowing through lines and equipment housing are caused by the internal transmitters of the system or, in some cases, by external transmitters, which are operating continuously or for short periods in the immediate vicinity of the equipment.

The most important feature is the screening attenuation resulting from metal structure of the system provided the place of equipment installation lies within this structure. In addition, the system's dimensions are of importance as large distances from the transmitting antennas may result, in some cases, in a significant reduction of the field strength produced.

Have a compilation of field strength data for various frequency sub ranges and for typical systems working in the setup. In addition also have field strength data for 'external equipment', i.e. equipment including associated cabling which is located within an antenna radiation field, and the respective screen attenuation levels for evaluating EMC management.

Immunity margins should be determined during system analysis and ensured by respective measures during the integration of the system of the setup. The voltages coupled into lines as a result of incident fields are dependent upon the ambient field, and the routing of the line.

The immunity to the disturbances at terminals of power supply lines, unscreened lines, and screens (microsecond and nanosecond pulses) are primarily dependent upon the power supply characteristics typical for the system (power supplies) and load connected. The peak values of disturbing pulses to be expected in a specific system, can only be determined with accuracy required for the exact specification of limiting values by means of a system power supply analysis.

Within the scope of this analysis, all characteristics of the system power supply (power generators, supply network and all inductively and capacitively coupled components of the control and signal network, load) which are essential to the generation, propagation and effects of the disturbing pulse, must be included and evaluated. Conductive couplings are difficult to predict and estimate. For the cases whose data are not available, fill in with a yes or no answer or a description of the signals on the information sheet?

A typical EMC management tabulation as shown on page; 30-31 of reference: 1, should be devised and used for analysing a system.

7.12 RF Shielding Material - Cable and Copper

These two items are most important and abundantly used in electrical and electronics equipments and components. Former one is used for interconnection of various sections of a system or equipment itself and second one for making components. In the following paragraph these will be described in details.

7.12.1 Cable Shielding Types and uses

There are various types of Shields in the form of continuously wrapped foil or woven multi-wire braid as per degree of shielding or immunity desired as well as for a particular type of application. These shields are always covered by an outer coating of PVC / rubberised or any other covering as per typical applications. In case of armouring, there is covering of PVC or any other over shield and then it is armoured with continuous armour foil or braided wires and is finally covered with outer covering of PVC.

(a) Foil Shielding
This is for protection against capacitive (electric field) coupling where shield coverage is more important than low DC resistance. This is used when probable sources of interference include TV signals, crosstalk from other circuits, radio transmitters, fluorescent lights or computing equipment. Also it is used for MATV, CATV, Video, networking, and computer I/O cables, industrial or commercial environments where ambient EMI levels are low.

(b) Braid Shielding
This is for superior performance against diffusion coupling, where low DC resistance is important, and to a lesser extent, capacitive and inductive coupling and is used when probable sources of interference exhibit low impedance characteristics, such as motor control circuits and switches which operate inductive loads, for computer to terminal interconnect for process, instrumentation or control applications.

(c) Spiral Shielding
This is for functional shielding against diffusion and capacitive coupling at Audio Frequencies only.

This is used when possible sources of interference are power lines and fluorescent lights, for applications when flexibility and flex life are major concerns, such as microphone and audio cables and retractile cords.

(d) Combination Shielding
This type of shielding is against high frequency radiated emissions coupling and ESD. This type of shielding combines the low resistance of braid and 100% coverage of foil shields. This is used when probable sources of interference include radio transmitters, TV stations, printed circuit boards, back planes, motor control circuits and computing equipment. Also it is used for Video, CATV, MATV, networking, computer I/O cables and computer aided manufacturing applications.

(e) Relative Cost Comparison

Relative cost comparisons are based on coaxial cable. Fig 7.11 shows relative shield cost as one component of the total cost of the cable. These cost ratings may change depending on the physical construction of the cable. For bigger size of the cables this may vary and may not hold good.

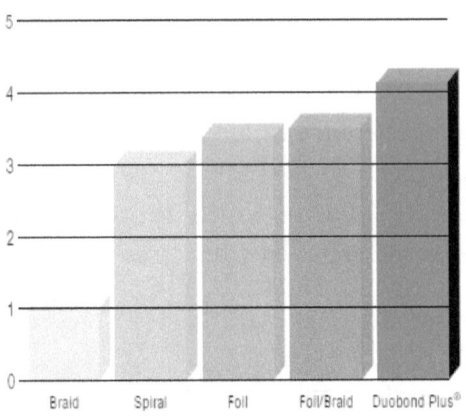

Fig 7.11: Relative cost of various Cable Shielding

(f) Performance comparison of various cable shields

Performance of various types of shield is not equal and varies for type of use it is put to. A cable shielding is chosen as per immunity desired and the budget available for project as well the acceptable tolerance level of interference. Table 7.8 details the shield performance comparison;

Table 7.8: Shield Performance Comparison Chart

Frequency Range & Parameters	Cable shield Ratings[1]			
	Braid (95% coverage)	Spiral	Foil	Foil/Braid
Frequency:		DC		
Capacitive	A	AA	AAA	AAA
Diffusion	AAA	A	C	AAA
Diffusion/inductive	NA	NA	NA	NA
Diffusion/Inductive/Capacitive	NA	NA	NA	NA
Frequency:		15KHz		
Capacitive	A	AA	AAA	AAA

Frequency Range & Parameters	Cable shield Ratings[1]			
	Braid (95% coverage)	Spiral	Foil	Foil/Braid
Diffusion	AAA	B	C	AAA
Diffusion/inductive	AA	C	A	AAA
Diffusion/Inductive/Capacitive	NA	NA	NA	NA
Frequency:	10 MHz to 1000 MHz			
Capacitive	A	AA	AAA	AAA
Diffusion	NA	NA	NA	NA
Diffusion/inductive	B	C	A	AA
Diffusion/Inductive/Capacitive	B	C	A	AA

1. Although ratings shown in table are based on shielded coaxial cable test results, these ratings also pertain to shielded multi-conductor and flat cable where shield types are available.

Legends: AAA —Best, AA — Better, A — - Good, B—- Satisfactory, C — Unsatisfactory, NA – Not Applicable

7.12.2 RF Shielding Material-Copper

RF shielding involves construction of enclosures for the purpose of reducing the penetration of electric or magnetic fields from one space to another. With the increase in sensitive electronic equipment, the issues of interference and protection from high voltages have become increasingly important.

All broadcasting transmitters Radio/TV and Radars radiate high RF field and affects the functional and qualitative performance of all working electric and electronics equipments and instruments in their vicinity. Similarly all electronic and computer systems also radiate certain frequencies of radio and magnetic waves. These signals can be received by special surveillance equipment, compromising the privacy of their source. In some cases, they can interfere with, or may be affected by, other equipment in the vicinity. RF shielding enclosures are used in these situations to reduce the levels of RF radiation that enters or leaves the enclosed space.

Copper is best materials among all, utilised for this purpose around the world. One of the characteristics of copper is its high electrical conductivity. This feature, combined with its other physical properties, ductility, malleability, and ease of soldering, make it an ideal material for RF shielding. Sheet copper can be formed into essentially any shape and size, and electrically connected to a grounding system to provide an effective RF enclosure.

RF shielding enclosures are usually designed to filter a range of frequencies under specified conditions. Properly designed and constructed enclosures can provide a high degree of effectiveness to suit practically any demand. **Types of Copper and Properties;** Most commonly used copper for sheet and strip applications complies with ASTM B370. It consists of 99.9 percent copper, and is available in six tempers designated by ASTM B370 as: 060 (soft), H00 (cold rolled), H01 (cold rolled, high yield), H02 (half hard), H03 (three quarter hard), and H04 (hard).

Soft tempered copper is extremely malleable and best suited for applications such as ornamental work. It was historically used in building construction. Because of its low strength, heavy gauge material was required. As a result, the use of soft tempered copper is not recommended for most building applications.

With the development of cold rolled copper many years ago, the gauge of the material could be reduced without compromising its low maintenance and long life. Cold rolled copper is less malleable than soft temper copper, but is much stronger. It is by far the most popular copper temper currently used in construction. The properties of cold rolled copper are summarized in Table 7.9;

Table 7.9: Properties of Cold Rolled Copper

Property	Value	Property	Value
Specific Gravity	8.89 – 8.94	Tensile Strength	32,000 psi min.
Density	0.322lb./cu. in. at 68°F	Yield Strength (0.5% Extension)	20,000 psi min.
Thermal Conductivity	226 BTU/Sq Ft/Ft/Hr °F at 68°F	Elongation in 2" – approx.	30%
Coefficient of Thermal Expansion	0.0000098/°F from 68°F to 572°F	Shear Strength	25,000 psi
Modulus of Elasticity (Tension)	17,000,000 psi	Hardness – Rockwell (F Scale) Rockwell (T Scale)	54 min. 15 min.

The significant properties of the six ASTM B370 designated tempers are summarized in Table 7.10.

Table 7.10: Mechanical Properties of cold rolled copper

Temper Designation Standard	Tensile Strength (Ksi) Min	Tensile Strength (Ksi) Max.	Yield Strength(Ksi) Min	Temper Designation Standard	Tensile Strength (Ksi) Min.	Tensile Strength (Ksi) Max.	Yield Strength (Ksi) Min
060 Soft	30	38	—	H02 Half Hard	37	46	30
H00 Cold-Rolled 1/8 Hard	32	40	20	H03 Three quarter Hard	41	50	32
H01 Cold-Rolled, high yield 1/4 Hard	34	42	28	H04 Hard	43	52	35

In general, cold rolled 1/8 hard tempered (H00) copper is recommended for most roofing and installations. Soft copper may be used where extreme forming is required such as in complicated thru-wall flashing conditions. However, it should be noted that cold rolled copper offers far more resistance than doe's soft to the stresses induced by expansion and contraction. Copper roof sheet of higher temper should be specified only if indicated for specific and engineering applications requiring such higher tempers. The yield strength of cold-rolled high yield (H01) copper is significantly higher than standard cold rolled (H00) copper upto 33,000 p.s.i. This allows the use of 12 ounce high yield copper in many applications where 16 ounce cold rolled copper is normally used.

Good resistance to corrosion, good electrical and thermal conductivity, ease of fabrication coupled with strength and resistance to fatigue are criteria by which copper or one of its alloys is selected.

Corrosion Resistance: Copper is a noble metal able to resist attack quite well under most corrosive environmental conditions. In the presence of moisture, salt and high sulphur pollution, copper quickly begins to oxidize and progress through the weathering cycle. Its high resistance to corrosion is due to its ability to react to its environment and reach weathering equilibrium.

Electrical and Thermal Conductivity: Copper and its alloys are excellent conductors of electricity and heat. In fact, copper is used for these purposes more often than any other metal. Alloying invariably decreases electrical conductivity and to a lesser extent, thermal conductivity. Coppers and high-copper alloys are preferred over copper alloys containing more than a few percent total alloy content for high electrical or thermal conductivity.

Ease of Fabrication: Copper and its alloys are generally capable of being shaped to the required form and dimensions by any of the common fabricating processes. They are routinely rolled, stamped, drawn and headed cold; they are rolled, extruded, forged and formed at elevated temperature.

Copper and its alloys are readily assembled by any of the various mechanical or bonding processes commonly used to join metal components. Crimping, staking, riveting, and bolting are mechanical means of maintaining joint integrity. Soldering, brazing and welding are the most widely used processes for bonding copper metals. Selection of the best joining process is governed by service requirements, joint configuration, thickness of the components, and alloy composition(s).

References

1. ELECTROMAGNETIC COMPATIBILITY –Dec 1995, Classification Notes No. 45.1, DET NORSKE VERITAS CLASSIFICATION AS Veritasveien 1, N-1322 Høvik, Norway
2. Analog Devices (MT-095 Tutorial) – EMI, RFI, and Shielding Concepts (pp-1 to 16)
3. Technical Information – 16.13, Shielding and Armouring –Belden Cables Beldon Electronics Division website: www. belden. com
4. Bureau of Indian Standards (BIS) – IS14700 (Part 1/Sec 1):2000 Part-1 to 6 Electromagnetic Compatibility (EMC)

CHAPTER 8
SAFETY PRACTICES IN RADIO STATIONS

8.1 Introduction

This section has been prepared as a guide for staff to safe working in radio installations, in particular in broadcast radio and television stations and to assist management faced with the need to draft or update rules and regulations applying to their technical facilities and staff. It outlines rules and recommended safe practices which help engineers, technicians and other staff to perform their work safely and to avoid accidents and injuries.

Every station management should have a safety policy prepared by keeping local safety rules in mind. The primary object of the rules must be to ensure the personal safety of the station staff and of third parties. All staff should be required to observe these with breaches being treated seriously, even if no accident occurs. Station supervisors should play a major role in safety training, counselling, and checking and by setting a good example.

Safety includes both the safety of staff who work on technical facilities and the safety of the equipment or plant. Equipment or plant malfunction, failure or damage may in turn result in personal injury to the staff working on the facility so the interactions between safety to staff and safety of equipment or plant are significant.

All technical staff must read station safety rules carefully, ask questions about their work and train themselves to form the habit of carrying out their work safely. It is important that they develop an appreciation of the potential dangers which exist in their station. An unsafe worker is a danger to himself, his fellow workers and the equipment with which he works. Safety comes first in any operation, and time and thought must be given to doing the job in a safe way.

Most accidents at radio stations are due to human fallibility such as the failure to use safety equipment and safe methods. Not many are due to imperfections in the radio equipment or plant. Elimination of all unsafe conditions and unsafe acts is the only sure way to eliminate accidents.

8.2 Safety Engineering Philosophy

Some basic factors such as adequate training by experienced and veterans personnel, supervision by properly qualified and aware personnel, correctly installed equipment, safe access to all equipment for maintenance purposes, the availability of the correct instruments, apparatus and tools and the knowledge of as to how to use them are all important safety practices in most fields of engineering. A full and proper appreciation of the various hazards involved in performing work is vital. There are, however, additional hazards in the field of radio engineering. These include high voltages (ac / dc) used with high power transmitters, the radio RF voltages encountered on transmission lines and antenna systems, electrically charged components, burns from hot water and vapours from vapotron and steam cooling systems of transmitting tubes, the biologically hazardous situation created by intense electromagnetic fields, the risks of high winds, rain and ice associated with work on tall masts and towers and the danger of working on compressed air operated switches in transmitters and transmission line matrix systems. Wherever possible work on unsafe apparatus and live circuits should be avoided but where there is no alternative then the work should be recognised as constituting an especially dangerous situation for which special safety precautions are necessary.

Well defined adequate maintenance programmes, frequent inspections and, above all, the education of workmen must be recognised as essential prerequisite for safe work. These are the responsibility of the organisation at all management levels. Supervisors must exercise constant vigil to ensure that rules are obeyed and best practices followed. The roles of supervisors being at that level are of prime importance and they should be encouraged to suggest new safe guards wherever necessary. It is a well established and recognised practice that cost must always be taken into account when expenditure on equipment and working procedures are under consideration, but the safety of the workmen must remain a prime objective.

In addition to persons at management level in an organisation, who is considered responsible for preventing the accidents, every person of the organisation must share the responsibility. Even those engaged in work not associated with the technical equipment, such as, labourers, cleaners, and the office workers, should carry out the safety instruction and see to it that neither they nor their fellow workers are injured due to an unsafe act that may be attributed to their own intentions, negligence, or thoughtlessness.

The basic philosophy should be to concentrate on the detection of hazards and then to eliminate them as far as possible. This applies to all phases of the work from planning, to design, construction, installation, testing, operations and maintenance. Thus, engineers engaged on developing methods and establishing standard practices should endeavour to apply the eliminate-the-hazard-first approach, wherever possible. However, it is recognized that there are times when certain constraints, such as funds, exist and the engineer is expected to produce the most practical and the safest design within these limitations. If an engineer has processed a job, and has left an avoidable hazard in it, he has not fully met his responsibility. In many instances, accidents occurred on a job because the work had not been adequately studied or engineered.

Safety programmes should be aimed at preventing the workmen injuring themselves and others. These should generally be accomplished though the use of safe guards like, mechanical, electrical etc, supplemented by written or oral instructions at regular intervals. The continuing educational programme is necessary because, unfortunately, the absence of accidents lulls personnel working in the hazardous areas or they may be so used to the presence of dangers that they are no longer conscious of it. Ironically, the more a person knows about the hazards of his job, the more complacent his attitude becomes.

A reduction in the total number of accidents on a particular radio engineering project does not necessarily involve a proportionate reduction in the hazards for each individual workman. Sometimes there is an increase, since high powers and automation make the mistakes of the workman more serious. In fact, the nature of the hazard has altered. The deterioration that has occurred in some areas can be attributed to the following factors;

(a) Inadequate vocational qualifications of staff to cope with the complexity of modern radio and electronic equipment. Ignorance of the presence of dangerous potentials on components may lead to a serious shock. For instance, many electric shocks and also other injuries from equipment have occurred when operators were unaware of the manner and sequence by which circuits were energized. Adjustment and maintenance of equipment should be carried out by properly trained and adequately qualified as well as certified workmen.

(b) Careless installation, operation and maintenance practices, particularly in relation to protection facilities. Standards of safety require that all intrinsically dangerous components or parts should be so protected, wherever situated in the equipment or plant, that no operator can perform unsafe act in relation to them. This of course is setting a high objective, but examinations of some modern broadcast transmitters, made by safety conscious manufacturers having the foregoing principles in mind, show that such a standard is not unobtainable.

There is unfortunately a tendency on the part of some designers to minimize safety features on relatively low voltage equipments, therefore there may be increased possibility that an accident will happen to a skill but careless worker. So a worker must keep in mind that no safety device can be 100% reliable and to be safe, he must observe all the safety rules.

8.2.1 Typical Safety Measures in Broadcasting set ups

Many large organizations and particularly Government departments operating broadcast and television services have a code of general principles which lays down the measures necessary to safeguard the safety and health of its employees while they are at work. Typical management provisions include the following;

(I) Obligations of the Organisation

The organisations have typical duties as mentioned below;

(a) To issue policy on safety and responsibilities and their review regularly.
(b) To adopt arrangements for joint consultation with employees on safety matters.
(c) To appoint safety coordinators and oversee their performance on continual basis.

(d) To provide of safe work places and a safe working environment.
(e) To provide of safe premises, environment, plant, machinery and equipment.
(f) The provision of safe work methods and appropriate training and placement of employees.
(g) The adoption of occupational hygiene principles and control of harmful chemicals and physical agents.
(h) The adoption of measures to minimise the risk of and harmful effects of fire and explosion.
(i) Provision of appropriate personal protective equipment and the adoption of measures to ensure its proper use.
(j) For establishment of medical, health and first aid services.
(k) To maintain records of injury and accident and arrangements for accident investigation.

(II) Obligation of Employees

The employee also has an obligation under the code as summarised below;

(a) Each employee shall have responsibility for safe working consistent with the extent of his control over or influence on working conditions and methods.
(b) Each employee shall take such action as is within his competence and responsibility or report or make such recommendations to a higher level as he deems necessary to avoid, eliminate or minimize hazards of which he is aware in regard to working conditions or methods.
(c) Each employee shall observe all instructions issued to protect his safety or safety of others or equipments.
(d) Each employee shall make proper use, or to the extent of his responsibility ensure that proper use is made, of all safeguards, safety devices, personal protective equipment and other appliances provided for safety purposes.
(e) No employee shall, or shall cause another employee to, interfere with, remove, displace or render ineffective any safeguard, safety device, personal protective equipment or other appliance provided for safety purposes, except where necessary as part of an approved maintenance or repair procedure.

8.3 Interpretation of Safety Rules

Safety practices may vary considerably in details from one station to another station, and yet both stations may have good safety records. The heart of the matter is that safe and consistent working procedures for a particular location must be prepared, implemented, and the workforce should be thoroughly familiar with, and willing to abide by these procedures. However, knowledge of the procedures is not enough – there must be keen interest in safety on the part of both worker and management. Hence ways and means of creating and maintaining interest in safety rules are required to be developed by keeping the necessary vigilance to prevent accidents.

Supervisors should ensure that no job is done more easily or quickly at the expense of safety, and that no man is ever urged to greater output if there is any question of a seriously decreased safety factor. In planning every job, safety for personnel is a major consideration and often the principal one, particularly when working on high voltage equipment or on live or low powered antennas and towers.

Supervisors must be constantly on the alert for problems resulting from the transfer of employees who have been working on low power equipment and antenna systems to jobs involving high powers. Where such transition occurs, supervisors must ensure specific job safety instructions update.

A careless worker is a potential hazard not only to himself, but also to his fellow workmen. The majority of the accidents, including electric shock, occurs because someone was careless. Carelessness by supervisor and qualified technical workmen cannot be excused, because on a large radio station complex, many unskilled workers may be employed and in many situations they depend greatly on the knowledge and skill of the technical people for their safety.

All new employees should be carefully instructed as to the hazards of the work and be issued with a copy of station safety rules. The immediate supervisor of the workmen should ensure that the rules have been read and correctly understood by the worker and at any time that worker should be called upon to show and demonstrate his knowledge of the rules. Workmen should not only comply with the rules but also use all safety devices, work carefully, and co-operate in activities having as their object of accidents.

Supervisors should warn off the staff / workers of the hazards involved if they keep wearing their personal accessories like metal rings and wrist watches etc while working on RF circuits and live voltages. Such accessories have been found responsible for many electric shocks, and burns. The area of contact on the skin of wrist watches and metal bands is considerable; pressure is usually firm and the skin beneath it usually moist and offers very good conductivity because of the salt in the perspiration. Survey has revealed that 20% had been the victim of shock or severe burns resulting directly from wearing of the metal accessories.

8.4 Classifications of Accidents

Accident statistics points to a certain number of hazards that can be identified in the installation, operation and maintenance of radio equipments and plant. An examination of several hundred accidents of all category, degrees of seriousness, including fatal accidents at broadcasting stations, have enabled the classification shown in Table 8.1 to be made.

More than 85% of the accidents were due to the human error; such as failure to follow the standard operating procedure (SOP), use of safety devices and prescribed safety methods for the setup. For the purpose of the Table, accidents are defined as those in which it was found necessary to use first aid facilities on the victim either at the workplace or in the hospital. Of fatal accidents included in the statistics,

Table 8.1: Classifications of Accidents of a Radio Station

S. No	Accident Cause	Percentage
1	Handling equipment and material on the ground	25.2
2	Tripping over objects or slipping on floor	12.6
3	Contacting sharp objects or bumping into objects	9.6
4	Contact with live components and conductors	9.0
5	Falls from structures, platforms and ladders	7.5
6	Falling or moving objects	6.7
7	Hand tool operations	4.4
8	Mechanical Aids and machinery operations	4.2
9	Erection of structures and antennas	3.4
10	Test Equipment operations	3.2
11	Working on tube hot water systems	2.2
12	Use of welding equipments	2.0
13	Operations of explosive powered tools and explosives	1.5
14	Handling harmful substances	0.8
15	Compressed air operations	0.7
16	Other cases	7.0

three contacted transmitter EHT supplies, eight were killed during mast erection operations, one was crushed by a mechanical aid and fracturing of leg/arm etc while shifting /placing equipments. There are many more non-fatal cases, which were not even reported by engineer-in-charge for fear of punitive action by authorities

By comparison with the number of accidents in factories, homes or on the roads, the number of accidents at radio stations is relatively small. However, some lives are lost and there must be continual effort to reduce this loss.

8.4.1 Findings of the accident statics

From an examination of many of the reports associated with the accidents in the table some action which needs to be considered for incorporating in the standard operating procedure (SOP) is:

(a) It is important to detect and eliminate hazards before they cause accidents.
(b) The cause of each accident should be ascertained and steps taken to eliminate the hazard and prevent the recurrence by updating the maintenance procedures.
(c) The prevention of accidents is not only good practice but also a wise economic investment.
(d) In the majority of cases, the accidents are preventable and can be prevented.
(e) The majority of accidents are due to similar reasons/causes, irrespective of the size of the organization.
(f) There are one or more causes for every accident, they do not just happen.

8.4.2 Important lessons learnt

Some important lessons, which can also be learned from the design, installation, operation or maintenance of the equipment or plant involved in some of the accidents, may be summarized as below:

(a) All equipment must be properly earthed. In the case of high power transmitting equipment, reference and protective earth system of low impedance, as per the standard norms, at the operating frequency is essential.

(b) Appropriate interlocking of control circuits and equipment as per standard safety norms must be provided.

(c) Equipment should have adequate built-in short circuit capability and be provided also with protective devices such as fuses, relays, vacuum switches and the like.

(d) High voltage equipment must be enclosed in earthed metal or insulated type enclosures, where practicable with interlocked access door.

(e) No equipment or plant should be operated without authority or warning. Typical unsafe acts include closing switches without authority, failure to place warning signs or signalmen wherever needed and failure to block or guard equipment against unexpected movement.

(f) Protective safety gears must be worn and used by all whosoever works on the equipment. Failure to use requisite rubber gloves, insulating mats or sleeves around energised equipment, failure to use protective equipment for the eyes, ears, head, feet and body where necessary have contributed to many of the accidents in broadcasting installations.

(g) The components comprising the equipment should be of high quality established and recognized standards, and correctly installed in the equipment.

(h) Equipments and plant must be adequately and effectively maintained to the standard norms all the time without any deficiency in the performance, safety and interlock features.

(i) Correct operating, installation or erection practices and procedures must be followed. Periodical review and refreshing practices / mock drills must be held to update the skill of the staff for prompt and correct action to avoid accident and loss to man, machine as well loss of service.

8.4.3 Bad Housekeeping -a reason for Accidents

Proper housekeeping is very important for any size of radio station as lot of accidents takes place for this nontechnical cause. Therefore there should be adequate properly documented and standardised guidelines for housekeeping jobs. A good housekeeping should be promoted by each radio station by demarcating proper place for keeping all the materials,

tools, and equipments when not in use. Accidents which have taken place in radio stations due to poor and bad housekeeping like;
(1) Slipping on floor as a result of oil leakage from a transformer tap as drip tray was not replaced back.
(2) Tripping over the transmitting tube kept on floor during tube replacement operation instead of keeping it in the socket in tube trolley.
(3) Tools dropped while working on the mast during erection or repair as these were not kept in tool bag.
(4) Injury to hands while handling packing crate lid as the nails was not removed before discarding the crate.

8.5 Emergency Plans of the Organisation

No radio station can be made completely immune from the accidents, loss or disasters. Fire, explosion, mast collapse, flood or cyclone may strike even the most carefully protected station. When the emergency comes, proper action can make the difference between a minor incident and a major catastrophe. If a serious loss does occur, a restoration plan prepared in advance may mean, especially in the case of a large commercially operated radio station, the difference between returning to normal operation with a minimum of delay or going out of business.

Management must also see that responsibilities are clearly defined, that clear, crisp and direct communications are maintained, and that all the staff on the station periodically receives proper training, to cope with possible emergency situations. The more decentralised and diversified the operation of an organization, the better its ability to recover from a disaster. It is essential that far more authority and responsibility be delegated to certain station staff during the time of disaster than is normally given. Management has to determine which members of the staff are to take over in these circumstances and must ensure that they receive all necessary help and support to prepare them for this work. Because most large station establishments are operated on shift basis, the station manager and other key personnel may not be immediately available on the site to take over the exigency. Organizations disaster plan should therefore include a clear succession list of the persons to take over the responsibilities as per situation at site.

As most of the staff might have had little experience in emergency situations on the station, it is understandable that at some levels there could be a degree of disinterest, even ranging on scepticism, about the risk of says a large fire, the threat to life and the problems of re-establishment. Management has to provide the leadership, direction and support necessary for a good emergency plan. It must show the way and by precept and example establish the interest and co-operation of the station staff.

8.5.1 Development of Emergency Plans

In the development of an emergency plan by an organization, some of the important actions which management must take may be summarised as:

(a) Train and periodically check all the station personnel in emergency procedures, including station shut-down, raising the alarm, fire fighting, and First aid and also salvage including the use of waterproof covers to protect equipments from water.
(b) Assign and periodically check the specific responsibilities to ensure their adherence.
(c) Post / display emergency instructions in appropriate locations throughout the station.
(d) Establish close liaison with local fire department, first aid services and civil defence organization.
(e) Prepare a written detailed plan covering the use of the alternative facilities, including the necessary arrangements and procedures for the transportation of replacement equipment or spare parts, and the support personnel.
(f) Familiarise personnel who would carry out the emergency plan with the alternative facilities, and test all the details of the plan in advance.

An emergency can happen any time whether the station is staffed or not. Therefore it is essential that always some person be nominated to take over the charge. Where the station is continuously staffed, an officer in each shift should be appointed to take the charge. In practice, this generally becomes the responsibility of the shift–in–charge as per the shift duty chart. In any case he should be:

(a) Trained in proper operation of all the fire fighting equipments, fire hydrants and appliances in the station.
(b) Familiar with every part of the station equipment and plant, and with existing and potential hazards.
(c) Trained in imparting first aid and artificial resuscitation and emergency for cardiac failures.
(d) Capable of effectively taking complete charge of all the staff, irrespective of the trade groupings, should the need arise. He should have the complete authority of the management to take whatever steps he considers necessary at the time to deal with the situation, particularly where the safety of the staff is involved.

Management is faced with the problems of assessing the damage and restoring the situation for normal working after the disaster has struck. So to minimize the effects of disaster, the followings should be considered;

(a) Evacuation plan, which is fully understood and known to the trapped staff, will help minimize accidents and may save the lives.
All doors of the building, transmitter hall and other technical areas should be easily openable from inside, even though they may be locked from outside. Pad locks, Mortise locks or similar locking devices, which need a key to open, should not be used on any emergency exit door or gate unless a key is placed in a glass cover near the exit. Bells and other alarm warnings for fire or emergency should be so placed that they can be heard in every room, even inside studios, frequented by the staff.
(b) The nature of work at some large stations involving work on high power equipments, tall structures, operational of mechanical aids, operation of high voltage generating plants etc, makes it essential that the organisation should include a first aid rescue crew in their emergency plan.
This type of arrangement is important and needs to be considered for the worker like, rigging crew erecting an AM tower or an antenna on a TV tower. By some happening a rigger is hurt on the tower, the winch man on the ground raises an alarm and first aid rescue crew arrives in the field immediately.

(c) All vital documents like records and copies of critical circuits and equipments should be safe guarded, as the loss of drawings / diagrams and other important records of the station for plant/ equipments /property could have more crippling effects than the physical destruction. So duplicate copies of all the vital documents to ensure availability for these in revival plan should be kept at city centre of the station or studios centre in case of a transmitter and vice-a-versa.

(d) During the conceptualisation, planning and system design of a new station or at the times of upgrading of old facilities, the provision of standby power plants, transmitters, antennas, and other critical plant and equipments, in minimising the effects of disaster should be considered. Standby facilities also provide high service reliability and flexibility during normal conditions.

(e) Alternative means of communications should be provided so that in an emergency some form of communication can be established. At big, important, and strategically located broadcast facilities, public address system, radio communication with fire tenders and civil defence etc should be maintained to summon aid immediately in case of emergency.

(f) To ensure restoration of facilities in the event of disaster in the quickest way, a pre-arranged and pre-designed plan with key men of various groups should be developed and kept ready for implementation without loss of time. The items like; to keep ready normal antenna matching arrangements in case of failure of one tower of DA system or failure of any component of phasing or power divider unit or sectionalising insulator or guy insulator or even the plan for arrangement to restore broadcast from a single tower in Omni mode in case of the failure of the base insulator or any section of the tower due to truncate of the top portion of the tower in gale or gust wind.

(g) The prompt restoration of the service aspect after an emergency has struck should be fully covered in the emergency plan of the station. By starting immediate salvage operation plenty of equipments can be salvaged for reuse or damage can be minimised. Say quick removal of water from the equipment cubicles, cable ducts, feeder hut etc after fire can restrict the damage to a minimum.

(h) Management is often mainly concerned with the safety of its own station staff, and the security of its property and the uninterrupted sustenance of the services. In addition there is a requirement of cooperation with other authorities for general community protection. This involves working closely with various local and national authorities such as police, civil defence etc.

(i) In addition one more aspect which is important for the management is the public relations in the event of a disaster. The public, and especially the families, of the staffs working in the station, are entitled to prompt information on the well being and safety of their bread earner and kins. An officer who has direct approach to the top management should be authorised to deal with the reporters, local authorities, civil defence organisation and others who can quickly come to disaster site.

8.5.2 Development of Shut down Procedures

As part of normal preparation for dealing with the disasters and emergencies, plans should be developed for emergency shut-down procedure of the station by the shift in-charge on the duty. A standard operating procedure (SOP) depending on the largeness and complexity of the station is required to facilitate the shift in-charge to sequentially and systematically shut down the transmitter and all other systems. The plans should be prepared for all type of situations like;

(a) Standard, routine or normal shut-down procedure when advance warning time of an impending disaster is sufficient to accomplish the shut-down in an orderly way without danger to personnel or damage to the equipment, plant and other property.

(b) An established alternate crash shut-down procedure when no advance warning was possible. To deal with such situations, emergency Mains isolators or switches should be located in a prominent, easily accessible position and clearly identified so that all the staff is able to disconnect the power supply to the equipments.

(c) Each possible type of emergency shut-down operations, which may involve safety of workers or damage to the plant, should be thoroughly examined to have automatic built-in safety measures or fail-safe devices.

(d) In emergency plan shut-down procedures, considerations should be given to the special requirements of availability of some form of lighting arrangements, means of ensuring the working of fire-alarm facilities, water pumps, telephone and other communicational facilities.

(e) Emergency can happen any time, so if in day time it is likely there are sufficient staffs on duty to deal with the emergency. However if it happens during night time then only minimum staff may be available, so this aspect must be kept in view while preparing the emergency plans.

8.5.3 Measures against failure in communication Networks

Breakdown in national telecommunication network may occur in unexpected places or in unexpected situations as a result of many causes. The causes can be manmade in the form of vandalism, major cable rupture by earth moving machinery or excavators or explosion as a result of gas leakage into ducts, collision by aircraft with tower main structure or Guys, or normal disaster like flooding, fire, lightning stroke, cyclonic or typhoonic wind, heavy snow fall, earthquake and landslide etc.

Although many public telecommunication networks including broadcasting and television networks have in-built flexibility and equipment redundancy it is not possible to cater to every situation and in the event of disaster well planned provisions against disaster in the form of mobile type radio communication, broadcasting or television systems of various capacities and frequencies will do much to ensure the speedy restoration of the network. In the case of natural disaster when public are also directly affected because of loss of life, damage to property and building, loss of transport and other essential services, social reliance on the telecommunications network is very intense and speedy restoration, even if only by temporary facilities, will do much to ease peoples mind at a time when they anxiously await information on plans for assistance or other vital news related to welfare.

Fixed radio communication, broadcasting and television facilities now forms part of most of the permanent public telecommunication services. So much can be done during the design and installation stage to ensure that they are physically strengthened particularly against high wind, sand or thunder storms, lightning strokes, earthquake, flood, landslide and heavy snowfall.

The measures which should be taken into account include; (a) Strengthening of the tower and antennas and support brackets to withstand violent earthquake shocks, or high wind loading, (b) reinforcement of equipment building, (c) improved lightning protection facilities,(d) automatic fire extinguishing system, installation of fire shutters, fire protection doors and the use of non combustible building materials,(e) security measures to prevent unauthorised entry to building and access to the tower,(f) protection against long duration failure of commercial power by installing fixed captive power diesel generators or a mobile unit or enhanced battery capacity,(g) protection of antenna system from built-up of snow by providing shrouds or in-built electric heating devices,(h) design of building near the tower to take into account falling snow and blocks of ice dislodged from tower, antenna, and (i) the decentralisation of radio terminals to improve network reliability so that all major network system do not pass through a single station. In case of transmission of a sound broadcast or television programme to a network of transmitters it is practice in some countries to put the programme into a loop circuit configuration for switching automatically or manually to another back-up route in case of disaster or catastrophic equipment failure at one or more stations.

On failure of telecommunication facilities, early re-establishment between isolated places or points should be ensured by having available a range of radio communication facilities which will allow the disaster situation to be handled quickly and effectively. The aim should be to restore normal operation as quickly as possible. The extent of back-up equipment facilities will depend on many factors including organisational policy, geographical size of the network, relative importance of traffic carried, density of traffic in various parts of the network, the extent of multi-routing of transmission circuits and physical security of the station building, towers and equipments.

The time taken to restore the service is dependent on many factors. These includes; site accessibility, the extent of damage, the condition and the age of the equipment, the availability and arrangement of spare parts whether components, modules, units etc. The availability of repair facilities in the local market, availability of manpower resources and the experience and the competency of the technical staff sent to the site for restoration of services. When the damage is extensive to permit onsite repair it may be necessary to transfer temporary replacement facilities to the site or to a suitable nearby site.

The complete restoration of service following breakdown requires several stages of action which might be spread over a considerable period of time. The time required for initial restoration of service has to be kept to an absolute minimum but the period of restoration to normal as a general rule is not of great consequence. Some type of failure like collapse of a mast or tower may require long time for manufacture and construction of a replacement but initial service restoration may be completed quickly by using a zip-up structure or a mobile tower.

8.6 Type of Disasters or Emergencies

There are many types of emergencies or disasters which can strike a radio station. So each need to be discussed in detail for tackling it and actions to be taken before its arrival, during its happening, and after it has passed.

8.6.1 Storm Emergencies

Storm warnings are usually issued well in advance by the meteorological authorities so that organisation, public can implement emergency precautions. Many large radio stations operating in tropical areas have pre-arranged plans to deal with these situations. Often the operating staffs are supplemented by other shift operators and maintenance workmen by recalling for the duty.

Following measures are put into operation after the receipt of advice on the approach of a cyclone, storm etc;

(a) Cover all large glass windows with shutters if provided for this purpose otherwise get made the arrangement. Particular attention is to be given to the windows in transmitter control rooms/halls, HT equipment enclosures, feeder hut etc.
(b) Fit special hoods over air inlet openings for transmitter ventilation to prevent water and debris being sucked into the air duct cooling system.

(c) Fit additional temporary bracing to the transmitter air outlet ducts, where theses project on the roof.
(d) Start emergency power plant for standby low power transmitter and its auxiliaries, and where this plant is capable of providing full backup power for the station, switch all the station plant to this service, and isolate incoming mains at the input of switchboard.
(e) Check that station emergency battery lighting system is functional.
(f) Check fire alarm system is functioning by operating special test facility provided for the purpose on the unit.
(g) Check and ensure that all staff on duty knows the fire drill.
(h) Check and ensure that all staff on duty knows the first aid and rescue drill.
(i) Check and ensure all tarpaulins and plastic covering required for emergency situations are readily available.
(j) Switch off the power to all non-essential buildings and areas.
(k) Check that automatic manhole drainage pumps, if installed are operational.
(l) Remove and keep in safe place all chemicals that will heat or produce flammable gases in contact with water.
(m) Check that mast light is working if the situations so demands otherwise inform the airport authorities.
(n) Check antenna counterweight systems are free to operate and any normal travel limit stops have been removed. Where automatic winches are provided they should be fully operational.
(o) Where station is prone to flooding; (1) Provide bulkhead and sand bagging to keep water out of ducts trenches, and chases and check the sealing of all ducts and pipes leading to building manholes and pits.(2) Ensure drums of fuel, oil (lube and transformer) and other flammable liquids are shifted and located above flooding level.
(p) In some stations standing instructions are there that tall radiators self or other type should be taken out of service and earthed during electric or sand storms and the standby radiators only should be used. In some stations during bad storms even complete transmission stops to avoid loss to life and property.

8.6.2 Bomb Threats

Explosives and incendiary devices have recently become powerful weapons in the hands of terrorists and other miscreants. Injury to staff and damage to the equipment have been extensive. It is an escalating threat to radio stations, particularly the less guarded tower in the field and also a city centre having studios where general public have access. So procedures for dealing with a bomb threat must be included in the emergency plan. Station staff must be appropriately trained to deal with the problem of bomb threat and the security should be tightened and good liaison should be established with police, ambulance, fire brigade and other Authorities dealing in disasters and relief like NDMA (National Disaster Management Authority) / NDRF (National Disaster Relief Force).

(I) Mast and towers being away from the sight of transmitter staff in the far field are popular targets of terrorists. Considerable damage has been caused at some transmitting stations as a result of the detonation of explosives. Areas where explosives and incendiary devices have been placed include;
  (a) Inside the transmitter cubicle by lowering the device through the air exhaust duct.
  (b) Inside the power substation enclosure by placing the device at a vulnerable point.
  (c) Near or on the base insulator by strapping it with tape or cord.
  (d) On the anchor block of guy wire by strapping it to turnbuckle or on the guy wire rope inside the enclosure.
  (e) Near the Antenna Tuning Hut at the base of the mast inside the mast base enclosure.
  (f) Inside the ground pits carrying the programme circuit to the transmitter near plinth protection of the building.
  (g) At the feeder outlet from the transmitter building near feeder line anchor frame on the wall.
  (h) On H-pole of the power supply line coming to transmitter site near substation.
(II) Studios are a frequently visited place of a radio station by general public and artists. It being in the city centre is prime target of subversive

elements to seek prominence and public attention. Most critical bomb location hides in a studios building includes;
(a) Air-conditioning ducts, under floor cable ducts, chases, and over false ceilings of studios.
(b) Announcers booth and network switching centre as well cable distribution frame.
(c) Film processing room, tape and records library as well Music Instrument room.
(d) Power supply H-pole, transformer enclosure, switching and battery room.
(e) Tape recording, replay rooms, and orchestral studios.
(f) Air-conditioning plant Air Handling Units room and compressor room.

Bomb threat warning may be received in numbers of ways as under;
(a) telephone call to a person in the organisation,
(b) Telephone call to police, or some other organisation.
(c) Written threat or a mail or sms or by throwing a pamphlet.

(III) Following action and measures should be put in place to deal with bomb threat;
(a) If call is received by telephone then person taking the call should extract as much information as possible from the caller.
(b) Have a bomb threat call check list ready placed near the telephone to deal with the situation and to facilitate collection of data after obtaining it from the organisation having such experience.
(c) The staff should be suitably instructed in the bomb threat situation so that they may be fully prepared to face it.
(d) The staff should be trained and made to be assured /believed that they have official support and backing and should offer high standards of realism and interest in case of a bomb threat call.
(e) An emergency control room officer should be nominated to coordinate and take decision for all emergency situations where large numbers of staffs are employed.
(f) After completing exercise of dealing with bomb threat, a full de-briefing to evaluate the operation should be held. This way the lessons learned can be translated to planning for future operations.

8.7 Staff Responsibilities

Large Broadcasting organisations employ numbers of technical staff to handle all major matters relating to system design, procurement of engineering equipments, safety, operation and maintenance of the engineering set up. In this section we will discuss the responsibility of various engineering staff of a radio station.

8.7.1 Designate a Safety Engineer

An engineer / engineering officers play very important role in the organisational set up of a radio station. They play an important role on the committees and one or more engineer may be appointed as part-time "safety engineer" with the task of establishing and implementation of an efficacious system of accident prevention. Safety engineer so designated is considered a specialist to deal with the matters concerning the prevention of accidents and as need arises; he should be trained in all aspects of safety of a radio station to enable him discharge his functions of safety engineer efficiently and effectively.

Some of the responsibilities of a safety engineer are;

(a) To do regular safety audit of the equipment and plant installation, maintenance and operating procedures, specialised tools, and testing facilities to ensure they are working properly and staff are using them correctly and in using them they are not exposed to hazards.
(b) To counsel, advice and assist management to fulfil its responsibility for safety by development, promotion, and maintenance of a continuing safety education programme for the staff.
(c) To direct or supervise for such catastrophes as fire, storm damage, collapse of structures, explosions, staff transport accidents and the organisation of the rescue service.
(d) To maintain, analysis and interpretation of accidents and injury statistics, accident cost data, and follow up investigation into accidents.
(e) To liaison with national safety bodies and other bodies like national disaster management authority.
(f) To disseminate safety knowledge by preparation of safety booklets, manuals, bulletins etc for use by the first line supervisors.

(g) To recommend procedures for introduction in new installations, plant, and apparatus in so far as operational safety is concerned.
(h) Inspection and clearance of all new safety devices and procedures.
(i) Checking the design of all temporary facilities such as anchorages, guys, scaffold, structures, equipment rearrangement and power feeds for safety.
(j) Liaison with design engineers to ensure that facilities are free from in-built hazards and operational difficulties. The facilities should be made to fit the people who operate and maintain them and not as so happens, the people having to fit the facility. Example: in some installations valves associated with the water return equipment of vapour cooling systems have been installed in inaccessible location underneath major components.

8.7.2 Designate a station Manager

A station manager being in charge of the operation and maintenance of the station is directly responsible for its safe operation. So a station manager should be properly qualified and adequately experienced. For this purpose a local officer-in-charge, regardless of his local title, should be designated as station manager.

In a large station complex, involving more than one site, and several different trade disciplines, the duties of station manager may be delegated for any particular section of the station work to a foreman, supervisor or other senior staff member, who should report to the station manager as required on all aspects of safety.

When station is under normal operation and maintenance, the shift-in-charge should be responsible to ensure that the directions of the manager are carried out and that in all other respects, the safety rules are observed.

When a station equipment or plant is undergoing test by the installation or specialist staff and is not under the control of official shift-in-charge, the senior specialist officer should be responsible for the safety of the equipment and staff working on it, and for any nonstandard conditions that may be caused by other operational equipment.

Duties of a Station Manager
The duties of a station manager for the safety of a station plant and equipments are;

(a) By keeping in view local conditions to create and issue local safety instructions.
(b) To hold safety meetings with supervisor and other key personnel, and to instruct them in safety procedures.
(c) To have safety inspection of work areas at regular periodic intervals including uninformed.
(d) To arrange to get issued safety tools and devices like torches, hard hats, gloves, and protective clothing.
(e) To take charge of the storage of flammable liquids and explosives.
(f) To arrange for ambulance, doctor, or hospital service as required.
(g) To investigate all accidents and prepare report for submission to the responsible engineer.
(h) To ensure that no workman is permitted to operate, adjust or maintain any equipment unless he is suitably qualified to do so. Also has the necessary consent and that adequate safety precaution are being observed.
(i) To ensure that all staff working on the station, and the including those employed by the contractors, are fully aware of such matters as the location of fire extinguishing plant, safety equipment, first aid equipment, building emergency exits, and the presence of live conductors in the vicinity of the work.
(j) To keep a station log book showing all changes in plant or equipment conditions. He should read and sign the log book when assuming the duty and sign again on relieving the duty.
(k) To ensure fire extinguishing equipment and services adequately maintained, readily available at all the times and that every operational shift includes at least one person trained or instructed in the use of these equipments.
(l) To ensure first aid equipments and safety aids are properly maintained and readily available all the times.
(m) To keep in safe custody key or device forming part of any safety facility, where this is required as per rules.
(n) Not to accept responsibility of the contractors staff working on the station plant or equipment yet to be handed over. But, he should ensure nothing is done that might endanger the safety of theses workmen.

(o) To, promote good housekeeping by demarcating proper place for keeping all the materials, tools, and equipments when not in use.

(p) To ensure that complete and up-to-date circuits of all the equipments on the station, together with all necessary instructions, are available all the times to staff in the course of their work.

(q) To ensure that all concerned staff are advised immediately of any changes in the critical circuit setting or adjustment, and are given all corresponding operating instructions.

(r) To ensure understanding with every workman at the station that the violations of the established safety rules will not be tolerated.

8.7.3 Supervisory staff

Supervisors at all levels are responsible for the completion of work efficiently, economically, and safely through their efforts and those of the staffs they control. As maximum efficiency, economy, and safety cannot be achieved without the earnest support and cooperation of all the staff, the role of the supervisor is most important and responsible. Supervisors are in key position to help in achieving these objectives. They know the work and are familiar with risk and the hazards involved in the work. They are continually and constantly at the workface and have a degree of intimacy with it, unequalled by anyone else in the organisation. Further they know the individual workmen so well with their abilities, sense of cooperation, safety habits and reaction to the authority.

Main Responsibilities

The main responsibilities with respect to safety, of supervisor of each trade group may be classified as;

(a) Before commencing work in which potential hazards exist, he should hold a short safety session meeting with the workmen involved in the job to discuss the hazards and the safety practices to be adopted. This meeting should be compulsory for all the workmen involved in the job. This means, he should take all such precautions, which are within his power to prevent the accidents, and to ensure safety rules are observed by workmen.

(b) To ensure that all new employees are carefully instructed as to the hazards of the work, and how to avoid them, are issued with a copy of

the station safety rules, have read the rules and correctly understood them. Failure to ensure this should be treated as negligence on the part of supervisor.
(c) To take steps to prevent unauthorised persons from approaching locations where works of hazards nature is being carried out.
(d) To prohibit the use of any machine, tool, test equipment or other device unsuited to the work in hand or which has not been properly tested by the competent person.
(e) To encourage and enforce wearing of safety aids, such as safety glasses, safety belts and, hard hats in the area where their use is essential requirement and mandatorily.
(f) To ensure that workmen when working in isolated places like aloft a mast or tower and where a foreseeable event could result in serious injury, are provided with a means of communication (a walkie-talkie).
(g) Always ensure to obtain the permission and clearance of station manager before permitting any member of his staff to work on the equipment which is operational or under testing.
(h) When equipment or plant has failed in service and staffs is engaged in repair of the fault, he must see that under no circumstances life is endangered to reduce the duration of the breakdown.
(i) To ensure that safe handling and appropriate lifting methods are practised by the workers under his control.
(j) To extend full support to all safety activities, procedures and programmes.
(k) To ensure that always there is more than one person when the work is being carried out on energised or could be expected to be energised like broadcast transmission lines or towers, or a broadcast transmitter or a job involving keeping assistance at hand for the worker.

So the role of a supervisor is very vital and crucial for the safe and secure working of a broadcasting setup.

8.8 Safety Rules

Safety rules should be established to ensure their enforcement in letter and spirit to take care of security and safety of life, equipments / plant of station and the property. Please make it sure that no safety rule should be made unless it is implementable in the field by workers. No disciplinary action against any worker should be initiated until management has discharged its safety obligations. Disciplinary action for failure to follow a rule must be taken or otherwise rule should be withdrawn forthwith. Management must make it clear beyond doubt that certain safety rule violations cannot be tolerated. Firmness in securing conformity with such safety rules would be consistent with the wishes of the majority of the staff of the station.

Safety rule book should be given to each and every workman by the station manager and the receipt of such including their undertaking to abide by the safety rules and procedures be obtained as soon as they join the station. Suggestions to the management for changes in, and additions to the safety rule book, should normally be called for periodically and originated from the supervisor but should be sought from subordinate staff also, as they actually works in the field.

Main Criteria for framing safety rules

Main considerations while framing safety rules should be;
(a) Should be less in number, easily understood, implementable and practical.
(b) Should be clearly explained to each staff / workman while giving him the rule book.
(c) Supervisors should set example in following the safety rules themselves.
(d) Safety rules should be made applicable to each section equally without any relaxation.
(e) No rule should be made which cannot be enforced and implemented.
(f) Prompt corrective action should be taken for failure to comply with safety rules.
(g) The emphasis should be to educate the staffs for need to observe the rules for their own, equipment and property safety, rather than punishment of defaulters.
(h) Each rule should be covered by general instructions by the station engineer /manager.
(i) Each hazard covered by a safety rule is definitely demonstrable as unsafe.
(j) Safety operating practices should not be called safety rules.

8.8.1 General Safety Rules for a broadcasting station

Safety rules should not be treated as restrictive measures but as guide to staff and workers in their cooperative efforts to prevent accidents. One of the most important points to keep in the mind about your job is the necessity for working safely by observing all safety regulations applicable to each operation.

Certain safety rules have a substantial impact on the safety of an individual or the entire workforce. A single violation might place the safety of many staff in jeopardy. So for this reasons, an individual violation of the following General Safety Rules should result in zero tolerance with immediate disciplinary action.

(a) No Smoking
Strictly prohibited in Flammable store rooms, battery rooms, and diesel generator rooms, control rooms of transmitters and in all rooms of studios centres.

(b) Observe all safety keys, locks, and tag Rules
These devices are used for safety of personnel and equipment and must be strictly observed. Unauthorised removal or replacement of keys or to use spare or duplicate keys and locks is serious offence requiring immediate disciplinary action.

(c) Use Safety Devices
For the personal protection, a workman must use the protective devices given to him and at no time he should make it inoperative or remove it. Also the equipment must not be operated if its safety device is not functioning properly.

(d) Wear Personal Protective Devices
For personal safety protective devices like safety glasses, hard hats, gloves and safety belts are provided for personal use for certain jobs like working on towers, feeder lines etc. These devices do not eliminate hazard but are must as second line of safety for protection from injury. Refusal to not wear these demands severe disciplinary action.

(e) Wear Proper Clothing
Personnel working on rotating machines and on towers/masts should not wear casual and loose clothes. Where heavy material is being handled do

not wear canvas shoes as well as slippers but wear proper safety and hard top shoes. No open toe or loose footwear should be permitted for rotating machinery work, high voltage and tower works.

No metal finger rings, wrist bands, and watches should be worn while working on radio and electrical equipments.

(f) Do not operate unless authorised

No workman or staff of the radio station should operate any equipment or plant unless he is qualified and authorised to do so by his supervisor. To operate most of the equipment and plant of a radio station in a safe and proper manner, special skill and training are required. So failure to observe this could be detrimental to the safety of equipment and life of fellow, yourself as well as other workers.

(g) Seek first-aid promptly

Always seek prompt first aid for any injury, even though of minor nature. Treatment prevents complications and will enable your supervisor to take steps to prevent similar accidents.

(h) Fire

Make yourself aware with the local arrangements for giving an alarm and the procedure to be followed in case of fire breakout. Instruction for this should remain displayed at prominent locations in the station.

(i) Good Housekeeping

Neat, tidy, clutter free and systematic working methods should be employed in your work to ensure jobs are carried out safety and efficiently. Do not leave any materials and tools lying on the floor or at work location where staff may trip over them.

(j) Good engineering Practices

Always follow good engineering practices with proper and specific tools, safety devices and with due defined procedure for a job. Always use correct tool for the work.

8.8.2 First-Aid Rules

The first-aid rules should be and is mandatory part of the safety rules manual of a radio station as to reduce the seriousness of an accident first-aid is very vital. Followings should be the part of rule book;

(a) A list containing all the details like names, addresses and telephone numbers of ambulances, doctors, fire brigade, civil defence units, members of management, engineering managers who are to be called in emergency.
(b) A copy of rules for first-aid with chart, procedural details, methods of resuscitation, anti-snake bite kit application and use of fire extinguishers.

The first-aid rules should be issued to every employee and also displayed at prominent location throughout the station and in work vehicles. In fact each employee must be got trained in first-aid and its various procedure for electric shock treatment etc through special agencies.

Radio stations particularly transmitters as general rules are in a remote or isolated location, so immediate medical assistance may not be available and possible. Certain accidents like electric shock or those accidents occurring on the structure needs immediate attention, and the chances of survival mainly depends on the promptness with which such attention and first-aid is given. In case of electric shock where asphyxiation is caused by muscular contraction or paralysis of the respiratory system there is an excellent chance that the victim can be revived if artificial respiration is promptly applied. There is a 95% chance of saving the injured person within the first minute, but only 25% chance from the fifth minute onwards.

In addition, there are certain mistakes to be avoided in the minutes immediately, succeeding the accidents, such mistakes can reduce the chances of successful medical treatment. It is very important to know what should not be and what should be done in the event of an accident. So it is very important to have a staff trained in first-aid in each shift. Names of all trained person/staff/workmen should be displayed near each first-aid box. Also the first-aid box always should remain fully stocked and replenished including validated anti-snake bite kit.

8.9 Induction of staff / employee

Each new staff joining a radio station must be imparted induction course by the supervisor of the unit immediately after joining the duty and he should be clearly instructed that he will be required to observe safety rules. He should be made aware and warned of the hazards of his work and advised how to reduce the danger of accidents for his and fellow worker's safety. He should be given a copy of safety rules and his acknowledgement that he has received, read and understood the rules, taken. All safety procedures and rules established by the station must be followed and adhered without any ifs and buts' by staff / employee.

Every organisation must have definite programme as induction course for training of personnel for the duty and safety rules formally or informally which should begin on the first day itself. The technical staff and their helpers should be trained to select the safe as well as proper tool, taught to get into safest position and to make proper use of safety equipment. All the safety rules, procedures should be available in writing and readily available to every worker as well as displayed at all prominent locations throughout the station for ready reference.

In addition special training courses with theory and demonstration should be conducted through qualified instructor / trainers' in electric shock treatment, emergency resuscitation, first-aid, use of fire extinguishers, fatal accidents due to fall from towers /masts, RF burns, and other casualties which happens in a radio station.

8.9.1 Qualifications and Fitness of Employees

Radio station through its organisation should have every reasonable means to ensure that each technical employee, particularly holding position of responsibility, is technically qualified, mentally fit, and physically capable of doing his work in accordance with the safety rules, codes or regulations or any other practices.

For radio transmitter tower / mast riggers and foreman who are required to work on masts / towers, station or the organisation should ensure that only those workmen who are certified medically fit for this type of work are employed. To ensure continued fitness for such type of employee, they should undergo periodical examinations. In addition all such employee should be insured against life risk and accidents with a govt insurance company, if already they do not have any other insurance policy as a mandatory welfare policy. This rule should be part of safety rules and policy of the organisation.

8.9.2 Visitor's policy of a Radio transmitting station

Even though more important for a transmitting station but this is equally applicable to a studios centre also. Visitors should be prohibited and not be permitted to have access to any live radio or electronic equipment or plant or other hazardous equipment unless accompanied by a qualified or authorised staff member of the station. He must take all necessary steps with regard to safeguards and ensure the safety of the visitors and ensure safety rules applicable to visitors are observed.

No visitor or even technical trainee should be permitted to touch or operate any equipment or plant. Visitors should be kept away and not be permitted access to strong EM fields, because of the possibility that some visitor may have implant or cardiac pacemakers which can malfunction. Even station should have a practice of issuing a precaution and safety instruction brochure or card for the visitors as part of safety practices. A typical safety brochure or card should include the followings;

(a) Brief description and purpose of the station. (b) RF power and peak voltages that may exist in the compound during transmission. (c) Plan, locations and type of danger of the area considered dangerous from the point of view of power, voltage, clearance etc. (d) Instructions regarding movement of pedestrians, vehicles. (e) Movement and clearance for other utility staff and vehicles. (f) Reasons for insistence on escorts for access to forbidden areas.

To keep pace with time and technology the safety procedures, rules and instructions should be regularly updated.

8.10 Safety Aids and facilities

An adequate and updated supply of first-aid material, equipments, protective clothing, safety aids, and fire extinguishing equipments and firemen aids to enable the employees to meet the requirements of safety rules should be provided in every station. First-aid items, charts and equipments should be as per ambulance practices or the advice of the district medical officers of govt departments. Similarly fire fighting equipments and their locations in the station should be as per recommendation of district fire officer.

The nature and extent of medical attention and the first-aid facilities should be sufficient for the needs of the employees, keeping in view the hazardous nature of the work and the distance of medical help or hospital from the station.

Many minor injuries can be treated satisfactorily at site if adequate facilities are maintained. All staff of the station should be trained in imparting first-aid, and every supervisor must ensure that each worker has been instructed as to how to get first-aid when needed.

8.10.1 Safety Aids

The type and quantities of safety aids, facilities, devices and equipments will depend on many factors, which may be power of transmitter, distance from the city, transportation facilities available etc. Typical list of safety aids for a Radio station is as below;

(a) First-aid kit including stretchers as per standard stipulated by Safety authorities or the station manual which also should comply with the occupational hazard and safety authorities.
(b) Insulating items like gloves, insulated appliances like pliers, screw drivers, and tongs for any necessary handling or testing of live equipments or lines as well as insulating shields, covers mats, and platforms.
(c) Protective goggles and helmets of suitable materials and appropriate construction.
(d) Portable signs such as, "Men at work", "Danger", "Do Not Touch", "keep Away", "keep at a safe distance" etc
(e) Fire extinguishing devices either designed for safe use on live parts or simply marked that not to be used.
(f) Earthing hooks, rods, barrier devices to make protective earthing.
(g) Portable communication devices and lighting system.
(h) Sufficient safety belts, ropes and ladders.
(i) Test equipments and probes.
(j) Hooters, Panic alarms and siren for warning.
(k) Ear muffs and ear plugs for use near blowers.

All the safety, protective devices, tools, and equipments should be inspected and testing at regular intervals of not more than six (6) months. Any safety device or testing equipment found unfit for ensuring standard desired of it should be withdrawn immediately and replaced by a new one. Each fit and functional equipment or device should be certified for use by affixing in writing by the project /station shift-in-charge, "Fit and certified for Use till—(date)".

Protective devices such as guards over belts of blower motors, other guard nets, access barriers should always remains in position when plants or machines are running.

8.10.2 Noise Attenuators

The noise level of high capacity air blowers of air conditioning plants or transmitter general ventilation or heat exchanger units or diesel generating plants may cause hearing damage but significant reduction of noise levels may be both extremely expensive and difficult. Measures to house noisy units behind partitions / in insulating / acoustic enclosures is effective but in many cases it create lot of other problems. Where it is not practicable to eliminate high noise levels, ear plugs and muffs should be provided to operating staff.

An external ear muff gives an attenuation of 25 to 40 db whereas a waxed cotton wool ear plug gives about 20 db attenuation. It is the responsibility of the management to ensure that operators are protected against any noise of sufficient level and duration which can cause damage to hearing. So safety rule relating to use of ear protection from noise should be enforced in all such transmitter premises and made compulsory to be worn by the operators as experience has shown that many operator are reluctant to use these either due to discomfort, inborn carelessness or other reasons.

8.10.3 Emergency Exit

The layout, demarcations, recognitions of emergencies exits and their easy thoroughfare is the most important part of safety rules and practices. All emergency exits should be marked as per standard practice so as to even be recognisable to fire officials and other visitors. The path of escape in case of fire, explosion and other emergency should be unobstructive and unmistakable. The escape route and emergency path must be provided with glow path signage as during emergency lighting may not be available and it may be dark all around.

Each room or area or working space of the transmitter equipment and the plant should have suitable exit clear of obstructions. At all times the correct key of emergency exit must be kept next to the door lock. The arrangement of keeping the key in thin glass cover near the exit serves the purpose but it will be preferred to have a keyless door. If the plan of the room is such that an accident is liable to block the emergency exits, like in a long narrow room then a second exit should be provided. As a basic rule, particularly where high power equipment is installed, a worker must have at least two safe exits to escape. If feasible the exits should be in opposite directions.

The layout of equipment and plant, particularly in high power transmitter's Antenna Tuning Huts, Antenna Switching Matrix rooms, enclosures and other high power areas should be such as to provide access and egress in a safe manner. A workman in both process of entering and leaving should enter and leave a safe area. All doors and gates should open in the direction of egress with the exception of sliding doors, which are permissible in exits through fire walls between such areas as the transmitter and transformer vault.

8.10.4 Working Alone

No person should work alone on a transmitter, transformer, power line or any other radio or electrical equipment, or carry out testing or experimental work on such equipment where the working voltage AC or DC or R F exceeds 350 volt peak except in routine operation or testing where live parts are suitably guarded or segregated or fully mechanically or electrically interlocked.

In some organisation it is practice and norms to have at least three operators in each shift where high power transmitters are in remote locations are involved. The main consideration is safety rather than equipment requirements, especially in case of modern transmitter equipments. In the event of an operator being hurt, the second will be there to give the first aid immediately leaving the third to raise alarm and seek medical help.

The continual watch of a worker engaged in testing, adjustment or maintenance of high voltage or RF equipments must be made mandatory because in the event of a person being the victim of severe electric shock or paralysed, he is virtually helpless to summon aid. So immediate aid and help can be given by those who were in the attendance, and who recognize the problem and know what to do.

The practice of allowing the workmen to remain inside the enclosure with live high voltage equipment should not be permitted, other than in exceptional cases and only after the station engineer is fully satisfied that the workmen is in a safe position, clear of all hazardous electrical and mechanical plant, and that he is under constant observation by another workman who is outside the enclosure and who can shut down the plant if necessary.

8.10.5 Fire extinguishing facilities

Adequate clearly marked fire extinguishing equipments and facilities as per the recommendation of local fire officer should be placed throughout the transmitter plant. All fire extinguishing equipments and facilities should be checked regularly by workmen on daily basis and on weekly basis by supervisors and on annually basis inspected in details by the engineer and got refilled as per requirements. This activity must be logged for future records and references.

While placing the unit ensure they are not too close to the equipment they are to protect lest they become inaccessible if the equipment catches fire. Also ensure extinguishers are placed inside as well as outside of room.

Also ensure Workmen trained in proper use of fire fighting equipments are kept in each shift. All the supervising staff should be got trained about fire fighting and in use of fire fighting equipments.

Equipment and rooms fitted with automatic co_2 or similar system should be fitted with an interlock system to protect staff when they are working on the equipment or in the room. When the interlock is disabled it should be displayed so. When remotely controlled, a notice should be placed on the remote control panel.

Action to be taken when fire starts should be in following sequences;

(a) Isolate the power from the equipment involved in fire.
(b) Raise the alarm in accordance with the local instructions and rules.
(c) Shut down all ventilation and forced air cooling devices.
(d) Attack the fire as per the procedure and availability of fire fighting facilities.
(e) Report the matter to station Manager/engineer or head as well as local fire officer if beyond your capability and means say in case of field fire and other major fire outbreak.

8.11 Safety standard requirements for Radio Transmitting Equipments

It is very essential that radio transmitting equipment designer's understand the underlying principles of safety requirements so that they can engineer safer equipments. Designers should take into account not only normal operating conditions of the equipment but also the fault conditions, foreseeable misuse and external influences such as temperature, altitude, pollution, moisture and over current's /over voltage's on the mains supply. Dimensioning of insulation spacing should take account of possible reductions by manufacturing tolerances, or where deformation could occur due to handling, shock ageing and vibration likely to be encountered during manufacture, transportation, installation and normal use.

There are two types of person whose safety need to be considered; SKILLED and UNSKILLED. Requirements for safety and protection should assume that UNSKILLED persons are not trained to identify the hazards, but will not intentionally create a hazardous situation. Consequently, the requirements must provide protection for the helpers, cleaners and even casual visitors and all also other UNSKILLED persons. In general, UNSKILLED persons should not have access to hazardous parts whether or not such parts are marked or provided with barrier. A person, having the necessary knowledge and practical experience of electrical and radio engineering to appreciate the various hazards that can arise from working on radio transmitters including its auxiliaries, and to take appropriate precautions to ensure the safety of personnel, is a skilled staff. A fresh new staff, having no or little knowledge and no or almost negligible practical experience of electrical and radio engineering to appreciate the various hazards that can arise from the working on radio transmitters including its auxiliaries, and to take appropriate precautions to ensure the safety of the personnel and also the equipment, is an unskilled staff.

A safety Standard should specify requirements intended to reduce the risks of; *electric shock, skin burns, high temperature, fire, injury from harmful radiation, and mechanical or any other hazard*, to the persons who may come in contact with the equipment covered by the Standard and also the equipment. A Standard is intended to reduce such risks with respect to installed equipment, whether it consists of a system of interconnected units or independent standalone units, subject to installing, operating and maintaining the equipment in the manner prescribed by the manufacturer. Design and construction requirements and, wherever appropriate, test methods should be specified covering the followings:

a) The safety of a SKILLED personnel when operating, carrying out routine adjustments, and as far as practicable, during fault finding and repairing the equipment; b) the safety of personnel, including UNSKILLED personnel directed by SKILLED personnel, when the equipment is operating normally, and also when operating under certain specific fault conditions which may arise in the normal use; c) The prevention of fire and its spread.

These requirements do not necessarily ensure the safety of UNSKILLED personnel working on the equipment when it is not in normal operation. Tests are specified, where appropriate, for checking that the equipment meets the safety requirements of the Standards when operating normally and also under the specified fault conditions. The tests should be carried out on a representative set of equipment in order to determine whether the design meets the requirements of the safety Standards as established for the territory. The tests are neither mandatory nor limiting and may be modified by agreement between the manufacturer and purchaser. The use of such Standard is not, however, intended to be restricted to type tests. It may also be used for acceptance tests after installation of the equipment, for tests after modifications to parts of the equipment, and for tests at appropriate intervals to ensure the continuing safety of the equipment throughout its life

The International Electrotechnical Commission (IEC) is a leading global organization that prepares and publishes International Standards for all electrical, electronic and related technologies. Bureau of Indian Standards (BIS) is the Indian body under the Ministry of Consumers Affairs Govt Of India that prepares and publishes National Standards with code: IS, for all such type of products, mechanical, civil, electrical, electronic products and related technologies with inputs from concerned experts of departments using those products.

IEC 60215 Edition 4.0 2016-04 is the latest International Standard for; Safety requirements for radio transmitting equipment and BIS have equivalent IS: 10437. This Indian Standard (2nd Revision to IS: 10437) which is identical with IEC 60215: 2016 'Safety Requirements for Radio Transmitting Equipment – General Requirements and Terminology' issued by the IEC, is adopted by the BIS on the recommendations of the Transmitting Equipment for Radio Communication Sectional Committee after approval of its Division Council.

This standard was originally based on IEC Pub 215 (1978)-Safety requirements for radio transmitting equipment, issued by the International Electrotechnical Commission and was subsequently revised in 1987 and has now been taken up for second revision to align it with the latest ISO/IEC Publication i.e. IEC 60215: 2016

The technical committee of BIS has reviewed the provisions of following International Standards referred in this adopted standard and has decided that they are acceptable for use in conjunction with this standard. For undated references, the latest edition of the referenced document applies, including any corrigenda and amendment.

International Standard	Title
IEC 60068-2-1	Environmental testing – Part 2-1: Tests – Test A: Cold
IEC 60244-6	Methods of measurement for radio transmitters – Part 6: Cabinet radiation at frequencies between 130 kHz & 1 GHz
IEC 60695-1-10	Fire hazard testing–Part1-10: Guidance for assessing the fire hazard of Electrotechnical products-General guidelines
IEC 60695-1-11	Fire hazard testing–Part 1-11: Guidance for assessing the fire hazard of Electrotechnical products-Fire hazard assessment
IEC 60825-12	Safety of laser products – Part 12: Safety of free space optical communication systems used for transmission of information
IEC 62232	Determination of RF field strength and SAR in the vicinity of radio communication base stations for the purpose of evaluating human exposure
IEC 62311	Assessment of electronic and electrical equipment related to human exposure restrictions for electromagnetic fields (0 Hz – 300 GHz)
ISO 1999	Acoustics – Estimation of noise-induced hearing loss

For the purpose of deciding whether a particular requirement of this standard is complied with, the final value, observed or calculated, expressing the result of a test or analysis, shall be rounded off in accordance with IS 2: 1960 'Rules for rounding off numerical values (revised)'. The number of significant places retained in the rounded off value should be the same as that of the specified value in this standard.

"The International Standard; IEC 60215: 2016; applies to radio transmitting equipment, operating under the responsibility of SKILLED persons. It also applies to auxiliary equipment and ancillary apparatus, including combining units and matching networks and cooling systems where these forms an integral part of the transmitter system.

The requirements of IEC: 60215 may also be used to meet safety requirements for cognate equipments that could be within the scope of this International Standard as shown in Table 8.2.

Table 8.2: Examples of cognate equipment

Generic product type	Specific example of generic type
RF amplifiers	High power RF amplifiers used for industrial, medical or scientific applications
High-voltage power supplies (HVPS)	DC HVPS based on PSM technology or any cognate technology

Table 8.2 is not intended to be comprehensive, and equipment that is not listed is not necessarily excluded. When the equipment is to be manufactured and/or to be installed in territories that have safety standards covering the scope of this International Standard that are more stringent, then those standards apply.

Antenna systems, associated feeder lines and matching networks, not forming an integral part of the transmitter, are excluded.

This International Standard does not apply to transmitters of safety-insulated construction using DOUBLE INSULATION or REINFORCED INSULATION and without provision for protective earthing. This type of equipment is designated CLASS II EQUIPMENT and is usually marked with a symbol as shown in 3.2.2 b). Refer standard IEC: 60215:2016 for detailed provisions.

Note: This International Standard does not apply to battery powered transmitters or to radio base stations and fixed terminal stations for wireless telecommunication, as this equipment is covered by other standards."

Symbols relating to safety

As per provision of IEC: 60215: 2016 some symbols related to safety standards are listed below. You will find these symbols on each electrical electronics and radio transmitting equipments.

(a) Safety earth IEC 60417-5019 (2006-08) ⏚
(b) Equipment of safety insulated construction IEC 60417-5172 (2003-02) ▢
   (CLASS II EQUIPMENT)
(c) Dangerous voltage IEC 60417-5036 (2002-10) ⚡
(d) Ionizing radiation ISO 7000-0907 (2004-01) ☢
(e) High temperature IEC 60417-5041 (2002-10) ⚠

References

1. Handbook for Radio Engineering Managers-J F Ross: Section-3 Safety Practices Part-1, Chapter 20, 21 and 22 (pp-312-346)
2. IEC 60215, Edition 4.0 2016-04 INTERNATIONAL STANDARD – Safety requirements for radio transmitting equipment – General requirements and terminology and its equivalent Bureau of Indian Standards IS: 10437 of BIS

CHAPTER 9

FIRST-AID AND OCCUPATIONAL HEALTH

9.1 History of First-aid

Little recorded information is known about the first-aid during pre-historic era. The instances of some recorded first aid were provided by religious personnel accompanying the armies. But, it was only in mid-19th century that the 1st International Geneva Convention was held and Red Cross was created to provide "aid to sick and wounded soldiers in the field." Soldiers were trained to treat their fellow soldiers before the medics arrived. A decade later, an army surgeon proposed the idea of training civilians in what he termed, "pre-medical treatment." Many developments in first-aid and many other medical techniques have been driven by wars, such as in the case of the American Civil War, which prompted Clara Barton to organize the American Red Cross. St. John Ambulance was formed in 1877 to teach first-aid, and numerous other organizations joined them. The term "first-aid" 1st time appeared in 1878 as a combination of "first treatment" and "National Aid."

9.1.1 History of Red Cross Movement

About 150 years ago, a battle in northern Italy sparked an idea that has since changed the world. On 24th June 1859, Henry Dunant, a young Swiss businessman from Geneva, witnessed horrifying suffering and agony following the battle of Solferino, Italy in 1859 during the Franco-Austrian war. He mobilized the civilian population, mainly women and girls, to care for the wounded irrespective of their role in the conflict. He secured them with the necessary materials and supplies and helped in the establishment of temporary hospitals. His book "A memory of Solferino" inspired the establishment of International Committee of the Red Cross (ICRC) in 1863.

The book 'Memory of Solferino' suggested that a neutral organization be established to aid the wounded soldiers in the times of war. Just a year after the release of this book, International Red Cross Movement was established by Geneva Convention of 1864. The name and the emblem of the movement are derived from the reversal of the Swiss national flag, to honor the country in which Red Cross was found. Providing first-aid services to injured people was one of the first services provided by the Red Cross Red Crescent volunteers for over 100 years. Now, almost all 187 Red Cross Red Crescent (RCRC) National Societies (NS) have first-aid as their core activity. RCRC National Societies are the major first-aid educator and provider in the world.

9.1.2 History of First-Aid in India

During the 1st World War in 1914, India had no organization for relief services to the affected soldiers, except a branch of St. John Ambulance Association and by a Joint Committee of the British Red Cross. Later, a branch of the same Committee was started to undertake the much needed relief services in collaboration with St. John Ambulance Association in aid of the soldiers as well as civilian sufferers of that Great War. A bill to constitute the Indian Red Cross Society, Independent of the British Red Cross, was introduced in the Indian Legislative Council on 3rd March, 1920 by Sir Claude Hill, member of the Viceroy's Executive Council who was also Chairman of the Joint war Committee in India. The Bill was passed on 17th March 1920 and it became Act-XV of 1920 after assent of Governor General on the 20th March 1920.

On 7th June 1920, fifty (50) members were formally nominated to constitute the Indian Red Cross Society and the first Managing Body was elected from among them with Sir Malcolm Hailey as Chairman. Indian Red Cross Society (IRCS) is a member of the International Federation of Red Cross and Red Crescent Movement.

9.2 Occupational Safety and Health

Occupational Safety and Health (OSH) of the staff working in the establishments is as important as the efficient output, smooth working and growth of the industry. So along with the growth of the industry, as professed in the National Policy for the working population everywhere in the country and also for the public in general to ensure safe and healthy working conditions good business practices should include sound labour practices which provide for healthy and safe working environment for workers at large.

Occupational health scenario has undergone a paradigm shift due to rapid industrialisation. Inadequate attention to developing the shift by concerned bespeaks of indifference to the very basic human resource which contributes to growth. Productivity at work is directly influenced by the health status of the workers.

An unhealthy workforce is a drag on productivity; affecting overall national productivity. Poor occupational health and reduced working capacity of the workers may cause an economic loss of up to 10-20% of GNP. WHO estimates that only 10-15% of workers have access to basic occupational health services. The burden of disease attributed to the occupational diseases is high and it is estimated to be about 11 million cases annually, with about 7, 00,000 deaths. According to a World Bank estimate, two-thirds of the occupationally determined loss of disability adjusted life years could be prevented by occupational health and safety programmes.

With over 40 million working population, India has a very large base engaged in industrial activity. The health needs of this population also differ according to the industry of work. Knowledge and orientation for diagnosing of occupation-specific cases are evolving globally in the form of speciality health care.

Most workers in India (about 90%) work in the vast informal sector. The variable and insecure nature of the work means that more and more workers are pushed into taking up hazardous and precarious employment both in the informal as well as formal sector. Informal workers give low priority to OSH, as having work is more important than the quality of the job.

In most places, occupational safety and health (OSH) invariably means prevention of accidents, very little attention is paid to occupational diseases. An accident free workplace by no means implies a safe workplace.

9.2.1 Legal Provision

Articles 39(e), 41, 43, and 48A of Indian constitution mandate "to make the life of the workman meaningful and purposeful with dignity". So the State for labour reforms shall direct its policy towards securing followings:

a) The health and strength of workers, men and women; b) That the tender age of children is not abused; c) The citizens are not forced by economic necessity to enter vocations unsuited to their age or strength; and d) Just and humane conditions of work and maternity relief.

Besides, various judgements of Supreme Court under Article-21, Right to life-have upheld the right of employees' health. The Court has noted that, "occupational accidents and diseases remain the most appalling human tragedy of modern industry and one of its most serious forms of economic waste."

The basic aim of concerned law making authorities is to devise rules which provide safety standards to protect basic needs of workers and their welfare. These laws should be flexible enough to create rather than destroy jobs, and increase overall wellbeing of workers without harming their efficiency, work output and establishment.

9.2.2 Objectives of Occupational Safety and Health

The main objectives of the Occupational Safety and Health related legislation should be: (a) Providing a statutory framework including enactment of a general enabling legislation on OSH in r.o all sectors of economic activities, and designing suitable control systems of compliance, enforcement and incentives for better compliance; (b) Providing administrative and technical support services; (c) Providing a system of incentives to employers and employees so that they achieve higher health and safety standards; (d) Establishing and developing research and development capabilities in emerging areas of risk and effective control measures; (e) Reducing the incidence of work related injuries, fatalities and diseases; (f) Reducing the cost of workplace injuries and diseases; (g) Increasing community awareness regarding areas related to OSH.

There are at least 18 ILO conventions that are targeted at addressing the issue of OSH. So far, India has ratified only three such conventions.

India is yet to ratify important conventions like Convention 155 on occupational safety and health and the working environment, Convention 161 on occupational health services, Convention 167 on safety and health in construction, Convention 176 on safety and health in mines, Convention 184 on safety and health in agriculture, Convention 187, the promotional framework for OSH.

Legal framework for the protection of workers in the formal sectors which employ only 10% of the workforce, has been in existence for long but the implementation has been lax. Number of safety officers, factory inspectors and medical inspectors has remained below optimal level. According to a recent assessment, there are 21 institutions across the country capable of training 460 specialists. This number obviously is inadequate considering the size of India's working class. There are around 1,000 qualified occupational health professionals in India and only around 100 qualified hygienists. At present, the need for occupational health specialists in the country is much higher and there is a significant gap in the demand and supply of this specialist service.

Occupational diseases-including cancers caused by various materials in workplace, including asbestos, carcinogenic chemicals, silica, cotton, dust, and radiation, job stress and work shifts-usually take a long time to develop (sometimes more than 10 years). Given the changing work practices, most of the establishments tend to hire workers on short-term contract. By the time they develop a disease, it becomes impossible to link the same to their working environment. Non communicable diseases result in more deaths than communicable diseases. Overall, people are more likely to die of work-related diseases than childhood or infectious diseases.

9.2.3 The Factory Act 1948

This is an Act to consolidate and amend the law regulating labour in factories and consists of 113 sections, detailing all nitty gritty concerning definitions, provisions, and how to implement it for providing standard OSH. It came into force on 1st of April, 1949 as the Factories Act 1948 and extends to the whole of India.

The legislation for labour welfare, known as Factories Act 1948, was enacted with the prime objective of protecting workmen employed in factories against industrial and occupational hazards. With that intent it imposes upon owners and occupiers certain obligations to protect unwary and negligent workers and to secure employment for them which is conducive and safe. The Act's objective is to protect human beings from being subjected to unduly long hours of bodily strain and labour. It provides that employees work in healthy and safe conditions as far as the manufacturing process will allow and that precautions be taken for their safety and for the prevention of accidents. In order to ensure that the objectives are carried out, local govts are empowered to appoint inspectors to call for returns and to ensure that the prescribed registers are duly maintained.

The Act provides for the health, safety, welfare and other aspects of OSH for workers in factories. It is enforced by the state governments through their factory inspectorates. It also empowers the states to frame rules, so that local conditions prevailing in the state are appropriately reflected in the enforcement, and that opportunities are taken advantage of to make certain other amendments found necessary in the implementation of the Act.

9.2.4 Section 45 of the Factory Act 1948

Section 45 of "The Factory Act 1948" of Govt of India deals with First-aid appliances in factory, Establishments, offices, projects and Institutions and it states that; (a) There shall in every factory be provided and maintained so as to be readily accessible and available during all working hours' first-aid boxes or cupboards equipped with the prescribed contents, and the number of such boxes or cupboards to be provided and maintained shall not be less than one (1) for every 150 workers ordinarily employed at any one time in the factory. Nothing except the prescribed contents shall be kept in a first-aid box. (b) Each first-aid box shall be kept in the charge of a separate responsible person, who holds a certificate in first-aid treatment recognised by the state government and who shall always be readily available during the working hours of the factory. In every factory where more than 500 workers are ordinarily, there shall be provided and maintained an ambulance room of prescribed size, containing the prescribed equipment and in the charge of such medical and nursing staff as may be prescribed and these facilities shall always be made readily available during the working hours of the factory.

Usually the only first-aid certificate which is accepted is that which is issued by the St. John's Ambulance. The basic certification training curriculum involves 3 to 8 days while the recertification usually involves one day training. Given that, it is a legal requirement to provide first-aid at workplace, all the organizations employing people are biggest customer for the first-aid services.

Though, the legal requirement for trained first-aiders at workplace is 1 per 150 workers, it would be prudent to train 1 in every 10 workers in first-aid, in-order to ensure availability of the first-aiders in all the work-shifts allowing for absence from work by the trained persons due to any reason.

9.3 What is First-Aid?

First-aid is the immediate care to an injured or ill, usually undertaken by a lay person, and performed with a limited-skill. First-aid is normally performed until the injury or illness is satisfactorily dealt with (as in the case of small cuts, minor bruises, and blisters) or until the next level of care, such as an ambulance or a doctor arrives.

First-aid includes identifying a life-threatening condition, taking action to prevent further injury or death, reducing pain, and counteracting the effects of shock.

9.3.1 Primary purpose of the First-Aid

The primary purpose of first aid is to: a) Care for life-threatening situations; b) Protect the victim from further injury and complications; c) Arrange transportation for the victim to a medical facility; d) Make the victim as comfortable as possible to conserve strength; and e) Provide assurance to the victim.

9.3.2 What is First-Aid Program

A workplace first-aid program is part of a comprehensive safety and health management system that includes;

a) Management Leadership and Employee Involvement b) Worksite Analysis c) Hazard Prevention and Control d) Safety and Health Training

The basic elements for a first-aid program at the workplace include; a) Identifying and assessing workplace risks that have potential to cause injury or illness; b) Designing and implementing a workplace first-aid programme that; 1) Aims to minimise the outcome of accidents or exposures, 2) Complies with essential provisions of regulations, and 3) Includes sufficient quantities of appropriate and readily accessible first-aid supplies and first-aid equipment, such as bandages and automated external physical and medical support; c) Assigns and trains first-aid providers who, receive first-aid training suitable to the specific workplace and receive periodic refresher courses on first-aid skills and knowledge; d) Instruct all workers about first-aid programme, including what they should do if a co-worker is injured or ill. Putting the policies and programme in writing is recommended to implement this; e) Provides for scheduled evaluation of first-aid programme to keep them current and applicable to emerging risks in the workplace, including regular assessment of the adequacy of first-aid training course.

9.4 First-Aid Rules and Flow Chart

First-aid rules are part of safety booklet of a Radio station and have already been described in section 8.8.2 of the Chapter-8 on safety Practices in AM Radio Stations, so please refer to same.

9.4.1 First-aid Rules

Thumb rule of first-aid keeping guiding principle in view is a word, "RAPABCH", as explained below;

"R" is for Responsiveness, which means that check the followings: (1) is the victim conscious? (2) Touch their shoulder, ask if they are alright; (3) Ask if he need help; (4) If he say no, then proceed no further; and (5) If yes, or no response, then proceed to next letter "A"

"A" is for Activate Emergency Management Services (EMS). In India recently the number 108 has been started jointly with health and police department.

Four things to remember when making an EMS call: (1) Your name and contact number; (2) Type of emergency and what has happened; (3) Location & condition of the victim; and (4) Check injured for responsiveness.

If victim does not responds or tells that he needs help? Stay till EMS arrives to take care of victim.

"P" is for Position and it's very important and is not to be disturbed. So re-position the victim only if the victim is in further danger in his present location. And / or there does not seem to be spinal injury and additional care requires moving the victim.

Next four letter **"ABCH"** stands for:

"A" is for airways and check to see if the airway is blocked by: Use your finger to sweep the mouth to remove any seen foreign object. If this fails, then perform the Heimlich manoeuvre or abdominal thrusts.

"B" is for Breathing; look, listen and feel by watching the chest and placing your cheek a few inches above the mouth of the victim to sense any movement of air. If the victim is not breathing, he may need his head to be repositioned. If he is still not breathing then he need rescue breathing. Give it yourself if you are trained.

"C" is for Circulation: If there is no pulse, then the victim needs **CPR**. Provide it yourself or by someone trained if you are not trained. Best place to check a pulse is carotid artery along the side of the neck along the windpipe.

"H" is for haemorrhaging: If victim bleeding; provide necessary care; If not; begin a secondary assessment.

9.4.2 Principles and Flow Chart

The key guiding principles and as to how the first-aid is to be given is as per the mnemonic "3 Ps" shown in flow chart in Fig 9.1. These 3 points govern all the actions undertaken by a first-aider Viz:

a) Prevent further injury b) Preserve life c) Promote recovery

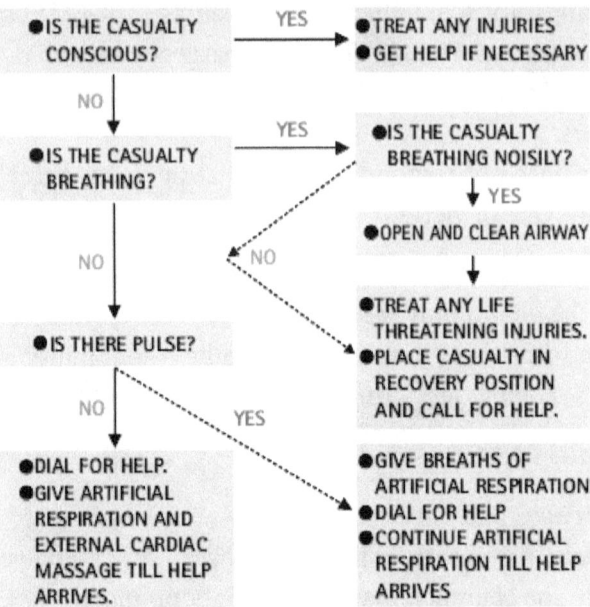

Fig 9.1: Flow Chart of a First-Aid

9.5 Protective Precautions in First-Aid

The first thing that any first-aider should be aware of when entering a situation is the potential danger to them. So in this section some protective precautions shall be described.

9.5.1 Awareness of Danger

Dangers while providing first-aid can consist of: (a) Environmental/surrounding danger – such as falling masonry, broken glass, fast vehicles or chemicals; (b) Human danger – Danger from people at the scene (including the victim) which can be intentional or accidental.

9.5.2 Protective Devices

Keeping protected should be the first priority of any first-aider. The key skill for this is awareness of surroundings and changing situations. Once you are aware of the hazards, you can then take steps to minimize the risk to yourself. One of the key dangers to a first-aider is bodily fluids, such as blood, vomit, urine and faeces, which pose a risk of cross contamination. The protective devices to be used while administering first-aid are:

(a) Gloves

The main tool of the first aider to avoid this risk is a pair of impermeable gloves. Gloves protect the key contact point (the hands) with the victim and allow you to work safely. They protect not only from bodily fluids, but from any dermatological infections or parasites that victim may have and are generally of three types:

(1) Nitrile – These Nitrile gloves come often in purple or blue color and are completely impermeable to bodily fluids. These are the gloves most recommended for use during victim contact. This material is also rated for dealing with chemical spills. To deal with chemical burns, use these gloves (you can brush off dry chemicals with gloved hands if you use Nitrile). Nitrile gloves, however, are also the most expensive.

(2) Latex – Usually white gloves often treated with powder to make them easier to put on or off. These are not used as widely as they once were due to a prevalence of allergies to latex. Latex allergies though are rarely life-threatening; if you must use latex gloves, ask the victim if they have a severe allergy to latex.

(3) Vinyl – Vinyl gloves are found in some kits, although they should not be used for contact with body fluids, though they are far better than nothing. They should primarily be used for touching victims who do not have external body fluids due to the glove's high break rate.

(b) CPR Adjunct

Another key protective equipment that should be in every first-aid kit is a CPR mask for helping to perform safe mouth-to-mouth resuscitation. With mouth-to-mouth resuscitation, there is a high probability of bodily fluid contact, especially with regurgitated stomach contents and mouth borne infections. A suitable mask will protect the rescuer from infections the victim may carry (and also the victim from the rescuer). It also makes the performance of CPR less onerous (not wishing to do mouth to mouth is a key reason cited for some bystanders not attempting CPR). CPR adjuncts come in a variety of forms, from small key rings with a Nitrile plastic shield to a 'pocket mask' such as shown in Fig 9.2.

(c) Other equipment

Large first-aid kits or those in high risk areas could contain additional equipments such as:

(1) **Safety glasses** – Prevents spurting or pooled fluid which could splay from coming in contact with the eyes.
(2) **Apron** – Disposable aprons are common items in large kits, and help protect the rescuers from contamination.
(3) **Filter breathing mask** – Some large kits, especially in high risk areas such as chemical & Gas plants, may contain breathing masks which filter out harmful chemicals or pathogens. These can be useful in normal first-aid kits for dealing with victim who are suffering from communicable respiratory infections such as tuberculosis.

Often all of these will be included as a part of a larger kit. The kit should have a list of instructions on how to properly put on / off the equipment. Follow these instructions to prevent an accidental exposure.

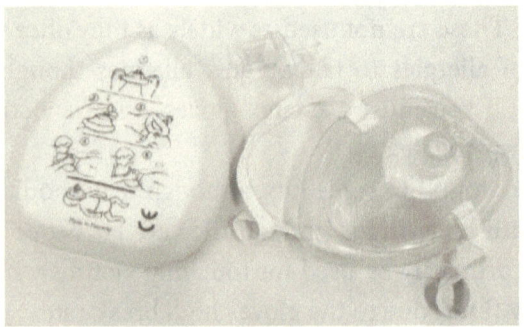

Fig 9.2: CPR Pocket Mask with carrying case

Fig 9.3: Improvisation for Bleeding in Leg

9.5.3 Improvisation for First-Aid

Many first-aid situations take place without a first-aid kit readily in hand and it may be the case that a first aider has to improvise materials and equipment. As a general rule, some help is better than no help, especially in critical situations, so a key first-aid skill is the ability to adapt to the situations, and use available materials until more help arrives. Fig 9.3 is a very good improvisation for bleeding in leg. Some common improvisations are:

(a) Gloves: plastic bags, dish gloves, leather work gloves (wash hands with soap well after using these); (b) Gauze: clean clothing (but no paper products); (c) Splints: straight sections of wood, plastic, cardboard or metal; and (d) Slings: Victim's shirt's bottom pinned to center of their chest will immobilize a forearm nicely.

9.6 Primary assessment & basic life support for First-Aid

First-aider never should place himself in a situation which might put him in danger. Remember, you cannot help a victim if you become a victim yourself. When a first-aider is called upon to deal with a victim, he must always remember to safeguard himself in the first instance and then assess the situation.

9.6.1 Initial steps

(1) Protect yourself
When called to a scene, remember that personal safety is paramount. Before you enter a scene, put on personal protective equipment, especially gloves. As you approach a scene, you need to be aware of the potential dangers to you as a first aider, or to the victim. These can include live electrical items, gas or chemical leaks, buildings on fire, traffic, or falling objects.

There are also human factors, such as bystanders in the way, victim not being co-operative, or an aggressor in the vicinity who may have inflicted the injuries on the victim. If these factors are present, have the police called to control the situation. Once you have made your first assessment for danger, you should continue to be aware of changes to the situation or environment throughout your time with the victim.

(2) What has happened?
As you approach, try to gain as much information as possible about the incident. Try and build a mental picture to try and help you treat the victim. Assess the Scene – Where are you? What equipments, structures, public buildings, etc. are nearby?

Also ensure to listen – While working on victim you may overhear information from witnesses in the crowd. Everything should be taken into account should no witnesses want to become involved or you cannot ask questions.

(3) Responsiveness of Victim

The next key factor is to assess how responsive the victim is. This can be started with an initial responsiveness check as you approach the victim. This is best as a form of greeting and question, such as: "Hello, are you alright?"

The best response to this would be a victim looking at you and replying. This means he is Alert. Victims can be quickly assessed and prioritized on the **AVPU** scale, and this will help make decisions about their care. This stands for **Alert, Voice, Pain, and Unresponsive**. If the victim looks at you spontaneously, can communicate (even if it doesn't make sense) and seems to have control of his body, he can be termed as Alert.

Key indicators on the victim are:

- **Eyes** – Are they open spontaneously? Are they looking around? Do they appear to be able to see you?
- **Response to voice** – Do they reply? Do they seem to understand? Can they obey commands, such as "Open your eyes!"?

If the victim is not alert, but you can get him to open his eyes, or obey a command by talking, then you can say that he is responsive to Voice. If a victim does not respond to your initial greeting and question, you will need to try and get a response to pain from him.

Any of the responses A, V or P, mean that the victim has some level of consciousness. If he is not alert, you should always summon professional help – call an ambulance. If he is only responsive to Voice or Pain, then consider using the Recovery position to help safeguard him.

If he does not respond to voice or pain, then he is Unresponsive and you must urgently perform further checks on his key life critical systems of breathing and circulation (ABCs).

9.6.2 Next Steps

If the victim is unconscious immediately call an ambulance. Waiting would endanger the victim's life unnecessarily. If you are alone with an adult victim, call immediately, even if you must leave the victim. Placing him in recovery position will safeguard his airway against aspiration if he should vomit while you are calling the ambulance. If you are alone with a child, continue your primary assessment; you will call once you have confirmed that the victim is breathing, or after 2 minutes of CPR. If you are not alone, have your bystander call the ambulance immediately while you continue your assessment and care of the victim.

If there is more than one person injured the rescuer must determine the order in which victims need care. In general, rescuers should focus on the victim with the injury that is the greatest threat to life..

9.6.3 Basic Treatment

The last step is to actually provide care to the limits of the first aider's training-but never beyond. Treatment should always be guided as indicated in Principal of First-Aid by the **3Ps:**

Preserve life **P**revent further injury **P**romote recovery

Treatment depends on the specific situation, but all victims must receive some level of treatment for shock.

(a) First–Aid: A for Airway

The airway of the human body is one of the more important parts to be checked when providing first-aid.

An unconscious person's airway may be blocked when his tongue relaxes and falls across the airways. The technique used to open the airway is called the "head-tilt chin-lift" technique. As shown in Fig 9.4 the victim must be supine (lying on their back). With one hand on the, forehead and the other hand under the chin, the victim's head is tilted back, and his chin lifted. The victim's jaw line should be perpendicular to the ground.

You may also check the airway for visible, removable obstructions in the mouth, which you could remove with a finger. You can remove any item in the mouth which is removable, but should not waste time trying to remove lodged items such as dentures. If a conscious victim's airway is obstructed by a foreign object, the object must be removed. Abdominal thrusts are the standard method for conscious victims.

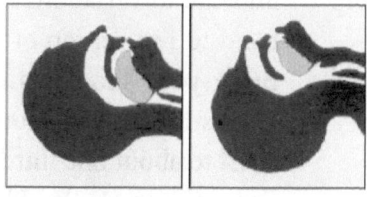

Fig 9.4: Head-tilt Chin-lift position

(b) First–Aid: B for Breathing

(1) Checking the respiration – After opening the victim's airway, check for breathing. To do this, place your cheek in front of the victim's mouth (about 3-5 cm away) while looking at his chest. You can also gently place a hand on the centre of victim's chest.
If there is no breathing, you must start CPR.

(2) Call for Help-If a bystander has not already summoned assistance, and then you must at this point call the Emergency Medical Service (EMS) or Ambulance Service.

(3) Rescue Breaths-Rescue breaths must be provided to victims in a state of respiratory arrest; do not provide them to a weakly breathing victim. If you cannot detect the breath of the victim, begin CPR. If you have a CPR mask, use it to protect yourself and the victim from exchange of body fluids. – Start by giving 2 rescue breaths.

(c) First-Aid: C for Compressions

 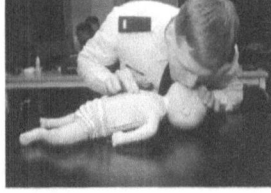

Fig 9.5: CPR-correct position   **Fig 9.6:** CPR Infant -2 fingers

(1) Technique-The aim is always to compress in the center of the chest, regardless of the victim. This means that compressions are performed on the sternum or breastbone of the victim, approximately in line with the nipples on males and children as shown in Fig 9.5. Compressions for infant CPR are done with two fingers as shown in Fig 9.6.

> For adults (>8) – Place the palm of one hand in the centre of the chest, approximately between the nipple line (on adult males – for females, you may need to approximate the ideal position of this line due to variations in breast size and shape). Bring your other hand to rest on top of the first hand, and interlock your fingers. Bring your shoulders directly above your hands, keeping your arms straight. You should then push down firmly, depressing the chest to about one third (1/3) of its depth.

> For children (1-8) – Place the palm of one hand in the centre of the chest, approximately between the nipple lines. Bring your

shoulder directly above your hand, with your arm straight, and perform compressions to one third (1/3) the depth of the chest with one arm only.
- For infants (<1yr) – Use your forefinger and middle finger only. Place your forefinger on the centre of the child's chest between the nipples, with your middle finger immediately below it on the chest, and push downwards using the strength in your arm, compressing the chest about one third (1/3) of its depth.

Give 30 compressions in a row, and then two (2) rescue breaths. Then restart your next cycle of compressions.

(2) Making compressions effective – Allow the ribs to come all the way back out after each compression, followed by a brief pause. This allows the heart's chambers to refill. Spacing compressions too close together will lead to them being ineffective.

(3) Obstructed Airway-If your ventilations don't go in, try adjusting the angle of the head (usually tilting it further back) and re-attempt ventilation. If the breath still doesn't go in, then do your compressions, and check the airway for obvious foreign obstructions after the compressions. If you see a foreign obstruction, remove it with your fingers if possible. Do not discontinue CPR because the airway is occluded.

9.7 Secondary assessment for: First-Aid

The purpose of a secondary assessment (composed of a head-to-toe, history and vitals) is to continually monitor the victim's condition and find any non-life-threatening conditions requiring treatment. This assessment should be done for any victim requiring ambulance intervention, or if there is a concern that the victim's condition may deteriorate. In some cases, you may do a quick secondary survey – use your best judgment.

9.7.1 What is Head-to-toe?

Head-to-toe assessment is a technique used by lay rescuers, first responders, and ambulance personnel to identify an injury or illness or determine extent of an injury or illness. It is used on victims with following criteria:

- Victim of trauma injuries (except minor injuries affecting peripheral areas)
- Unconscious victims
- Victims with much reduced level of consciousness

If a victim is found unconscious, and no history is available, initially assume that the unconsciousness is caused by trauma, and where possible immobilize the spine, until you can establish an alternative cause.

The secondary assessment should be performed on every such victim (especially trauma) regardless of gender of rescuer or victim. However, be sensitive to gender issues here (as with all aspects of first-aid), and if performing a full body check-up on a member of the opposite sex, it is advisable to ensure there is an observer present, for your own protection. In an emergency however, victim care always takes priority.

(a) Priority of ABCs

You should always make ABCs a priority when dealing with victims who are appropriate for a secondary survey. In case of trauma victims, where victim is conscious and able to talk, keep talking to him throughout. This not only acts to reassure them and inform him what you're doing, but will assure you that they have a patent airway and is breathing.

For unconscious victims, if you are alone, check the ABCs between checking entire body area, or if you are with another competent person, make sure they check ABCs continuously whilst you perform the survey.

Remember that if the person is unconscious and if you know or suspect it to be a trauma injury (evidence of blood, fall etc.) than you MUST treat it as a potential spinal injury in the first instance. This is because in trauma, any blow to the head sufficient to cause unconsciousness is also sufficient to cause spinal injury. In this case immobilization of the head, neck and spine takes priority over the secondary survey. If you have a second rescuer or bystander, then have them immobilize while you perform the head-to-toe.

(b) What is being looked for?

The head-to-toe is a detailed examination where you should look for abnormality. This can take the form of asymmetry; deformity; bruising; point tenderness (wincing or guarding – don't necessarily expect them to tell you); minor bleeding; and medic alert bracelets, anklets, or necklaces.

It is important to remember that some people naturally have unusual body conformation, so be sensitive about this, but don't be afraid to ask the conscious victim or relatives if this is normal for them. It is always worth looking for symmetry – if it is the same both sides, the chances are, it's normal.

(c) The six areas of the body
Divide the victim body into 6 areas; after you examine each area (with gloves on), you reassess ABCs.

(1) Head and neck
Important areas to assess, and take time and care to look for any potential problems.

Head – Using both hands gently run your hands across the skull, pressing it gently but firmly, starting at the forehead and working around to the back of the head. Feel for indentations, look for blood or fluid and watch the victim for signs of discomfort. If it is a trauma injury, check both ears for signs of blood or fluid.

Neck – The neck is an important area. Start at the sides of the neck and gently press in. Watch carefully for signs of pain. Continue until you reach the spine, moving as far down the neck as possible without moving victim, if he on his back. If there is pain, tenderness or deformity, then stop the survey and immediately immobilize the neck, placing one hand on each side of the head, with the thumb around the ear. This is most comfortable done from 'above' with the victim lying supine on their back, although you should support the victim in the position you find them.

(2) Shoulders, chest and back
This area of the body contains many of the vital organs, so it is important to look for damage which could indicate internal injury

Shoulders – Try and expose shoulders if possible, looking for obvious deformity, especially around collar bones. Try pressing along the line of the collar bone, watching for deformity or pain. Then place hands on each shoulder, and gently push down, to ensure that one side does not move more than the other.

Chest – The chest is ideally done exposed, although you should be aware of the sensitivity of females to this, and if you are able to keep breasts covered, it is advisable to do so. You should be looking for sections of the chest which are out of line with the rest of it, or which are moving differently to the rest of the chest whilst breathing. You should also look for obvious wounds. You can then gently press on the chest.

Back – If the victim is lying on their side, or front, you can also feel down their spine. If they are lying on their back, then skip this part of the check, and leave it for the ambulance crew.

(3) Arms and hands

Run both your hands down one arm at a time, looking for deformity or pain.

(4) Abdomen

The abdomen contains the remaining critical organs of the body, which need to be checked for potential damage. The abdomen checking is mostly done by gentle pushing, using the flat of your hands. Again use symmetry, and push both sides simultaneously.

(5) Pelvis

Pelvis (hips) is a large bone, with potential for a fair amount of damage. The main diagnostic test is to place a hand on each hip and first gently compress the hips together with both hands (should be very little movement and little to no pain). If the patient has moderate to severe pain when the hips are compressed, or the hips move when compressed, do not rock the hips from side to side. If there is no pain or movement, gently push down on the hips in a "rocking" motion to see if there is any movement.

(6) Legs and feet

As with arms, use both hands at the same time, running them down the inside and outside of each leg simultaneously (avoiding the groin area on the inside). You should also look for any shortening or rotation of one leg compared to the other. Finally, you take each foot, check that it has normal motility (can be moved normally) and has no obvious injuries

9.7.2 Victim's History

Taking a victim's history is a crucial step. If an ambulance needs to be called and the victim is conscious, taking history before the victim's condition worsens will assist the responding paramedics and the emergency department to better help the victim and be aware of medical conditions of the victim. Some common things to ask for in a history can be remembered using the acronym CHAMPION;

Chief complaint - What is the problem?

History of chief complaint - How did this happen? -Has it ever happened before?

Allergies- Are you allergic to anything?

Medical history and medications
- Do you have any medical conditions (angina, high BP, diabetes…)?
- Do you take any medications?
- Do your medications help when this happens?
- What is the name of your normal doctor?

Pain assessment
- Pain location, Quality of pain (sharp/dull, squeezing…), Radiating pain?
- Severity of pain (on a scale from 1 to 10)
- Timing (Constant? How long?)
- Also try to find out what makes it feel better/worse

Important Information
- Name, date of birth, age, sex, address…

Onset
- When did the symptoms start?
- What were you doing?

9.7.3 First-Aid: Vital signs

(a) Purpose

As part of ongoing assessment of the victim, and in preparation for the arrival of any assistance you have called, it is important to keep a check on victim's vital signs. If possible, write down these so that you can keep a record of any changes, and hand this over to the ambulance crew who take over the victim from you. Ideally, it should be recorded on a report, which should form part of every first-aid kit.

(b) Assessments

The vital signs you are looking to record relate to the body's essential functions. It starts with the airway and breathing already covered in basic life support and continues with circulation, look of the skin, level of consciousness and pupil reaction.

(1) **Breathing:** While maintaining open airway, ensure that the victim is breathing and count the rate of breathing. The easiest way to do this is to count the number of breaths taken in a given time period (15 or 30 seconds are common), and then multiply to make a minute.

In addition to rate, note if the breathing is heavy or shallow, and importantly if it is regular. If it is irregular, see if there is a pattern such as breathing slowly, getting faster, then suddenly slower again. Note whether breathing is noisy (wheezing could be a sign of asthma, rattling – a sign of fluid in the throat or lungs)

(2) **Circulation:** In primary survey, we did not check the circulation of the victim to see if the heart was beating (we assumed that if the victim was breathing, their heart was working and if he was not breathing, his heart was also stopped), it is important in monitoring the breathing victim to check the circulation. Two main checks are:

Capillary Refill – The capillaries are the smallest type of blood vessel, and are responsible for getting blood in to all the body tissues. If the blood pressure is not high enough, then not enough blood will be getting to the capillaries. It is especially important to check capillary refill if the victim has suffered an injury to one of their limbs. Check capillary refill by taking the victim's hand, lifting it above the level of the heart, and squeezing reasonably hard for about a second on the nail bed. This should move the blood out, and the nail bed will appear white. If the pink colour returns quickly A normal time for the pink colour to return is less than two (2) seconds. If it takes longer than 2 secs for colour to return, this indicates a problem and seeks medical advice.

Pulse check – As a first aider, you can also check a victim's heart rate by feeling for their pulse. There are three main places you might wish to check for a pulse:

Radial pulse – This is the best pulse to look for on a conscious victim, as it is relatively easy to find. It is located on the wrist (over the radial bone). To find it, place the victim's hand palm up and take the first two fingers of your hand and on the thumb side of the victim's wrist you will feel a rounded piece of bone, move in from here 1-2cm in to a shallow dip at the side of the bone, and press your fingers in (gently), where you should be able to feel a pulse. Should there be no pulse in a victim who is pale and unwell, you are advised to seek medical assistance urgently.

Carotid Pulse-This is in the main artery which supplies blood to the head and brain, and is located in the neck. This is best used on unconscious victims or those victims where you are unable to find a radial pulse. To locate it, place your two fingers in to the indentation to the side of the windpipe, in line with the Adam's apple (on men), or approximately the location an Adam's apple would be on women.

Pedal Pulse – The pedal pulse can be found in several locations on the foot, and this is used when you suspect a broken leg, in order to ascertain if there is blood flowing to the foot. When measuring a pulse you should measure the pulse rate. This is best achieved by counting the number of beats in 15 seconds, and then multiplying the result by four. You should also check if the pulse is regular or irregular.

(3) **Skin**: Changes in circulation will cause the skin to be of different colours, and you should note if the victim is flushed, pale, ashen, or blue tinged. It should also be noted if the victim's skin is clammy, sweaty or very dry, and this information should be passed on to the ambulance crew.

(4) **Level of Consciousness:** You can continue to use the acronym **AVPU** to assess if the victim's level of consciousness changes while you are with them. To recap, the levels are:
Alert – Voice induces response – Pain induces response – Unresponsive to stimuli

(5) **Pupils:** Valuable information can be gained from looking at victim's pupils. For this purpose, first aid kits should have a penlight or small torch in it. Ideally, the pupils of the eye should be equal and reactive to light, usually written down as; **PEARL** — Pupils — Equal — And — Reactive to — Light

9.8 Application and Categorization of: First-aid

Some of the most probable emergency situations under various categories of medical classification being described from the First-aid point of view are:

9.8.1 First-Aid for: Circulatory System emergencies

Bleeding (internal or external) is a common reason for application of first-aid measures. The principle difference is: external bleeding can be seen as blood leaves the body, whereas in internal bleeding, no blood can be seen.

(1) First-Aid: External Bleeding

There are many causes of external bleeding, which falls in to six main categories, which are:

- **Abrasion / Graze** – a transverse action of a foreign object to the skin, usually does not penetrate the epidermis
- **Excoriation** – Same as Abrasion, a mechanical destruction of skin, usually has an underlying medical cause
- **Laceration** – Irregular wound due to blunt impact to soft tissue overlying hard tissue or tearing as in childbirth
- **Incision** – A clean 'surgical' wound, caused by a sharp object, such as a knife
- **Puncture Wound** – by an object penetrating the skin and underlying layers, such as a nail, needle or knife
- **Contusion** – Also known as a bruise, this is a blunt trauma damaging tissue under the surface of the skin
- **Gunshot wounds** – caused by a projectile weapon, this may include two external wounds (entry and exit) and a contiguous wound between the two

Recognition

Recognizing external bleeding is usually easy, as the presence of blood should alert you to it. It may be difficult to find the source of bleeding, especially with large wounds or (even quite small) wounds with large amounts of bleeding. If there is more than 5 cups of bleeding, then the situation is life-threatening.

Treatment

As with all first aid situations, the priority is to protect you, so put on protective gloves before approaching the victim.

All external bleeding is treated using three key techniques, which allow the body's natural repair process to start. These can be remembered using the acronym mnemonic '**RED**':

Rest Elevation Direct pressure

Rest: In all cases, less the movement the wound undergoes, the easier the healing, so rest is advised.

Elevation: Direct pressure is usually enough to stop most minor bleedings, but for larger bleeds, it may be necessary to elevate the wound above the level of the heart (while maintaining direct pressure as shown in Fig 9.7). This decreases and slows the blood flow to the affected area and assists in clotting. Elevation only works on the peripheries of the body (limbs and head) and is not appropriate for body wounds. You should ask the victim to hold their wound as high as possible.

You should assist them to do this if necessary, and use furniture or surrounding items to help support them in this position. If legs are affected, make the victim lie on his back (supine), and raise his legs.

Direct Pressure: Most important of these three is direct pressure. This is simply placing pressure on the wound in order to stem the flow of blood. This is best done using a dressing, such as a sterile gauze pad (in emergency, any material is suitable). If the blood starts coming through the dressing, add additional dressings to the top, to a maximum of three. If you reach three dressings, you should remove all but the one in contact with the wound itself (as this may cause it to reopen) and continue to add pads on top. Repeat this again when you reach three dressings.

In most cases, during the initial treatment of bleeding, apply pressure by hand in order to stem the flow of blood. In some cases, a dressing may help you do this as it can keep pressure consistently on the wound. If you stop the flow by hand, you should then consider dressing the wound properly, as below in fig 9.7.

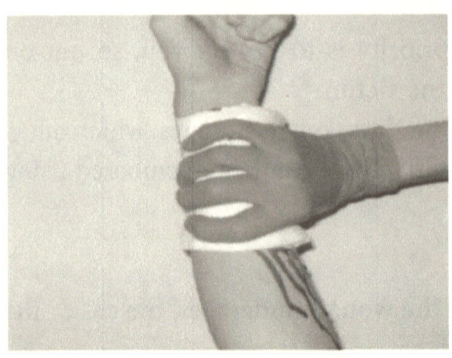

Fig 9.7: Direct Pressure and Elevation of Arm

Dressing: Once the bleeding is slowed or stopped, or in some cases, to assist the slowing of the blood flow you should consider dressing the wound properly. To dress a wound, use a sterile low-adherent pad, which will not stick to the wound, but will absorb the blood. Once this is in place, wrap a crepe or conforming bandage around it firmly. It should be tight enough to apply some pressure, but should not be too tight to cut blood flow off below the bandage. A simple check for the bandage being too tight on a limb wound is a capillary refill check; see section 9.7.3 (b) (2)-circulation. If blood starts coming through the dressing, add another on top, to a maximum of three. If all these are saturated, remove top two, leaving the closest dressing to the wound in place. This ensures that any blood clots that have formed are not disturbed; otherwise, the wound would be opened a new.

Special cases
(a) **Nosebleeds (epistaxis):** If victim has nosebleed, pinch the soft part of nose firmly between thumb and forefinger, just below the end of the bone. If necessary, you do it, but it is preferable, ask victim if able, to do it. The victim should lean their head slightly forward and breathe through their mouth. Tilting the head forward ensures that blood isn't ingested (as it can cause vomiting) or inhaled (choking hazard).
If bleeding does not stop after 10 minutes of direct pressure, you could consider using an ice pack on the bridge of the nose to help stem the flow.
If the nose continues to bleed profusely, seek medical assistance, probably from the ambulance.
(b) **Embedded Objects:** If something is embedded in the wound, don't remove it. Instead, apply pressure around the object using sterile gauze

as above. Rolled bandages are perfect for this. Be careful not to disturb the object, as moving it may exacerbate the bleeding. This doesn't apply to superficial splinters and such.

(c) **Stab / gunshot wounds in the body:** Such wounds are life threatening, so after assessing the **ABCs** of the victim call an ambulance immediately. As with all embedded objects, do not remove the item from the body. If possible, make the victim sit (as blood go to the lowest point, allowing the heart and lungs to work as efficiently as possible). Also lean the victim on the injured side, keeping the healthy side free from incursion by blood. Assess the victim for open chest wounds or abdominal injuries, and treat accordingly.

(d) **Amputations:** If a body part has been amputated, immediately call ambulance, and treat the bleeding as above. Cover the amputated part with a moist dressing and get it into a clean plastic bag, and place this bag into a bag of ice and water, sending it with the victim to the hospital. (Label date & time, what body part it is i.e.: Right finger) You should avoid putting the part in direct contact with ice, as this can cause irreparable damage, meaning that surgeons are unable to reattach it. *Note:* If the body part is partially amputated, do not detach.

(2) First-Aid: Internal Bleeding

Sometimes blood will leak from inside the body through natural openings. Other times the blood stays inside the body, causing pain & shock, even though you cannot see the blood loss.

Causes

Internal bleeding can be caused in many ways. Any time someone could have internal bleeding, no harm by treating him for internal bleeding, but not treating the victim could lead to death. Some causes are: Falls, Accidents, Vehicle accidents, Gun shot wounds, Injury from Explosions, Impaled Objects, and stab wounds.

Recognition

A person may be bleeding internally if one of these things happens; Blood out of the nose or mouth, Blood or clear fluid out of the ear (both occurs from severe head trauma), Blood is in the stool or urine or vomiting, pain over vital organs or Fractured femur.

But remember, a person may be bleeding inside the body, even though you cannot see the bleeding. If you see the signs of shock and no apparent injuries, always suspect internal bleeding. In cases of internal bleeding the skin may become pale and cold and cyanosis may be present.

Treatment
As with any victim, before treating, put on disposable gloves and take other necessary body substance isolation precautions;

- Check the victim's **ABC's**; Call an ambulance; Treat for shock; If the victim has ABC complications; treat those first – ABC s always take priority; Assist the victim into the most comfortable position; and Monitor ABC's & vitals until the ambulance

(3) First-Aid: Heart Attack & Angina

Heart attack (myocardial infarction-MCI) is caused when blood supply to the heart or its part is cut off partially or completely, which leads to death of the heart muscle due to oxygen deprivation. Heart attacks usually occur after periods of rest or being recumbent, and only rarely occur after exercise (despite popular portrayal). Heart attack can be caused by blockage in arteries supplying blood to the heart.

Angina: is a 'miniature heart attack' caused by a short term blockage. Angina almost always occurs after strenuous exercise or periods of high stress for the victim.

The key differentiation between a heart attack and angina is that, in line with their typical onset modes, angina should start to relieve very shortly after resting (a few minutes), whereas a heart attack will not relieve with rest.

(a) Recognition
Chest pain: tightness in the chest or between the shoulder blades, often radiating into the left arm, and the jaw; Nausea or indigestion; Pale, clammy skin; Ashen Grey skin; Impending sense of doom and Denial.

(b) Treatment
Assist the victim with medication, if he has any. People with angina will often have medication to control it; either as pills or a spray. The pills should never be touched with bare skin by the rescuer, as they may cause a migraine headache, and they are placed under the tongue for absorption.

The spray should be taken on the bottom of the tongue. Only the victim should administer his medication. If he is unable to do so, then the rescuer should not do it for him. Helping to take the lid off or handing the bottle to the victim is fine; this should be documented if patient is transferred to other rescuers.

- Call ambulance if don't have medication, or it doesn't help
- Loosen tight clothing, especially around the neck
- Assist the victim into a recumbent position, with the body leant back at about 45 degrees, with feet on the floor, but knees raised – this puts the patient in a 'W' position.
- If the patient is not on any anti-coagulant medicine such as heparin or warfarin, then help him take one dose of aspirin if they decide to do so.
- Continue monitoring **vitals** be prepared to do **CPR** should the victim go into cardiac arrest

Do's during Heart Attack;
- Make patient to sit down, rest and try to keep calm
- Loosen any tight clothing
- Ask if the patient takes any chest pain medication for a known heart condition such as nitro-glycerine and help him take it
- If the pain does not go away with rest or within 3 minutes of taking nitro-glycerine, call for **EMS**
- If the person is unconscious and unresponsive, call for **EMS** and begin CPR

Don'ts during Heart attack
- Do not leave the person alone
- Do not allow the person to deny the symptoms
- Do not wait to see the symptoms to go away
- Do not give the person anything by mouth expect a prescribed heart medication (such as nitro-glycerine)

(c) CPR (Cardiopulmonary Resuscitation)
CPR (Cardiopulmonary Resuscitation) an emergency life saving procedure and is a combination of;
- Chest compression to keep Patients blood circulating
- Rescue breathing to provides oxygen to Patient's lungs

(d) Chest Compressions
As explained in 9.6.3. (c) Administer it as below;
- Place the heel of one hand on the lower half of the person's breastbone
- Place the other hand on top of the first hand and interlock the fingers
- Press down firmly and smoothly (compressing to 1/3 of the chest depth) upto 30 times
- Administer 2 breaths as described below in mouth to mouth
- The ratio of 30 chest compressions followed by 2 breaths is the same, whether CPR is being performed alone or with the assistance of a second person
- Aim for a compression ratio of 100 per minute

Effective chest compression will be tiring. It is important to get help from others if possible, to allow change over for rest and to keep the compression effective.

(e) Mouth to Mouth Respiration
- If the victim is not breathing normally, make sure he is lying on his back on a firm surface
- Open the airways by tilting the head back and lifting his chin
- Close his nostrils with your fingers and thumb
- Put your mouth (better use a mouth barrier) over victim's mouth and blow into his mouth
- Give 2 full breaths to the victim (called rescue breathing). Make sure there is no air leak and the chest of victim is rising and falling, if his chest doesn't rise and fall then check tightly pinching of his nostrils and sealing of your mouth to his mouth. If still no breathing, check airways of victim for any obstruction.
- Continue CPR, repeating the cycle of 30 compressions the 2 breaths until professional help arrives.

(f) When to stop CPR
Generally CPR is stopped when;
- Victim revives and starts breathing on his own
- When professional medical help arrives – Or when the person giving CPR needs some rest due to exhaustion

(4) First-Aid: A Stroke

A Stroke is a small blockage in a blood vessel of the brain, which causes oxygen starvation to that part. This oxygen starvation can cause a loss of function, related to the area of the brain affected. Depending on duration of time the area is blocked, the damage may become irreparable. The blockage is usually caused by a small blood clot, although incursions such as air bubbles can have the same effect.

There are two main types of stroke – a **CVA** (Cardiovascular Attack – sometimes called just a stroke or major stroke) and a **TIA** (Transient Ischemic Attack – sometimes called a mini-stroke). The difference between a **CVA** and a **TIA** is simply the duration of the symptoms.

Recognition

Key recognition signs for a stroke can be remembered with the acronym FAST, which stands for:

- **Facial Weakness** – Can the person smile? Has their eye or mouth drooped?
- **Arm Weakness** – Can the person raise both arms and hold them parallel? If they squeeze your hands can they exert equal force?
- **Speech problems** – Can the person speak clearly and understand what you say

Test all three symptoms. Also patient may experience additional symptoms, which may not indicate a stroke. These include:

- Sudden blurred, dim or patchy vision; Sudden dizziness, and Sudden, severe, unusual headache.

Treatment

- Call for an ambulance, and Reassure the victim
- Encourage and facilitate the victim to move in to a position of comfort if possible. If they have significant paralysis, they may be unable to move themselves, so you should make them as comfortable as possible where they are. If possible, incline them to the unaffected side (if there is one), as this will help you relieve some symptoms such as a feeling of floating.
- Take vitals, history and regular observations

Unconscious victim
- Call for an ambulance
- Assess the victim's ABCs (attempt CPR if not breathing)
- Assist victim into recovery position on his unaffected side where gravity may assist blood reach injured side of the brain, which is below the unaffected side of the brain.

(5) First-Aid: A Shock
Shock is a range of related medical conditions where the delivery of oxygen and nutrients is insufficient to meet the body's needs. The main carrier of oxygen and nutrients in the body is the blood, so most causes are related to the blood, with the most common first-aid consideration being loss of blood. Shock is a life-threatening emergency; it should not be confused with a feeling of extreme surprise.

Types of shocks:
- **Hypovolaemic shock** – This is caused by the loss of blood from the circulatory system (not necessarily from the body, as it may be the result of internal bleeding)
- **Cardiogenic shock** – This is where the blood is not pumping effectively through the body – usually caused by heart problems, such as a heart attack.
- **Anaphylactic shock** – Caused by an allergic reaction that causes air passages to swell, blocking the flow of oxygen, and causing a lack of oxygen in the blood.

Recognition
Signs of shock can range greatly, some common signs are;

Early Phases
- A rapid pulse
- Pale, cold, clammy skin
- Sweating

Developing phase
- Cyanosis – Grey/blue skin, especially in areas such as the lips. Sometimes known as 'bluing'
- Weakness and dizziness, Nausea and possibly vomiting, Thirst
- Rapid, shallow breathing, Weak, 'thready' pulse

Advanced phases
- Absence of pulse from the wrist (radial)
- Restlessness and aggressiveness
- Yawning and gasping for air
- Unconsciousness

Final phase; Cardiac arrest
Treatment; Check the carotid pulse of the victim in the "recovery position". The most important treatment for shock of any type is to try and maintain the blood flow to the body's core (thorax and head). To do this, lay the patient flat on the floor and raise his legs about 6-12 inches off the ground.

Other important factors in the treatment of shock can be remembered by the simple mnemonic **WART**:

Warmth ABCs (Airway, Breathing, Circulation) Rest & Reassurance Treatment of underlying cause

Unconscious patients: Should a patient become unconscious, call ambulance and take the following steps;

- Assess **ABCs**. Should any change occur, compensate with required treatment. (Ex. Patient goes into cardiac arrest, begin **CPR**.)
- As airway takes priority over other treatment, place victim in recovery position to ensure a patent airway.

9.8.2 First-Aid: Respiratory related emergencies

Respiration means breathing in and breathing out. It helps in supply of oxygen (of the air) to all parts of the body. Breathing consists of three phases: –

(1) Breathing-in (inspiration),
(2) Breathing-out (expiration), and
(3) Pause

Respiratory rate: Average adult-16 to 18 times (per minute) Children – 20 to 30 times (per minute)

Note; These rates vary during stress, exercise, injury, illness, etc. Heart rate will accordingly increase to carry the extra oxygen around the body.

(1) First-Aid: Asthma & Hyperventilation

Asthma is a medical condition which causes swelling of airway, constricting airflow. Hyperventilation is simply breathing at an inappropriately high rate.

Recognition

Asthma is characterized by difficulty in breathing, wheezing, increased secretions in the airway, and a history of asthma. Hyperventilation can be recognized by fast breathing which is inappropriate for the circumstances, a feeling of not being able to catch one's breath, and light headedness.

Treatment

(a) For Asthma

Asthma inhalers come in several types. Does not use slow type of inhaler during attack?

- If the victim has a fast-acting inhaler, ask him to use it.
- Have the victim match your breathing patterns – calm the victim while slowing his breathing rate
- Help victim to sit in a position which relieves pressure on his chest. Tripod position is ideal – sitting up, leaning slightly forward, and supporting body weight with his arms either on knees or on a table or like in front of them.
- Call EMS if the victim's condition does not improve or if the victim's level of consciousness is lowered

(b) For Hyperventilation

Aim to calm the victim down, to reduce his rate of breathing, and if possible to increase concentration of carbon dioxide in air they breathe, perhaps by getting him to breathe into a paper bag.

(2) First-Aid: Obstructed Airway

(a) Conscious Victims

Abdominal thrusts are used to clear the obstructed airway of a conscious victim. It is an effective life-saving measure in cases of severe airway obstruction.

A person performing abdominal thrusts uses his hands to exert pressure on the bottom of the diaphragm. This compresses the lungs and exerts pressure on any object lodged in the trachea, hopefully expelling it.

This amounts to an artificial cough. (The victim of an obstructed airway, having lost the ability to draw air into the lungs, has lost the ability to cough on their own.)

Even when performed correctly, abdominal thrusts can injure the victim. Abdominal thrusts should never be performed on someone who can still cough, breathe, or speak – encourage victim to cough instead.

Signs of a severe airway obstruction
- The person's face turns blue from lack of oxygen
- The person desperately grabs at their neck
- The person cannot speak or cry out

Obstructed Airway for Adults & Children
Abdominal thrusts are only used on conscious adult or child victims with severe airway obstructions.

Before attempting abdominal thrusts, ask the victim "Are you choking?" If the victim can reply verbally, you should not interfere, but encourage the victim to cough.

If the victim's airway obstruction is severe, then perform abdominal thrusts:

- The rescuer stands behind and to the side of the victim and wraps their arms around the victim's sides, underneath the victim's arms
- One hand is made into a fist and placed, thumb side in, flat against the victim's upper abdomen, below the ribs but above the navel
- The other hand grabs the fist and directs it in a series of upward thrusts until the object obstructing the airway is expelled
- The thrusts should not compress or restrict the ribcage in any way.
- If you're not able to compress the victim's diaphragm due to their size or pregnancy, then perform the thrusts at the chest. If the victim loses consciousness, call for an ambulance.

(b) Unconscious Victims
If a victim has become unconscious as a result of an occluded airway, you should immediately call for assistance from the emergency medical services, and commence a primary assessment, starting with Airway and if required, commence CPR to keep the victim alive.

9.8.3 First-Aid - Soft-tissue injuries

First-aid for all soft tissue related injuries occurring from burns due to friction or frictional heat, heat due to sun, Electrocution, RF radiations, Chemicals, Radioactive and fire shall be described in this section.

(1) First-Aid: Burns

Burns are special types of soft-tissue injuries that can damage one or more layers of the skin and underlying tissues. A burn is defined as an injury to the skin or other organic tissue primarily caused by heat due to radiation, radioactivity, electricity, friction or contact with chemicals. Burns are also caused by extreme temperature (hot or cold) or RF voltage.

Type and Recognition

There are various types and 3 degrees of burns as shown in Fig 9.8:

(1) Superficial Burn – First-degree: Skin will be dry, reddening, may swell and will usually be painful.
(2) Partial thickness – Second-degree: Skin will be red and may swell, usually is painful, shall have blisters that may open and release a clear fluid. This will make the skin appear wet.
(3) Full Thickness – Third-degree: Destruction of deeper tissues such as fat, bones, nerves, and muscles and scaring scene.

Skin may be brown or black and will look charred. Tissue underneath may appear white and may be very painful or painless due to nerve endings being destroyed.

Burns may be caused by heat (thermal burns), chemicals (wet or dry), by electricity, or by radiation.

Treatment
(a) Thermal burns

Stop the burning by removing the person from the source of the burn. Check for any life-threatening conditions by checking the ABCs. Cool the burn with lots of water until the pain is relieved.

Cover the burnt part of body loosely with a sterile (preferably non-adhesive) dressing. Do not use butter, oils, creams, etc.; as they can trap heat and increase risk of infection. They will also need to be cleaned out by the hospital later, which only increases the pain of the patient. Also do not use antiseptics that may aggravate sensitive skin. Treat the victim for shock.

Burns cripple the body's ability to regulate heat. Ensure the person does not become over-heated or chilled. Aloe-Vera extract, silverdene (Silver Sulfadiazine), topical analgesics, and NSAIDs (such as ibuprofen or aspirin) are commonly used medications. Consult a doctor before use.

Fig 9.8: Degree of Burns

(b) Chemical burns

These are caused by strong acids like Sulphuric Acid, Nitric Acid, etc., strong Alkalis like Caustic Soda, Potash, Quick Lime, etc.

Don't

- Apply ice or iced water except on small first degree burns.
- Remove adhered clothing.
- Touch a burn with anything other than a sterile covering.
- Try to clean a severe burn.
- Use any kind of ointment on a severe burn.
- Break blisters.

  - If there is a dry chemical, brush it off the skin using paper, cloth, or with a gloved hand. Be sure not to get any on yourself or on the patient. Once the bulk of the dry chemical is gone, flush with running water.
  - Call EMS immediately.
  - If the burn is caused by a wet chemical, flush with plenty of water for 15 minutes and while flushing.

(c) Electrical burns

These are caused by high tension electric current. Electrical burns look like third-degree burns, but are not surrounded by first and second-degree

burns. They always come in pairs: an entry wound (smallest) and exit wound (larger).
- Call EMS immediately as electrocution shock can cause cardiac and respiratory problems.

1. Be prepared to give CPR or defibrillation. 2. Care is the same for thermal burns.

(d) Radiation burns

These are caused by exposure to RF voltage, Sun, X-Rays, and radiation after nuclear explosion, etc. Radiation burns, could also include ultraviolet radiation in the form of sunburn which should be treated as a thermal burn.

Burns caused by a nuclear source, though rare, are still possible. RF or Radiation burns cannot be treated by a lay rescuer. Individuals working in high-risk environments for possible radiation exposure are trained in the treatment of RF / radiation burns. The rescuer may unknowingly put him/her at risk of radiation exposure by treating someone with a radiation burn. For all nuclear radiation burns, call local emergency immediately.

Radiation burns also come in the form of snow blindness (or other intense light burns to the retina). Cover the eyes with sterile gauze, and contact EMS immediately. Do whatever you can to keep the victim comfortable, monitor ABCs, treat for shock, and keep the victim calm.

(e) Burns due to Fire

Burns caused by general fire should be treated judiciously to prevent loss of life or limb.

How to help a person whose clothes have caught fire?

- Put out flames using water or whatever suitable means available.
- Water is best since it quenches the flames & also cools the burnt area so that the damage is minimised.
- Don't allow the person to run about.
- Hold a wet rug in front of you, while approaching victim. Lay victim down & wrap wet rug tightly around victim so as to smother the flames.
- If the clothes in front have caught fire, lay victim on his back & vice versa, till suitable material is brought to smother the flame.

How to rescue a person from fire? As air at floor level is clean, so crawl along the floor to pull out the person.

- Tie a wet handkerchief or cloth around your face, to prevent inhalation of smoke.
- Act quickly.
- Don't open the windows.

(2) First-Aid: for Electrocution

Electrocution is a related set of injuries caused by direct contact with live electrical wires. The effects can vary from minor to causing cardiac arrest.

Actions and Treatment

- Be aware of Danger
- The real danger in this situation is the electrical supply.
- If the victim is still in touch with a live electrical source, either turn off the power to the source, or break the victim's contact with it. Find a non-conductive object (like wooden broom handle) and break the contact between the victim and the source. If the victim is in contact with a downed power line, do not attempt a rescue, instead, call power utility help and wait for professional rescuers to come and ensure the power lines are no longer live.
- Call ambulance immediately-all electrocution victims, conscious or not, require assessment in hospital.
- After ensuring the area is safe, begin a primary assessment
- check ABCs & begin CPR if required.
- Conduct a secondary assessment looking specifically for 2 set of electrical burns.
- Electrical burns look like third-degree burns, but are not surrounded by first – and second-degree burns. They always come in pairs: an entry wound (smaller) and exit wound (larger). You should cover the wounds with non stick, sterile dressings.

Electrocution causing unconsciousness; Serious electrocution may cause unconsciousness, at least for a brief period. If this is the case, conduct primary assessment by checking ABCs. If victim is not breathing, begin CPR. Airway swelling can occur from being electrocuted. Frequently check the victim's breathing. If the victim received a serious electric shock, do not put the victim in the recovery position. Head/neck/back injuries along with multiple fractures can occur from strong muscle contractions from electrocution. Begin a secondary assessment; look for 2 set or more electrical burns – one entrance wound and one exit wound.

Continually evaluate ABCs. Cardiac rhythm disturbances can cause the victim to go into cardiac arrest.

Electrocution not causing unconsciousness; Victims who are conscious are likely to feel unwell after the shock, and may complain of numbness or needles in the area from where electricity has passed. The victim must still be transported to a hospital for evaluation, as heart rhythm disturbances can lead to cardiac arrest.

(3) First-Aid: Chest & Abdominal Injuries
(a) Closed Chest Wounds

Chest wounds can be inherently serious as this area of the body protects the majority of the vital organs. Most chest trauma should receive professional medical attention immediately, so consider calling for an ambulance for any serious chest injury.

The most likely injuries that can be caused with a chest injury include broken ribs. A single broken rib can be very painful for the victim, and a rib fracture carries with it the risk of causing internal injury, such as puncturing the lung, which can lead to lung collapsing. Some specific, more complicated, rib fractures include:

- Flail chest—2 or more rib fractures along the same rib(s): Can cause a 'floating' segment of the chest wall which makes breathing difficult
- Stove chest —all ribs fractured: Can cause entire ribcage to lose its rigidity, causing great difficulty breathing

Recognition
- Trouble or and Shallow breathing
- Tenderness at site of injury
- Cynosis
- Deformity & bruising of chest
- Uneven expansion of chest
- Pain upon movement/deep breathing/coughing & may cough with blood
- Crackling sensation in skin if lung is punctured

Treatment;
- Assess ABCs and intervene as necessary
- Call for an ambulance
- Assist victim to a position of comfort
- Do a secondary survey
- Monitor vitals carefully

(b) Open Chest Wounds
An open pneumothorax or sucking chest wound — chest wall has been penetrated (by knife, bullet, falling onto a sharp object…)

Recognition
- An open chest wound – escaping air
- Trouble breathing, and coughing up blood
- Entrance and possible exit wound (exit wounds are more severe)
- Sucking sound as air passes through opening in chest wall
- Blood or blood-stained bubbles may be expelled with each exhalation

Treatment
- Call for an ambulance
- Assess ABCs and intervene as necessary
- Do not remove any embedded objects
- Flutter valve over wound, as described below
- Lateral positioning: victim's injured side down
- Treat for shock, Conduct a secondary survey, and monitor vitals

Making a flutter valve; get some sort of plastic that is bigger than the wound. Say plastic patch, Zip-lock bag, some first-aid kits will have a ready-to-use valve. Tape the plastic patch over the wound on only 3 sides. The 4th side is left open, allowing blood to drain and air to escape.

This opening should be at the bottom (as per victim's position).

(c) Abdominal Injuries
If a trauma injury has caused the victim's internal organs to protrude outside the abdominal wall, do not push them back in. Instead, have the person lie flat with their knees bent and cover the organs with a moist, sterile dressing (*not paper products – use gauze*). Do not allow the victim to eat or drink, though they may complain of extreme thirst. Call an ambulance treat for shock and monitor ABCs until the emergency medical team arrives.

If the abdominal injury does not cause an open wound, have the person lie flat with their knees bent and treat for shock until EMS arrives.

9.8.4 First-Aid: Bone & joint injuries

All bone & joint related injuries occurring due to falling from work places, accidents and other happenings shall be described in this section.

(1) First-Aid: Musculoskeletal Injuries – Sprain or Fracture

All sprains, strains, dislocations, and fractures normally have same symptoms. It is very difficult to determine what the injury may be. It is not necessary to know which injury the victim has as the treatment is same for all of them. If victim has any of the following symptoms, he should be treated for a possible muscle or skeletal injury.

- Bruising and swelling
- Deformity at the injury point
- No pulse below injury point
- Crepitus – A grinding or cracking sound when the affected part is moved (usually accompanied by extreme pain). (No test! Victim should tell)
- Inability to use the affected body part normally

Treatment

The treatment for any muscle, bone or joint injury follows the acronym **"RICE"**.

Rest – Rest is very important for soft tissue injuries, both in the short term and for longer term care.

Immobilize – Sprains, strains and dislocations can be slinged; but fractures should be splinted and slinged.

Cold – Ice should be applied periodically, for around 10-20 minutes at a time. Also take the ice off for same time it was on. In order to avoid problems, always place some fabric between the ice and the skin.

Elevation –Wherever appropriate, the injury should be elevated, as this may help reduce the localized swelling. Do not elevate if this causes more pain to the victim.

Immobilization

Proper method of slinging depends on the location of injury on the arm. After applying a sling, ensure circulation to the arm is not compromised by doing a distal circulation check. Remember that moving an arm into a position where you can put a sling on it may be painful for the victim. If that is the case, simply immobilize in the position found. You will have to improvise something based on the victim's position of comfort.

(a) The arm sling – for injuries to the forearm

A splint and sling should be applied to the forearm as shown in Fig 9.9. Note the second triangular bandage immobilizing the arm by holding it against the torso.

Fig 9.9: A Splint & Sling applied to forearm

- Support the injured forearm approx parallel to the ground with the wrist slightly higher than the elbow.
- Place an open triangular bandage between the body and the arm, with its apex towards the elbow.
- Extend the upper point of the bandage over the shoulder on the uninjured side.
- Bring the lower point up over the arm, across the shoulder on the injured side to join the upper point and tie firmly with a reef knot.
- Ensure the elbow is secure by folding the excess bandage over the elbow, securing it with a safety pin.

(b) Elevated sling – for injuries to the shoulder

- Support the victim's arm with elbow beside the body and hand extended towards the uninjured shoulder.
- Place an opened triangular bandage over the forearm and hand, with the apex towards the elbow.
- Extend the upper point of the bandage over the uninjured shoulder.
- Tuck the lower part of the bandage under the injured arm, bring it under the elbow and around the back and extend the lower point up to meet the upper point at the shoulder.
- Tie firmly with a reef knot.

➤ Secure the elbow by folding the excess material and applying a safety pin, and then ensure that the sling is tucked under the arm giving firm support.

(c) Collar and cuff – for upper arm or rib injuries
➤ Allow the elbow to hang naturally at the side and place the hand extended towards the shoulder on the uninjured side.
➤ Form a clove hitch by forming two loops – one towards you, the other away.
➤ Put the loops together by sliding your hands under the loops and closing with a "clapping" motion. If you can tie a clove hitch, simply tie it on the wrist.
➤ Slide the clove hitch over the hand and gently pull it firmly to secure the wrist. Extend the points of the bandage to either side of the neck, and tie firmly with a reef know.
➤ Allow the arm to hand naturally.
➤ It is especially important for this sling that you ensure that circulation to the hand is not compromised – do distal circulation checks often

(d) Femoral fractures
The femur is the largest bone in the body, and has a large artery, the femoral artery, directly beside it. Because a mechanism of injury which can fracture the femur is likely to also displace the fracture, it is possible that the femoral artery will be damaged internally. Call EMS immediately.

Damage to the femoral artery is likely to cause massive internal bleeding, so it is a major emergency.

Be sure to maintain as much immobilization as possible and monitor ABCs until EMS arrives.

(2) First-Aid: Head & Facial Injuries
(a) Head Injuries
Head wounds must be treated with special care, since there is always the possibility of brain damage. The general treatment for head wounds is same as that for other flesh wounds. However, certain special precautions must be observed if you are giving first-aid for a head wound. Victims with a head injury having decreased level of consciousness (even brief) require assessment by a physician and also for a potential spinal injury. Any mechanisms of injury that can cause a head injury can also cause a spinal injury.

(1) Concussion

If a person suffers blow to the head, the brain can be shaken inside the skull. This is called concussion. It results into a short loss of consciousness (a few seconds to a few minutes even). Most victims make a full recovery from it, but occasionally it may become serious. If you suspect someone have become concussion, then it is a medical emergency and act accordingly. Briefly it is described as below;

- Mild head injury that causes a brief "short-circuit" of the brain
- Essentially, the brain has been rattled within the skull
- No damage or injury to brain tissue

Symptoms of Concussion

- Anxiety & agitation
- possibly unconscious for a short period of time
- Dazed and confused for several minutes
- Visual disturbances (seeing stars)
- Vomiting, Amnesia (memory loss), Head pain
- Pupils unequal in size or unreactive to light

(2) Compression

- Pressure on the brain caused by a build-up of fluids or a depressed skull fracture
- The brain has been bruised, Damage to brain tissue is likely
- Symptoms are progressive, and will usually get worse over time Recognition; same as above for serial no: 1

Treatment

- Call Emergency Medical Service (EMS)
- Immobilize spine if required
- Treat for any bleeding, bruising or swelling (if you suspect a skull fracture, do not apply pressure – instead, use a thick dressing with as little pressure as possible)

Notes for head injuries;

- If the level of consciousness is altered, call EMS
- Do not use direct pressure to control bleeding if the skull is depressed or obviously fractured, as this would cause further injury by compressing the brain

(b) Injuries involving the eye

Wounds that involve the eyelids or the soft tissue around the eye must be handled carefully to avoid further damage. If the injury does not involve the eyeball, apply a sterile compress and hold it in place with a firm bandage. If the eyeball appears to be injured, use a loose bandage.

(**Note:** Never attempt to remove any object that is embedded in the eyeball or that has penetrated it; just apply a dry, sterile compress to cover both eyes, and hold the compress in place with a loose bandage).

A person who has suffered a facial wound involving eye, eyelids, or tissues around the eye must receive medical attention as soon as possible. Be sure to keep the victim lying down. Use a stretcher for transport.

Many eye wounds contain foreign objects. Dirt, coal, cinders, eyelashes, bits of metal, and a variety of other objects may become lodged in the eye. Since even a small piece of dirt is intensely irritating to the eye, the removal of such objects is important. However, the eye is easily damaged. Impairment of vision (or even total loss of vision) can result from fumbling, inexpert attempts to remove foreign objects from the eye. The following precautions must be observed:

- DO NOT allow the victim to rub the eye.
- DO NOT press against the eye or manipulate it in any way that might cause the object to become embedded in the tissues of the eye. Be very gentle; roughness is almost sure to cause injury to the eye.
- DO NOT use such things as knives, toothpicks, matchsticks, or wires to remove the object.
- *DO NOT Under Any Circumstances Attempt To Remove An Object That Is Embedded In The Eyeball Or That Has Penetrated The eye!*

If you see a splinter or other object sticking out from the eyeball, leave it alone! Only specially trained medical personnel can hope to save the victim's sight if an object has actually penetrated the eyeball. Small objects that are lodged on the surface of the eye or on the membrane lining the eyelids can usually be removed by the following procedures:

(1) Wash the eye gently with lukewarm, sterile water. A sterile medicine dropper or a sterile syringe can be used for this purpose. Have the victim lie down, with the head turned slightly to one side. Hold the

eyelids apart. Direct the flow of water to the inside corner of the eye, and let it run down to the outside corner. Do not let the water fall directly onto the eyeball.
(2) Gently pull the lower lid down, and instruct the victim to look up. If you can see the object, try to remove it with the corner of a clean handkerchief or with a small moist cotton swab. You can make the swab by twisting cotton around a wooden applicator, not too tightly, and moistening it with sterile water.

Caution: Never use dry cotton anywhere near the eye. It will stick to the eyeball or to the inside of the lids, and you will have the problem of removing it as well as the original object.

(1) If you cannot see the object when the lower lid is pulled down, turn the upper lid back over a smooth wooden applicator. Tell the victim to look down. Place the applicator lengthwise across the center of the upper lid. Grasp the lashes of the upper lid gently but firmly. Press gently with the applicator. Pull up on the eyelashes, turning the lid back over the applicator. If you can see the object, try to remove it with a moist cotton swab or with the corner of a clean handkerchief.
(2) If the foreign object cannot be removed by any of the above methods, DO NOT MAKE ANY FURTHER ATTEMPTS TO REMOVE IT. Instead, place a small, thick gauze dressing over both eyes and hold it in place with a loose bandage. This limits movement of the injured eye.
(3) Get medical help for the victim at the earliest opportunity.

(3) First-Aid: Suspected Spinal Injury
The spinal cord is a thick nerve that runs down the neck and back; it is protected by bones called vertebrae. If the spinal cord is injured, this can lead to paralysis. Since the vertebrae protect the spinal cord, it is generally difficult to cause such an injury. Note that only an x-ray can conclusively determine if a spinal injury exists. If a spinal injury is suspected, the victim must be treated as though one does exist.

Recognition
- Mental confusion (such as paranoia or euphoria)
- Paralysis, Dizziness, Head, neck or back pain
- Head or back injury, Resistance to moving the head

- Any fall where the head or neck has fallen more than two (2) metres
- Cerebrospinal fluid in the nose or ears
- Pupils which are not equal and reactive to light

Treatment

The victim should not be moved unless absolutely necessary. Without moving the victim, check if the victim is breathing. If not, CPR must be initiated; the victim must be rolled while attempting to minimize movement of the spine. If the victim is breathing, immobilize his spine in the position found by sandbagging. Means pack towels, clothing etc. around the victim's head such that it is immobilized. Be sure to leave the face accessible, since you'll need to monitor his breathing.

If you must roll the victim over to begin CPR, take care to keep his spine immobilized. You may need bystanders to help you. Hands-on training is the only way to learn the various techniques which are appropriate for use in this situation.

(4) First-Aid: Unconsciousness

If a person on duty becomes unconscious, his head should be tilted backwards to avoid tongue to fall backward and block the airways. Tilting the head backwards and pulling the tongue forward will help to clear the airways.

Back or Neck Injury – If it is felt that someone has a back or neck injury, it is advisable to move him to his side. The priority should be to keep him breathing. Try to keep his spine straight when turning him. Do not turn alone, if available get someone else's help to turn the victim otherwise call for help and wait.

Feeling Faint-If someone is feeling faint, advise him to lie down on his back and raise his legs to improve blood flow to the brain. Fainting is caused by a temporary reduction in flow of blood to the brain and can result in a brief loss of unconsciousness. A person who has fainted should quickly regain consciousness. If not, treat him as an unconscious and handle accordingly.

9.8.5 First-Aid: Environmental related illness & injury

This para will describe all environment related such as heat, weather, cold etc causalities as below.

(1) First-Aid: Heat-Related Illness & Injury

(a) Burns
Already discussed under soft tissue injury as "Thermal Burns", so please refer to that section.

(b) Heat Cramps
Heat cramps, usually occur when a person has been active in hot weather and is dehydrated. Treating heat cramps is very simple, do the following:

- Remove the victim from the hot environment, a shady area will suffice.
- Stretch the calf and thigh muscles gently through the cramp. This usually results in immediate relief.
- Hydrate the victim; use a small concentration of salt for best results. (Saltine cracker to eat while drinking)
- Have the victim rest.

Should the cramping continue, seek further medical advice.

(c) Heat Exhaustion
This is a milder form of heat-related illness that can develop after several days of exposure to high temperatures and inadequate replacement of fluids. Those most prone to heat exhaustion are elderly people, people with high blood pressure, and people working or exercising in a hot environment.

Symptoms of Heat Exhaustion
- Heavy sweating, Paleness, Muscle cramps, Tiredness, Weakness, Dizziness, Headache, Nausea or vomiting, Fainting

Treatment of Heat Exhaustion
- Loosen the clothing, apply cool wet cloths.
- Move the victim to either a cool or an air-conditioned area, and fan the victim.

The treatment priority for heat exhaustion is to cool the victim. Heat exhaustion is not life-threatening (unlike heat stroke), so EMS is not needed unless the victim's condition worsens to the point of entering heat stroke. If the victim's level of consciousness is affected, that is heat stroke.

(d) Heat Stroke

Heatstroke occurs when the core body temperature rises too far for the body's natural cooling mechanisms to function. It is a serious, life-threatening problem that can cause death in minutes. The treatment priority with heat stroke is to call EMS and cool the victim down. When you provide first aid for heat stroke, remember this is a true life-and-death emergency. The longer the victim remains overheated, the higher the chances of irreversible body damage or even death occur.

Symptoms
- Unconscious or has a markedly abnormal mental status
- Flushed, hot, and dry skin (although it may be moist initially from previous sweating or from attempts to cool the person with water)
- May experience dizziness, confusion, or delirium
- May have slightly elevated blood pressure at first that falls later
- May be hyperventilating
- Rectal (core) temperature of 105°F or more

Treatment
- Call EMS.
- Cool victim's body by dousing the body with cold water.
- Apply wet, cold towels to the whole body.
- Pack ice into the victim's heat-loss areas (underarms, groin, neck). Do not let ice contact the victim's bare skin as this may cause frostbite!
- Maintain an open airway.
- Wetting and evaporating measures work best. (Think, artificial sweating)
- Move the victim to the coolest possible place and remove as much clothing as possible (ensure privacy).
- Expose the victim to a fan or air-conditioner since drafts will promote cooling.
- Immersing the victim in a cold water bath is also effective.
- Give the victim (if conscious) cool water to drink.
- Do not give any hot drinks or stimulants.
- Never give an unconscious victim something to drink as it may obstruct the airway or cause vomiting.

➤ Get the victim to a medical facility as soon as possible. Cooling measures must be continued while the victim is being transported. Monitor victim's vital signs frequently. Be ready for CPR should the victim become unconscious.

(2) First-Aid: Cold – Related Illness & Injury

(a) Frostbite

There are three (3) types of frostbite as shown below in Fig 9.10. Frost bite is when tissues freeze. If the frozen tissue is more than skin deep, this is considered deep frostbite.

Treatment

➤ Call EMS immediately or be ready to transport victim to a medical facility, even after treatment of frostbite.
➤ Make sure there is no risk of re-freezing. Skin that re-freezes after thawing will have more damage.
➤ Remove victim from cold environment, ensure there is no possibility of hypothermia.
➤ Fill a shallow container with enough water to cover the frostbitten body part. Make sure the water is at room temperature. The water should be cool, but should not be too warm as warm worsen the pain.
➤ Immerse the injured area; ensure that the skin does not come into contact with anything!
➤ Repeat above step by refreshing the water as it cools until the skin is back to a normal color and texture. This may take several hours depending on the severity of injury.

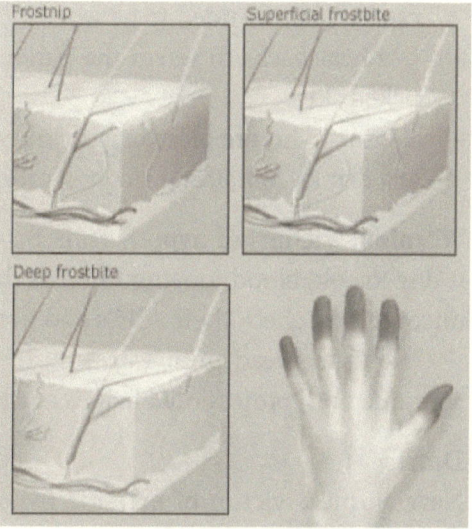

Fig 9.10: Types of Frost Bites

Note; transport the victim to a medical assistance for further assessment after the above steps.

(b) Hypothermia

Hypothermia is when the body's core temperature drops so low the body can no longer warm itself back up. Severe hypothermia is classified as when the body drops below 95 degrees Fahrenheit.

Treatment
- Remove the victim from the cold environment.
- Wrap victim in blankets.
- For cases of extreme hypothermia, where the patient is showing signs of confusion, slurred speech, fumbling hands, or go unconscious, call EMS.
- Remove wet clothing from victim and replace with dry clothing. (A dry cap is recommended to be worn.)
- Use heat packs to warm the patient. Do not allow the packs to touch naked skin.
- Victims who are alert may drink warm liquids, however, do not give any drinks containing alcohol, caffeine, or give a drink that is too hot.

Warning: Victims of hypothermia may become worse as they warm, this is due to cold blood moving towards the core of the body. If a patient goes unconscious, check their ABCs and notify EMS.

"Nobody's dead until warm and First-aid procedures must therefore continue until professional help is available.

Don'ts;
Never jostle a victim of extreme hypothermia as this may cause cardiac arrest!

9.8.6 First-Aid: Medical conditions & poisoning

This section describes first-aid relating to general medical conditions and the poisoning.

(1) First-Aid

Diabetes – A disease causing an inability to regulate the level of sugar (glucose) in the blood

Insulin – a hormone that allows glucose to travel from the bloodstream into the cells

(a) Hypoglycaemia (Insulin Shock)

Hypoglycaemia is a condition in which blood sugar levels are too low to power the body. The symptoms of hypoglycaemia will come on suddenly.

Causes
- Lack of food (low glucose), Excessive exercise
- Too much insulin, Vomited meal

Recognition
- Pale, cool, clammy, Dizziness, weakness
- Hunger, Seizures
- Confusion (like being drunk)
- Strong, rapid pulse (May be normal in some patients)

This may be confused with stroke or other cardiac disorders.

Treatment
If possible, have the victim tested for glucose level to correctly identify Hypoglycaemia or Hyperglycemias'.
- EMS, Monitor ABCs, Treat for shock
- Assist with glucose in any form (candy, juice, Monogel), but only if the victim is fully conscious
- Encourage any victim of a diabetic emergency to use their test kit if it is nearby.
- Giving glucose to a victim with insulin shock will help.
- Don't give glucose to an unconscious victim as it can easily become an airway obstruction.
- Some victims carry with them glucagon injections as a rapid treatment for severe insulin shock.

The victim should know how to administer it, and should administer it himself.

(b) Hyperglycaemia

Hyperglycaemia is a condition in which the body's blood sugar level is too high to maintain. This condition is less common and usually occurs very slowly, over the course of several days.

Causes
- Victim doesn't take enough insulin
- Eats too much (high glucose)
- has an infection

Recognition
- Flush/redness of skin, Deep or rapid respirations
- Weak/dizzy, Dehydration/extreme thirst/excessive urination
- Loss of appetite, Weak, rapid pulse

Treatment
- EMS, Monitor ABCs, Treat shock
- Encourage any victim of a diabetic emergency to use their test kit if it is nearby.
- Some victims carry with them insulin injections as a fast treatment for hyperglycaemia. Assist them if required.

(2) First-Aid: Seizures

Random, uncontrolled electrical activity in the brain causes seizures. A seizure occurs when the electrical activity of the brain becomes irregular. When the electrical activity is severely irregular, the result may be a seizure. A seizure is a medical emergency. Seizures may be caused by either an acute or chronic condition such as epilepsy.

Risk Factors for Seizures
- Head trauma, Infections of the brain or spinal cord
- Epilepsy, Stroke
- Drug use or withdrawal, Heat Stroke
- Hypoglycaemia (Low Blood Sugar), Fever in infants

Often before a seizure occurs, the victim may feel an aura, which is an unusual sensation that typically precedes seizures. Auras may come in many forms; often if the person is epileptic, they may be aware that a seizure is imminent and may tell others or sit or lie down to prevent injury.

Recognition
Typically seizures usually last no more than three minutes. Some common occurrences during a seizure include stopped or irregular breathing, body rigidness or convulsing, defecation, urination, and drooling.

Treatment
Don'ts
- Never try to restrain the seizure
- Never put anything in the mouth

Seeing a seizure may be a frightening experience which may cause you hesitation to act to aid the victim. However, it is very easy to care for the victim.

Never attempt to hold victim in any way to stop their seizure – the victim is unaware and is unable to control it. Attempting to restrain an individual having a seizure may result in injuries to both you and the victim.

Also, do not attempt to stick anything into the victim's mouth – the victim will not swallow their tongue and sticking something in their mouth can cause further injury or death. The tongue may obstruct the airway during the seizure, but this is normal.

Care for Seizures
Call EMS or ask someone to call for you

(1) Move anything the victim can injure away from themselves such as chairs or other objects
(2) Gently support the victim's head to prevent it from hitting the ground
(3) Request all bystanders to move away (persons having a seizure are often embarrassed of their seizure)
(4) After the seizure has ended, roll the victim into recovery position but only if you do not suspect a spinal injury

After the seizure, the victim will slowly "**awaken.**" Ensure that bystanders are away and offer reassurance for the victim. Victims who have a seizure in public are often self-conscious about their condition. The victim will be very tired after his seizure. Continue to reassure the victim until fully aware of the surroundings or EMS arrives.

(3) First-Aid: Poisoning
Poison, in sufficient quantity, can cause temporary or permanent damage. Specific information concerning treatment can be obtained from labels or written matter such as the MSDS (Material Safety Data Sheet). Expert advice (poison control) and rapid transport to advanced medical

care (EMS) is urgently needed in poisoning cases. A poisoning victim may require basic life support at any moment; monitor the victim's ABCs throughout.

(a) Absorbed poison

Absorbed poisons are taken into the body through unbroken skin. Absorbed poisons are especially dangerous as they may not only cause local damage, but they can enter the bloodstream and cause widespread damage. It is important to note that certain poisons such as agricultural chemicals or insecticides may enter the bloodstream through absorption while leaving the skin undamaged.

Treatment

As with any type of poisoning EMS should be immediately called and the rescuer should always start with the initial assessment and treat any life-threatening problems before continuing. Once all life-threats are taken care of, then the poison should be removed.

Removing

- Powders: Wearing gloves, brush the powder off the victim, and then irrigate the affected area with plenty of water for at least 20 minutes
- Liquids: Flush with clean water for at least 20 minutes
- Chemical in Eyes: Flush with clean water for at least 20 minutes

In case of all absorbed chemicals, it is crucial to remove the chemical immediately to prevent further damage to the victim.

Signs and symptoms

- Skin reactions, Itching
- Irritation of eyes, Headache
- Increased relative skin temperature, Anaphylactic shock.

Treatment

- Remove the patient from the spot.
- Brush off any dry material.
- Wash the area with soap and water
- Remove all contaminated clothing, shoes, jewellery, etc.
- Shift to hospital.

(b) Inhalation Poisoning

Inhaled poison can come from a variety of sources including the inhalation of Chemical vapours (From Sprays, cleaning fluids etc.) or Fumes from fire, stoves, exhausts, sewer gases etc. Shift victim to fresh air. Use caution in giving rescue breathing to a person overcome by hazardous chemicals, as you may be contaminated in doing so.

Signs and symptoms
- Unconsciousness or altered behaviour, Shortness of breath
- Cough, Abnormal pulse
- Burning eyes, mouth, nose, throat, chest, etc., Severe headache
- Nausea & vomiting, Reddening of lips. (In carbon monoxide poisoning)

Treatment
- Remove the patient from the spot.
- Open airway, CPR if needed
- Remove contaminated clothing, carefully
- Provide intensive First aid, if possible, Shift to hospital.

(c) Ingested Poison

It's by eating or drinking poisonous substances (Rat poison, poisonous plants, etc.). Internal poisoning may not be immediately apparent. Symptoms, such as vomiting are sufficiently general that an immediate diagnosis cannot be made. The best indication of internal poisoning may be the presence of an open container of medication or toxic household chemicals. Check the label for specific first aid instructions for that specific poison.

Call for help immediately as advanced medical care will be required. If possible contact a poison control center and provide information about the suspected poison. Depending on the type of poison, the poison control center may suggest additional first aid measures pending the arrival of emergency medical technicians. These might include dilution with water or milk, administration of syrup of ipecac or activated charcoal, or the use of other common household products as improvised emergency antidotes.

Signs and symptoms
- Burns and stains around the mouth, Abnormal breathing.
- Abnormal pulse, Sweating.

- Unusual odours of breath, on clothing or at the scene (From garlic smell, rat poison).
- Dilated or contracted pupils.
- Painful swallowing, Nausea & vomiting.
- Excess salivation or foaming from the mouth.
- Distension, pain & tenderness of abdomen.
- Diarrhoea, Convulsions.
- Altered state of consciousness.

Treatment
- Open airway, Dilute poison with water or milk.
- Induce vomiting with syrup ipecac, soap & water, etc. (except if patient is unconscious, ingested corrosives, gasoline products, etc.)
- Ensure patient does not aspirate vomitus.
- Give one tablespoon of powdered charcoal or white of egg.
- Antidote, if known & available, can be given.
- Transport to hospital in lateral recumbent position.

Do not apply such measures without the expert advice. Appropriate first-aid measures vary depending on the type of poison. Induced vomiting may do more harm than good, because the poison may harm the alimentary canal or oesophagus. Vomit may also block the airway. However, induced vomiting may be necessary with some poisons to save the victim's life.

(d) Injected Poison

An injection poisoning can occur from a variety of sources; Insect bites (From Spiders, wasps, Snake bite, drug abuse to animal bite etc.). Poison Control Centers will provide the best information for first aiders. Basic treatment involves monitoring the patients ABCs, treating for shock, observing the patient for an allergic reaction, and calming the patient.

To help EMS, gain as much information about the poison as you can. What it was, when it was injected, how it was injected, and if the person has any allergies to the injection. (E.g. a bee sting causing anaphylaxis)

Signs and symptoms
- Noticeable stings or bites.
- Localised pain or itching or burning sensation.

- Headache, Swelling or blistering at the site.
- Weakness or collapse, Dizziness.
- Abnormal pulse and breathing, Nausea & vomiting.
- Excessive salivation, sweating, etc.
- Muscle cramps chest tightening & joint pains.

Treatment
- Treat for shock.
- Scrap away bee or wasp stingers & venom sacs.
- Place ice bag over sting area.

(e) Snake bite
Signs and symptoms
- Noticeable bite on skin, Pain & swelling over the bite area.
- Rapid pulse & laboured breathing, Progressive general weakness.
- Vision problems. (Dim or Blurred), Nausea or vomiting.
- Convulsions, Drowsiness or unconsciousness.
- Bleeding (in viper bites)

Treatment
- Treat for shock.
- Clean the area with soap & water.
- Remove constrictive items on the bitten extremities.
- Immobilise bitten extremities.
- Keep bite area at the level or at a lower level from the heart.
- Apply constrictive band above and below the wound.
- Give antidote if available and sure of type of snake and expert is available at site to administer antidote.
- Avoid cutting over the wound and shift to hospital.

Important Points in poison cases
- Look for evidence.
- Get details from the patient or others.
- Retain any leftover poison, bottle or any container, etc to ease in identification.
- Retain the vomitus, stained clothing, dead snake, insects etc to help identify the poison.

9.9 Contents of a First-aid box

First-aid box must have medicines for common ailments. Whether for a headache, an allergic reaction, and a broken toe nail or have suffered a few burns, it is these items in your first-aid box that will provide instant relief. However, don't just stock it and forget about it. You must remember to clean and re-stock your medicines and ointments time to time. Do check expiry dates and throw out items that have outlived their life. In an emergency, nothing can be worse than having an expired medicine or tube of gel in your first-aid box.

9.9.1 Typical Items of a First-aid box

Even though the contents of a First-Aid box will vary as per the locations and the requirements of each establishment but following items generally are essential for a kit;

(1) Gloves – Have assorted size containing medium and large size, which should fit just everyone. Inspect them periodically and Replace if they age, becomes brittle and cracks.
(2) Mouth Barrier Device – Have a pocket mask type, the one that comes in a hard plastic case. A must as protective device for first-aid for victim's as well first-aider protection.
(3) Adhesive Bandages-To cover cuts need adhesive bandages. Have different sizes (1/2 inch to larger sizes like 2 or 3 inch) and assorted shapes to cover bigger cuts, medium and smaller cuts and scrapes. Once applied it keeps most dirt out, is flexible, and will stay on the skin even if wet. It's very useful for outdoor use.
(4) Ace-Bandages – Also called elastic bandages. Have 3 inch for wrists to 6 inch size for larger joints.
(5) Band-Aids-Have self-adhesive medicated band-aids of assorted sizes (small, medium & large) and shapes (circular, square & strips) and types (dry & wet use).
(6) Gauze Pads – Purchase the "semi-sterile" type that comes in bulk packages. Transfer them to plastic bags to put in the kit. 4x4 inch and 2x2 inch are useful sizes.
(7) Sterile Gauze and Tape – For bigger injuries, especially those that are bleeding profusely, you will need sterile gauze and medical tape

to create a larger bandage. Where a band-aid seems to be too small to cover the wound, use these two. Create padding with sterile gauze, apply a little antiseptic cream, and cover the wound. Then secure in place with the tape.

(8) Small splints- A few wooden tongue depressors or Popsicle sticks work as finger splints or tongue depressor.

(9) Triangular Bandages – They can be used as slings, dressing, and ties for splints in case of bone injury.

(10) Antiseptic Ointments, Creams and Lotions – Before you put on the bandage, you will need to thoroughly clean a wound. While soap and water work fine, but it is advisable to use a good antiseptic lotion to thoroughly rinse out any debris or particles in the wound that could infect it. Also, if the wound is large and could get pus formation, you will need to use an antiseptic regularly while you dress up the wound. So buy Dettol or Savlon to kill all germs and bacteria that can thrive in an open wound.

(11) Sanitizer – As a quickest way to sterile your hands in the field before and after first-aid treatment.

(12) Scissors – A must to have a pair of small medical scissors to cuts bandages, ties, tape, and clothing.

(13) Tweezers – A fine-tipped pair of tweezers to remove foreign objects lodged in the skin like splinters. Make sure you sterilise the tweezers by cleaning them with an antiseptic lotion before and after use.

(14) Nail Clippers & Safety Pins – A nail clipper in your first-aid box is needed to clip toe and hand nails and keep the nails clean as dirty nails are the biggest cause for fungal infections, bacterial growths and warts. A set of Safety pins to hold the normal dressings in position is a must for a first –aid box.

(15) Muscle Creams etc – Have pain relief spray, muscle creams and gels to ensure instant relief. However, use them in moderation as the ingredients do get absorbed directly in to bloodstream through the skin.

(16) Pain Relievers – Having a few pain relievers like Combiflam or Ibuprofen in your first-aid box is a must. Even if the pain is persistent and needs to be checked by a doctor, you can still pop a painkiller to help you bear with the trip down to the clinic and the long wait before you get the turn to meet with your doctor.

(17) Antihistamines – From sneezing to breaking out in a rash, antihistamine will take care of all allergies. While sinus and dust allergies might not need medical attention, a food allergy might need a trip to the doctor. In any case, the allergy medication will provide a little relief while you rush the patient to the emergency.
(18) Fungal Medicines –In a hot, humid climate there is risk of fungal infections, anywhere from genitals to feet, face and hands. These fungal infections can be itchy and embarrassing. So make sure to have anti-fungal medicines: gels, creams, powders and even pills in your first–aid box to provide relief from the persistent itch.
(19) Antibiotic – Carry the triple antibiotic type. Consult a physician before buying one. A pharmacist also can give you a practical advice on the latest one and the effective one for general use.
(20) Aspirin – Very useful for heart ailments' and it must be part of first-aid kits for cardiac related issues.
(21) Common Fever – Fevers are common, so keep Crocin or basic paracetamol to ensure to regulate normal body temperature while your doctor diagnoses the cause.
(22) Antacid –It helps for stomach cramps / disorder /bloating or tightness feelings.
(23) Thermometer – Any first-aid box is incomplete without a good thermometer, so have a digital thermometer.

Lastly have a good-sized approved colour & design box for first-aid to store all the items systematically.

9.9.2 Contents of a Professional First-aid box

Appropriate first-aid facilities should be available at all sites. A risk assessment should be completed to determine the number of First-Aid boxes and location. First-Aid kits should be clearly marked, easily accessible and should not be locked. In each shift there should be a trained first-aid provider. For first-aid kits in the workplaces should be as per legislation (factory act) which specifies what must be provided; which in turn will depend on the size and type of the workplace. Make sure you know where first-aid kits are located. There should be sufficient indication of the kit's location for those who are unfamiliar with its location to identify it.

First-aid kits must be kept well-stocked; supplies do expire, and must be replaced periodically. Consider creating a schedule for checking that the kit is stocked, and replacing any expired items as required. Quantities below are guidelines; you should determine what is required based on the kit's expected use. If possible use a bright-colored, Water tight plastic container. It may be worthwhile to purchase a kit made specifically for this purpose. Table 9.1 below details the recommended contents of a professional First-Aid box;

Table 9.1: Typical recommended contents of a professional First-Aid box

S. No	Items / Person	First-aid –box contents		
		Upto 10	Upto 25	Upto 50
1	Gloves – medium and large size, to fit just everyone.	3set	5set	10set
2	Adhesive plasters	20	20	40
3	Band Aids – Have both small and large sizes (Assorted)	3set	5set	10set
4	Sterile Eye pads (no16)	2	2	4
5	Triangular Bandage (Individually wrapped) – to be used as slings, dressing, and ties for splints	2	2	4
6	Assorted size Sterile unmedicated wound dressings (Individually wrapped)	10	15	20
7	Disinfectant wipes (Individually wrapped)	10	15	20
8	Paramedic Scissors	1	1	1
9	Tweezers	1	1	1
10	Safety Pins	5	5	10
11	Nail clipper	1	1	1
12	Digital thermometer	1	1	1
13	Instant BP meter	1	1	1
14	Sterile water or Sanitizer – If tap water not clear	1x500ml	2x500ml	2x500ml
15	Pocket face mask	1	1	2
16	Water based burn dressing large	1	2	2
17	Crepe bandage Assorted size (each)	1	2	2
18	Disinfectant like Dettol etc	1x500ml	2x500ml	2x500ml
19	Antiseptic Ointments, Creams and Lotions	Assorted	Assorted	Assorted
20	Muscle Creams and Sprays	Assorted	Assorted	Assorted

Contd...

S. No	Items / Person	First-aid –box contents		
		Upto 10	Upto 25	Upto 50
21	Pain reliever like Crocin and Combiflam	Assorted	Assorted	Assorted
22	Antihistamines for allergies & sneezing	Assorted	Assorted	Assorted
23	Fungal creams / lotions	Assorted	Assorted	Assorted
24	Fever medicines like Crocin Paracetamol	1 strap	2 strap	2 strap
25	Aspirin – for cardiac issues.	1 strap	2 strap	2 strap
26	Antacid – for stomach illness or disorder.	1 strap	2 strap	2 strap
27	List of emergency / ambulance service numbers			
28	Log Register for use/ replenishment of stock			
29	First –aid box of appropriate size(each)	1	1	1
30	Approved Anti snake bite kit (For Project / field works)	1	1	1
31	Artificial resuscitation / CPR chart	1	1	1
32	Instruction chart for Do's & Don'ts of First-aid	1	1	1
33	Instruction chart for each type of First-aid	1	1	1
34	A Chart for type of snake as per area for proper antidote	1	1	1

Any other specific item as per the requirement of a particular project on the recommendation of project in-charge also should be included in the first-aid kit.

9.9.3 Marking and Physical appearance of a First-aid box

Bureau of Indian standards (BIS) also have Indian Standard – IS 13115:1991 for First-Aid kits. As per this standard each first-aid kit shall be kept in a durable container indelibly and legibly marked: 'FIRST-AID KIT', in letters not less than 20 mm in height, First–aid symbol, and trade Mark if any, and manufacturer's name and address. Each box must have a booklet giving detailed instruction on first-aid treatment.

A typical First-aid box should be packed professionally for ease in use and locating the contents easily. Fig 9.11 shows exterior, inner views of a typical First–Aid box with first-aid symbol on the box in English and Hindi.

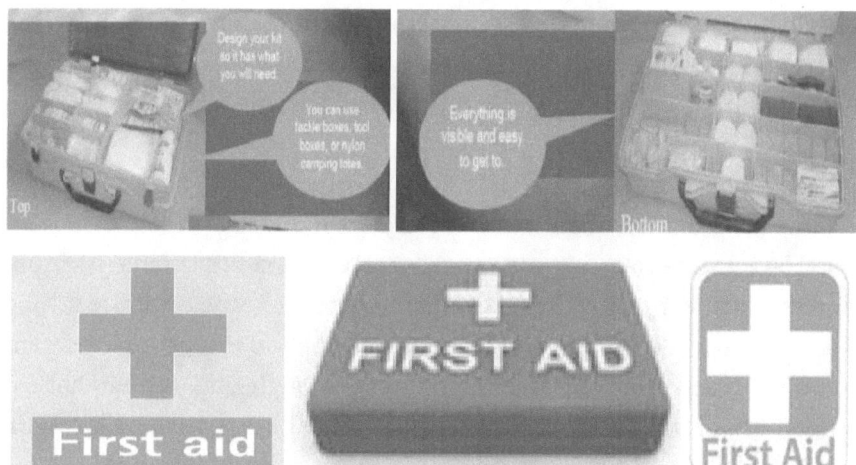

Fig 9.11: Typical Bottom and Top view including symbol in Hindi and English on a First-Aid box

9.9.4 Steps to make a first-aid kit

A first-aid kit is very important item for an establishment and it should meet certain prescribed criteria for the organization as per factory act for the occupational safety of its employees. So it should be planned meticulously. The following points need to be addressed before making a first-aid kit:

a) Identify how kit will be used and where it will be kept. Determine the size and type of container), b) Analyse what type of injuries are expected in the premises to determine what to keep in First–aid box, c) Select and obtain the bag, box, or container for the kit, d) Obtain the contents, repackage them if necessary, and pack the kit, e) Decide where the extra items are to be kept to restock the kit, f) Make up an inventory list. Keep one copy in the kit, and another copy with the in-charge, and g) Show every employee the kit, what is in it, and how to find and use the contents.

9.10 Elements of Training in First-aid

Training is very important element for first-aiders from each establishment. These trained, first-aider in turn are required to train the remaining staff of the establishment for extending the first-aid services independently in the absence of trained first-aider as well as during mass need of first-aider in a major accident, calamity or like. Each country through its National Health

Policy has specified agencies for imparting training in first-aid, in India its Indian Red Cross through Saint John Ambulance.

9.10.1 Training Methods

Training programmes should incorporate the following principles;
    a) Have appropriate first-aid supplies and equipment available; b) Expose trainees to acute injury etc settings and to appropriate the response through the use of visual aids; c) Include a course information resource for reference both during and after training; d) Allowing enough time for emphasis on commonly and frequently occurring situations; e) Emphasise skills training and confidence-building over classroom lectures; and f) Emphasise quick response to first-aid situations.

9.10.2 Assessing the scene and the victim(s)

The training programme should include instructions on the followings;
    (a) Assessing the scene for safety, number of injured, and nature of the event, (b) Assessing the toxic potential of the environment and the need for respiratory protection, (c) Establish the presence of a confined space and the need for respiratory protection and specialised training to perform a rescue exercise, (d) Prioritising the care as to who is to be attended first when there are several injured, (e) Assess each victim for responsiveness, airway blockage, breathing, circulation, and medical alert tags, (d) Taking a victim's history at the scene, including determining the mechanism of injury, (e) Performing a logical head-to-toe check for injuries, (f) Stressing the need to continuously monitor the victim's vital signs, (g) Emphasising early activation of EMS (emergency Medical Services),(h) Indications for and methods of safely moving and rescuing the victims, and (i) Repositioning ill/injured victims to prevent further injury means preserving the life

9.10.3 Responding to Life-Threatening Emergencies

The training program should be designed or adapted for the specific worksite and may include first-aid Instructions on the followings:

(a) ABC's
- Establishing responsiveness;
- Establishing and maintaining open and clear airway;
- Performing rescue breathing;
- Treating airway obstruction in a conscious victim;
- Performing CPR;
- Using an AED;
- Recognising the signs and symptoms of shock and providing first aid for shock due to illness or injury;
- Assessing and treating a victim who has an unexplained change in level of consciousness or sudden illness; Controlling bleeding with direct pressure;

(b) Poisoning

Recognising asphyxiation and the danger of entering a confined space without appropriate respiratory protection. Additional training required if first-aid personnel to assist rescue from the confined space.

(c) Responding to Medical Emergencies
- Chest pain, Stroke
- Breathing problems, Anaphylactic reaction
- Hypoglycemia in diabetics taking insulin
- Seizures, Abdominal injury
- Pregnancy complications, Impaled object
- Reduced level of consciousness.

9.10.4 Responding to Non-Life-Threatening Emergencies

The training programme should be designed for the specific worksite and include first aid instruction for the management of the followings:

(a) Wounds
- Assessment and first-aid for wounds including abrasions, cuts, lacerations, punctures, avulsions, amputations and crush injuries;
- Principles of wound care, including infection precautions;
- Principles of body substance isolation, universal precautions and use of personal protective equipment.

(b) Burns
- Assessing the severity of a burn;
- Recognising whether a burn is thermal, electrical, or chemical and the appropriate first aid;
- Reviewing corrosive chemicals at a specific worksite, along with appropriate first aid.

(c) Temperature Extremes
- Exposure to cold, including frostbite and hypothermia;
- Exposure to heat, including heat cramps, heat exhaustion and heat stroke.

(d) Musculoskeletal Injuries
- Fractures; Sprains, strains, contusions and cramps;
- Head, neck, back and spinal injuries;
- Appropriate handling of amputated body parts.

(e) Eye injuries
- First aid for eye injuries;
- First aid for chemical burns.

(f) Mouth and Teeth Injuries
- Oral injuries; lip and tongue injuries; broken and missing teeth;
- The importance of preventing aspiration of blood and/or teeth.

(g) Bites and Stings
- Human and animal bites; Bites and stings from insects;
- Instruction in first-aid treatment of anaphylactic shock.

9.10.5 Trainee Assessment, Skill and Program Updates

Assessment of successful completion of the first-aid training program should include instructor observation of acquired skills and written performance assessments.

Note: Keep provisions for annual critical skill up-dates, retraining and first-aid program updates as per changes or modernization in the workplace facilities & locations.

References

1. Certificate Programme, Occupational Health and Safety: Legal and Operational Guide, Unit 6, Occupational health and safety legislation in India ©2014 PRIA International Academy
2. Legal Provisions for protection of health and safety at work in India: Jagdish Patel: jagdish.jb@gmail.com
3. First Aid, Emergency care for the injured, Source: http:/ / en. Wikibooks. org/ w/ index. php? title=First_ Aid/ Poisoning & old id=1167235: Principal Authors: Mike. lifeguard, Nugger, White knight, SBJohnny, Rama, Alex S, Kernigh, Jomegat, Uncle G, MichaelFrey, Leonariso
4. Disaster Management Institute, Paryavaran Parisar, E-5, Arera Colony, PB No. 563 Bhopal-462 016 MP (India), www.hrdp-iDRM.in, Chief Editor: Praveen Garg, IAS, Executive Director, DMI, Bhopal, India
5. Guide to the Safety, Health and Welfare at Work (General Application) Regulations 2007 Chapter 2 of Part 7 First-Aid: Published in December 2007 by the Health and Safety Authority, The Metropolitan Building, James Joyce Street, Dublin 1.
6. ISID Discussion Notes – Health and Safety at Workplaces in India – M.M.K. Sardana (Author is a Visiting Fellow at the, "Indian Institute of Public Health", Hyderabad, India.
7. IS 13115: 1991, Indian Standard for PORTABLE FIRST-AID KIT FOR GENERAL USE — SPECIFICATION
8. Best Practices Guide: Fundamentals of a Workplace First-Aid Program, U.S. Department of Labor, Occupational Safety and Health Administration OSHA 3317-06N 2006
9. Handbook of Radio engineering Managers – J.F. Ross: Chapter-21; Staff Responsibility-First-Aid Rules (pp 337-338)

INDEX

A
Arc Detectors, 224

C
Calorimetric Measurements, 105
Capillary Refill, 518, 522
Carotid Pulse, 519, 529
Charge Separation, 201-202
CPR Adjunct, 507

D
Dosimetry Method, 102

E
Earth resistance, 13, 307, 309, 349, 352, 354, 357, 359-361, 366-368, 371, 373, 375, 377, 383-387, 396
Effect of peak power, 52-53
Egress System, 45
Electric Arc Radiation, 30

F
Fall of Potential, 377, 384, 387
Far Field, 67-71, 74, 77, 84, 408-409, 412-413, 475
Ferrite Cores, 228, 233
Fire Detection Devices, 112, 184
Fire Protection Requirements, 130, 146, 162, 167, 190
Fire Triangle, 31, 114
Flame Detectors, 132, 134
Flutter Valve, 537

G
Grid Earthing, 352
Ground Loops, 337, 343, 416, 431, 433, 438-439

H
Harmonic Distortion, 398
Hazard map, 45
Heat Detectors, 131, 134

I
Impedance method, 78, 80
Infra-red Detectors, 133, 185
Ionizing, 49, 51, 59-60, 66, 94, 204, 268, 496

L
Leader Formation, 202

M

Method of moments, 78-79

N

Near Field, 60, 67, 71-72, 78, 84, 86, 102, 408-409, 412
Non-continuous, 56, 109
Non-Ionizing, 49, 51, 60, 66, 94, 268

O

Ordinary Person, 281

P

PFAS, 41-42
Power Density, 51, 56-57, 67-69, 71, 76-78, 82, 84, 88, 92, 95-96, 98, 109, 408
Process map, 46

Q

Qualified Person, 33, 274-275, 281-283

R

Radiated, 34, 50, 58, 61, 67, 75, 92, 108, 110, 182, 220-221, 346, 397, 408, 415, 420-421, 432, 437, 441, 445, 451

RF Chokes, 223, 232
RF Voltage, 106, 180, 220-222, 458, 532, 534

S

SAR Measurements, 103
Shielding Effectiveness, 410-411, 413-415, 421-422
Smoke Detectors, 131, 134, 177, 184, 186, 192, 194
Soil Resistivity, 13, 250, 349-351, 369, 371, 373-374, 376, 388-389
Static Charge, 12, 216-220, 222-224, 249, 251-253, 276, 340, 344, 398
Static Charge Estimation, 220
Static Drain Resistors, 223
Surge Protectors, 12, 226-228, 233, 237, 246

T

Tetrahedron of Fire, 119-120
Touch, 107, 216-217, 222, 263-265, 276, 316, 398, 487-488, 504, 533, 535, 548

V

Varistor Operation, 239

www.ingramcontent.com/pod-product-compliance
Lightning Source LLC
Chambersburg PA
CBHW020718180526
45163CB00001B/21